DEDICATION

To my kids Samuel and Meghan Rose, and to my mother, Mary Anne.

World Peace

The Transition

Of an Automated Society

#WPProjects are our Generation's "Moon Launch" - and Greatest Legacy

Check out WP Magazine online at csq1.org/mag

- 100% Employment, Education, Safety Nets, Pensions
- Two Years and 200 Countries Ago
- The Meaning of Life, Accountability & Worthwhile Projects
- Human Rights and Aristotle's "Right Plan" for a Good Life and the American Dream
- Democracy's Guide to Success in Today's Manual Economy
- Includes the WP Projects TED Talk

From the Author who wrote the book on Common Sense in Economics, Technology, Business, Government, History and Transformation

Edward Tilley

Copyright © 2015 Edward Tilley

All rights reserved.

Hardcover ISBN: 978-1-987964-07-3

Book ISBN: 1-987964-04-2

Book ISBN-13: 978-1-987964-04-2

eBook ISBN: 978-1-987964-05-9

AudioBook ISBN: 978-1-987964-06-6

CONTENTS

Chapter 1 - Introduction 1

Level-Setting 4

Why is World Peace important? 5

Academic Discussion 7

Great Thinking 8

Top-Down vs. Bottom-Up vs. Sustainable Bottom-Up 11

Forward Thinking 12

Great Process 13

Step One – A Good Life at Home 17

Step Two – Technology Projects 21

Step Three – The Transition 22

Step Four – Rollout World Peace Worldwide 23

Offending Content and Topics 24

CSQ Certifications & Social Responsibility 25

WP-TV Documentary and News Programs 28

Chapter 2 - World Peace 29

World Peace through Economic Controls 31

National Wealth Distribution Policy 37

Economic Control Roadblocks 38

Sustaining a K-Wave Spring 41

The Old War-Distraction Reuse 46

Status of World Peace Economic Controls 47
A New Option 48
World Peace through Technology 48
Status of World Peace through Technology 50
Timeline for World Peace through Technology 51
Sustainable Economic World Peace 54
Three Teams 54
Voting and Growing Together 57
Chapter 3 – War 59
Why do Wars begin? 60
Non-Combatants 61
Death Rates in War 64
Democracies don't war with Democracies 66
War of Needs 66
The American Revolution 66
The Iraq War 70
War in the Congo 71
Wars over Oil 72
Replacing Oil 74
Rapid Charge Vehicle Battery Systems 79
Oil is not even needed really 80
Wars that Changed Society 82
The Two Faces of Every War 83

Rational Wars 84
Non-Rational Wars 84
Effect of Democracy on War 86
Chapter 4 - A Government Systems Primer 87
Government Leadership Categories 87
Common Misconceptions 91
Socialist Governments 91
Communist Government Misconceptions 92
Choosing a System 94
Do your needs of a society include 95
Right vs. Left 96
Systems within Systems 97
Election of a President or Prime Minister 97
Types of Government 97
Chapter 5 – Government Economic Policy 102
The Pitfalls of Managing Proactively 106
Economic Controls - Right and Left 107
Economic Sustainability Goals 109
Chapter 6 – Social Projects 111
Wealth Distribution 113
Patterns in Income Equality 116
Setting Targets – Minimum Incomes & Tax 117
Wealth 117
Income 118

The Lowest Quintile (20%)....... 118
Next Lowest Quintile 119
Middle Quintile....... 119
Second Highest Quintile 119
The Highest Quintile....... 119
Wage Stagnation 120
Minimum Wage....... 120
Graduated Tax 121
Wealth Creation 121
Socialistic Policies Mean Business....... 123
Reshoring Production & Engineering 126
Mitigating Multinational Extortion....... 128
Philanthropy....... 129
Billionaire Bequeathals....... 130
The Case for Distributing Wealth Quickly 131
Team One - Next Steps and Targets....... 132
Chapter 7 – Technology Projects 134
The "Worldville" Competitions 136
Science Fiction is just a Technology Project....... 140
World Peace Agenda Projects....... 142
Team Two Project Assignments v1.0 143
Many Hands make Light Work 144
Assigning Projects....... 145
Team Two – Automating our Production Economy....... 148

Adding World Peace Agenda Projects 157
Project and Operational Performance 158
Building Great Technology 159
Example Project - Industrial Robotics 159
Peer Review and Coaching 161
Chapter 8 – The Transition 163
Our Jobs are Automating 163
Current Social Safety Nets 166
Dawn of the Engineers 167
Wealth Distribution 168
Transitioning to a Good Life 168
Transitioning Countries One Part at a Time 168
Wealth Creation 171
Automate your production economy 171
Support and Sell 171
Running Projects that Solve Anything 173
Here we are, the Good Life. 175
Chapter 9 – World Peace TEDTalk 177
Cover - World Peace – The Transition 179
Step 1 + World Peace Starts at Home 185
Step One – 187
We Used to have this Good Life - in G8 Countries 189
What Happened to our Good Life? 191

What do Capitalist Cycles Look Like? 193

K-Wave Fall – 1982 to 2002 195

K-Wave Winters - the Great Depressions and Financial Panics 197

Controls for Sustainable Capitalism 199

Our Leadership Changed 201

The Time of the Engineers 201

Technology in 1917 204

By the end of World War II 204

1945 to 1969 - the Cold War 205

Business School Administrators Lead 210

45 Years Later 212

Miniaturization, Improvement, Profit 215

Emerging Technology 217

What Can Era Differences Teach Us 217

Team 1 – Recessionary Policy 221

Team 2 builds Step 2 - Technology Projects 224

Team 2 – One Project to Automate our Civilization 226

Each Country is Assigned a #WPProject 229

Automating Production Economies 231

Step 3 – Manage the Transition 233

Step 4 – World Peace Agendas - Rollout Worldwide 235

Next Steps – Vote and Choose Projects Well 237

GAPs in the UN's Global Goals 239

CSQ CSR Compliance and Certification 241
Chapter 10 – The World Peace Agenda 243
World Peace Thermometer and Clock 245
Communicating Status and Needs 246
Crowded Projects 247
Procedures to get Funding and Launch 248
Acknowledgments and Recognition 248
Operational Performance Measures 249
WP Dashboard 249
The World Peace Agenda Call Center 250
Project Metadata: 250
Form 1: Charter 251
Form 2: KPIs 251
Form 3: Project Documentation 251
Form 4: Status 251
Form 5: Approvals 251
Chapter 11 – Technologia in Voting 252
Think like a Founding Father 254
The Mayflower Compact 257
The Importance of Education in Democracy 258
Voting Right or Left 261
Splitting the Left or Right Vote 262
Which Policies to Prefer 263

Voter Comparison Charts 264

Educating Policy Makers 266

Chapter 12 - Accountability 267

Good Process Never Takes Longer 268

Designing Accountability 269

Insist on Forward Momentum 270

Business Accountability 272

Raising Successful Girls; Women in the Workplace 273

Chapter 13 - Land Ownership 279

Understanding lending costs 280

Economic Controls for Housing Bubbles 282

Government Land Ownership 285

Can we Switch - back and forth? 285

What sort of home would I get? 287

What is the right place for you to live? 289

Value & Performance 290

Housing KPIs (Performance Measures) 291

Corruption 292

Dante's Description of Hell 293

Chapter 14 – The United Nations 296

GAPs in the Plan 300

Functions of the United Nations 303

UN Departments 304

Statistics at the United Nations 304
United Nations Annual Agendas 305
The General Assembly discuss: 306
Economic and Social Council 306
Department of Development 307
Human Rights and Humanitarian Affairs 308
Peace and Security 308
Chapter 15 – Disarmament 310
Nuclear Arsenals 311
Weapons, Militia, and Conventional Arsenals 312
Armies 312
Religion in the Army 312
Chapter 16 – Rating your Government 314
Rating the Management Team of your Country 315
Human Development Index, GINI, and GDP 315
Area and Export – Wealth Creation 319
Top 10 Country Management Teams 320
Rating Elected Leaders 321
Conclusions 324
Chapter 17 - Building World Peace 329
Document 1 - Program Charter 331
2. Requirements & Inventory 334
3. The Design Document 336

4. Detail Implementation Plan Build, Test, Pilot, Release 337
Signoffs Requested 338
Run Book – Operators Guide 339
How-Tos 342
Life-cycle Managed Documentation 343
A Robotic Hand - Example Project 344
Performance Management Dashboards 345
Status Options and Stakeholder Duties 349
Sharing Knowledge 349
Everything is Solvable 350
Funding Technology Projects 352
The Productive Complaint 353
Corporate Stewardship Cautions 353
Chapter 18 - Diligence not Fear 358
Mitigating Security Risks 359
Responding to Terrorism 361
Putting an End to Terrorism 366
Global Crises 367
Global Warming 368
A Global Energy Crisis 368
World Population Control 370
Too Many Monkeys 374
Moderating Birth Rates 374

A Global Water Crisis ... 375

The Real Estate Super-Bubble ... 376

Who will Lead the Change ... 377

Chapter 19 – Performance & Merit ... 378

Maintaining Performance ... 382

Judging Merit ... 383

Chapter 20 – Human Rights ... 385

Respect for Life ... 387

Respect for Equality ... 388

Women's Rights ... 389

Children's Rights ... 389

Respect for Family ... 389

Right to Healthcare ... 390

Free Speech ... 390

The Right to a Good Life builds World Peace ... 390

CSR – Corporate Social Responsibility ... 391

Chapter 21 - World War III ... 392

Chapter 22 – The Meaning of Life ... 396

Worthwhile Projects ... 397

Less-Than Worthwhile Projects ... 398

Make Religion a "Good" Thing ... 400

At the detail level... 400

At the forest level ... 401

The 100 km Summary view ... 402

The Meaning of Life 403
All You Need Is Love 403
Chapter 23 – Causal Conclusion Case Studies 406
Case 1 – Working the Numbers 407
Case 2 – Building HOW back into our Schools 409
Case 3 – An Easy Case Study 413
Case 4 – Not as easy 415
Case 4 Conclusions 418
Militia and Organized Crime 418
Wealth Distribution in the U.S. 419
Case 4 Conclusion 420
Chapter 24 - Getting the Word Out 421
Beginning a Good Life in Law 423
Difficult Summaries 425
Fixing problems for our Young People 426
Productive Safety Nets 426
Context in Great Thinking 427
The Do Nothing Option 427
Social Planning 428
Business & Government 429
Leading the Cheer 430
Get the Word Out 431
Chapter 25 – Transition Economics 432
Important Data 434

Optimal Rate of Transition 435
Accelerating Trends in Industry Automation 435
Unemployment and Underemployment 436
Underfunded Retirements 436
Debt 437
Real Estate 438
Worldwide Markets 439
Financial Indicators Summary 439
Chapter 26 - Conclusion 441
Leadership with Results 443
ABOUT THE AUTHOR 445
Bibliography 447

ACKNOWLEDGMENTS

My most sincere thanks to Mariusz Sokolowski and Elena Schwartz, who provided insights in world governments, civilization, and history that cemented an outline for a very important plan that was broadcast to 200 World leaders this year. To friends, fellow authors, scientists, historians, futurists, Alan Ellis and Sharon Gibney who provided research, feedback and helpful academic challenges.

What a journey. This book is a Proof in the academic sense of the word; a consolidation of solutions from fields in problem-solving, leadership, project method, history, technology and economics research – all boiled down to a simple equation and a handful of steps. This plan for World Peace can work because it has worked before; the plan does work because it has begun already and its conclusion is inevitable.

Earlier this year, CSQ Common Sense 101 consolidated the lessons accumulated from a lifetime of building solutions in technology, engineering, business, community, family and individual process and wisdom. It dawned on me later that the projects needed to build a bright future in that course, are also the solution to a sustainable World Peace - I just needed to explain The Transition.

I test those theories here successfully. It took unlearning dozens of well-intentioned bits of nonsense that we have all been conditioned to believe since childhood, and I leave it to the reader now to decide that this is Aristotle's Right Plan of worthwhile actions to build a Good Life.

John F Kennedy, Aristotle, Hammurabi, Constantine, Alan Turing, Gene Roddenberry, William Hanna, and Joseph Barbera, all contributed to this plan for a bright future for us all. I cannot count all of the millions of World Peace builders to come – but thank you too.

Being ahead of your time is perhaps the most frustrating genius of all, we owe you such a debt.

Chapter 1- Introduction

The knock on your door is a delight as you have been waiting for this day for a lifetime. On the doorstep is a package with a few-days-supply of canned goods, fresh vegetables, water, a cell phone, a tablet computer and a universal charger. A sheet of paper provides instructions to call with questions, and it has directions on how to use your tablet or phone to order your groceries.

From this day forward, water, groceries, energy, and household supplies will continue automatically as needed without interruption for the rest of your life, for the lives of your children, and that is just the beginning.

The World Peace Transition Program, at the direction of your government, has recognized that the world is just one big living community and has replaced your manual economy with a sustainable automated production infrastructure – similar to how Audi builds almost an entire car without a single human being today.

More than groceries, this knock also brings the basic Human Right of a Good Life, in a family friendly community, without security concerns, with the highest respect for human life. A Good Life is a life with equal opportunity, strong family and community values, universal healthcare, advanced education, with advanced technology, 100-year longevity, and the assurance of liberty and the pursuit of happiness too.

If you are wondering, don't we already have this Good Life? You may be surprised to learn that you are among less than 20% of people on the planet that do – and you are probably over the age of twenty-five as well.

If this sounds like an unusual start for a discussion of the plan for World Peace, take a minute to realize that the first step to solving any problem is to recognize that it exists. Second, without our basic needs met, and personal happiness fulfilled, we can rarely look beyond our situations to address problems elsewhere. When 80% of the planet are not in a position to help build an infrastructure for a Good Life, chances are greatly reduced for World Peace.

In these two ways, World Peace has been, and will be, delayed – until finally a very special knock arrives at the door of a family that does not have a Good Life today.

As a student, I liked math classes because a right or wrong answer was what it was. If my biology teacher was having an off day, he could mark my handed in drawings of cell layers higher or lower. Or, if my phrasings bent to poetic license versus strict grammar, my language teacher's mark was higher or lower depending on the day, his sense of humor, and how many other papers that he or she had to mark at the same time. Math was easy, it was defendable; it was right or it was wrong, and there was nothing subjective about it.

This plan takes similar care to minimize subjective interpretations by supporting problems and conclusions with well-referenced research and hard numbers. As a math person, I wouldn't stand for less because it would not be credible to me as a writer nor a reader otherwise. I believe that expert opinions and verifiable statistics make interesting learning, and then conclusions all by themselves. I was personally surprised by a lot of the facts uncovered during researches in science, business, leadership, economics, history, philosophy and even - good old math. Writing this book caused me to rethink long-held personal opinions, and I hope you will enjoy the process as much as I did.

What happens when 49% of people within a society are not provided with the basic needs of life – needs like water, food, shelter, security, love and family? Historically, communities deprived of security will create militia; take a chapter from American, European, or Chinese History to confirm this. Regions deprived of food, healthcare, income and the basics of life, lash out as in a revolution; they raid neighboring communities, or create other social problems, and their lives are regarded as a hard life.

If this group lives in a pure democracy, and normal capitalist cycles continue, their votes will grow to exceed 51% in time, and democratic elections will ensure that wealth distributes and capitalism can begin over for everyone. To prevent half the people in society from starving while waiting for this majority vote, Constitutions give all citizens a definition of basic rights of food, shelter, healthcare, security, and then these minimum supplies are afforded them.

The challenge for some countries next becomes, what minimum levels of support, or Human Rights, should we grant to all members of society?

The best countries in the world for everyone to live within, were those that delivered the American Dream of the 1950s - a “Good Life” for all citizens - and many other G8 nations provided this too. The Soviet Union was the last to end support for a Good Life in its major cities in the mid-1980s when Perestroika converted its United Communist Republics to Independent Capitalist countries.

As this Good Life for all citizens shrunk from 100% to 90%, 80%, then to 51% over time in Capitalist G8 democracies, it started increasing toward 100% for thirty-eight million people in Socialistic-Capitalist countries like Sweden, Norway, Denmark, Finland, and the Netherlands.

Social values within these regions explain this difference. Where fathers in the harsh north and deserts, teach sons to feed and shelter travelers and others as a priority; fathers in lands of milk and honey, saw the assistance of others as unnecessary.

Countries with a Good Life for all began to show better economic performance too because 100% could afford to pursue commerce that generated revenue for their country's GDP. The average Dutch citizen generates almost 50% more export (wealth) than an American and 300% more than a Canadian citizen.

Capitalist, Socialist, and Monarchy nations maintain many policies that are capitalistic, socialistic and communal or communistic too.

Level-Setting

The US News makes it clear that "Socialism is bad"; and when I surf to Investopedia, places like North Korea, China, Vietnam, and Cuba are labeled Socialist Countries incorrectly. I confirmed after a bit of research that these countries are Communist states and that North Korea is a Fascist Regime as well. Socialistic countries like Switzerland, Japan, Sweden, Norway, the Netherlands, Denmark and most other G20 countries, set the highest standard for best places to live in the world; and so I realize that even an online encyclopedia can struggle to understand these terms consistently.

Clearly, to begin a discussion on how to build a Good Life and sustainable ommunities for seven billion people, we first have to wipe our slates clean of lessons provided in the news and by some schools, politicians, and special interest agendas. Next let's level-set with a shared understanding of the basic terms and definitions that define how we might prefer to live today, and perhaps more importantly, how we might like our children to live tomorrow too.

Our final level-set realizes that World Peace is a real and attainable thing, the tangible result of a series of well thought out and planned list of projects that Aristotle called the "Right Plan" 2500 years ago. World Peace is not a religion, although it has been a goal of many religions; it has also been the goal of teachers, governments, veterans and great leaders and great thinkers in time.

The Puritans' first democratic constitution - the Mayflower Compact;

Constantine, Mohammed, Buddha, and others, each set comparable goals for society, but planning the "How" was largely left to us.

World Peace – The Transition is a comprehensive Plan to return a Good Life to the G20 with straightforward explanations of the economic controls, safety nets and technology projects needed. The same technologies that have already automated the production of some of the biggest employers in the United States, U.K., Canada, and all other G20 nations.

If carefully transitioned, these tools deliver a Good Life and sustainable community to each of us, and can direct government and business investment that distributes existing wealth and also generates new wealth from profitable exports as a top priority. Once our technology automation begins working, money loses importance quite a bit as it begins to afford troublesome wants more than needs.

So much of our news is a distraction because we have no idea what to do next. This plan lets us see what our next steps are and then it lets us simply work the problem, solve new problems as they arise and get the work done quickly as well.

Along the way we are going to compare governments, countries, policy, best practices in economic controls, and we are going to build an automated production economy too.

Welcome to the Transition.

Why is World Peace important?

Can we stop building a Good Life at a point when it exists for everyone within just our community?

Both history and most of the great thinkers supported that we can only realistically begin to consider building a sustainable community for others, once we have met our basic needs first. And also, that if others see our lives as selfishness, then our life might also become unsustainable until our neighboring communities also share a Good Life as well. In CSQ terminology, the smallest unit of community is your country.

World Peace has been the passionate, safe and admittedly vapid wish of beauty pageant contestants since the 1920s. Early covers of this book, included Miss Russia 2014, Yulia Alipova for this reason and a couple of other reasons which I will explain throughout the book. Ms. Alipova is seen here in traditional 12th Century Byzantine Era Royal formal dress.

The Byzantine Empire was the Greek-speaking, more wealthy continuation of the Eastern Roman Empire. This Era was marked by the renaming of the City of Istanbul – then "Byzantium" - to Constantinople in 330 AD/CE in honor of one of our History's most impressive leaders, the Holy Roman Emperor Constantine. The relative peace of the Roman Empire had allowed great strides in engineering, philosophy, education, law, government and science that slowed dramatically after Constantine's time.

Constantine assembled the modern Bible and made a Christian-Pagan blended Catholic Religion and European Church. In 1453, the Muslim Ottoman Empire took control of Christian Byzantium and Hungary became the front line of Western-European Christianity for centuries to follow. I will chat about why religion was important later in Chapter 3 - War.

Why did the Roman Empire fall? In a word, Sustainability: The people of the cities held wealth and security while farmers and soldiers fed and fought-off growing insurgent armies from Muslim and Germanic tribes who were jealous of that wealth. At a point, the economy's producers realized that the cost outweighed the benefit of defending Rome. Over the course of two centuries, outlying farmers and soldiers became unwilling to feed nor defend the cities any longer - and Rome fell.

As the Roman Empire receded, constant warring between fiefdoms, feudal lords, and Christianity's pervasive belief in poverty, denied a Good Life and stalled human evolution in Europe for a thousand years. The End of the Byzantine Era marked the end of a thousand years of stagnation in European society and by the 14th Century, things started to get interesting again.

If only the beauty contestants' wishes for World Peace, had included a strong plan or explanation on how to achieve the thing that these young women wanted most. World Peace, most of us realize, is a wonderful aspiration.

And then there are the cynics too of course; those who feel compelled to remind us all that we can never accomplish World Peace. These voices are no better nor more impressive than the pageant contestants really, as they too are merely acute or frustrated individuals without the understanding of the strong process and frameworks needed to design and implement the sustainable solution of a large, complex, and very achievable goal.

Academic Discussion

Productive, intelligent, academic discussion is a process that improves with patience and practice. It is a reflection of good education both at home and in academic settings, and it is a reflection of good character. In the company of a romantic interest, it can be a second for poetry very nicely as well.

Prodigious reading is far from uninteresting, so strap in and look forward to the debate ahead.

Academic discussion looks at a thought as one might examine a table or a piece of wood - and it examines it critically, emotionally, and then sees the unemotional pluses and the minuses as they are too. Academic discussion thinks through, and asks will a solution meet the needs of its users; will it be scalable, flexible, and reliable; is it as simple, and therefore, as actionable as it could be?

My personal technique is to consider a thought and quantify its qualities. I assign percentage probabilities to the results of decisions made, and I mitigate the risk of things that could go wrong a significant percent of the time. When I use the word "mitigation" throughout this book, I just mean to say that I make a list of possible risks to a positive outcome - and then I think of a countermeasure or solution for each risk, threat, or negative thing should it occur. This mitigation list takes thirty seconds to compose in thought, and I don't have to worry emotionally about it after that because I have a plan for 99% of the negatives. It has taken me years of practice to get good at it, and this process has replaced fear and worry with logic in both personal and professional life because thinking about it, makes more sense than worrying about it.

It's an aside, but whenever you feel worry or fear, build a mitigations list. With practice; after you write it down a few times, you will be able to do it in your head. You will sleep easier and make better decisions too. I keep a notepad, or EverNote smartphone app, by my bed to capture thoughts that wake me up at night. John F Kennedy used a similar approach that I will describe two pages hence.

With this note, our first academic discussion has started.

Great Thinking

Great thinkers, individuals whose process of thought and reflection were so strong that they changed the way we live 3000 years later, did this - with just thought. I struggled to find contemporary contexts for the teachings of Plato, Aristotle, Socrates, Elizabeth I, and others before writing this book - but no longer. Much of their process is brilliant thinking that took them, and I, fifty years of life and career experience to recognize as profound wisdom. I simply had to translate relevant discussions into a contemporary context.

Like Alan Turing, inventor of the computer, we must realize that our next challenge becomes one of implementing great thinking into our physical lives using tools and processes that are simple, scalable and automatable. Mr. Turing realized that cracking Enigma needed an automatable machine that could solve any problem. Given available resources, he first made a machine that could solve anything, and then he used it to solve a specific puzzle. Initially, he built with limited electro-mechanical components and then improved his working design by leveraging fully digital components.

Soon, engineers will build onto his universal computer, the ability to solve all the other puzzles needed to automate our production economies – in just the same way.

We are going to practice academic discussion in this book. There will be a time to listen and learn, and there will be a time and place to argue point by counterpoint and provide feedback.

The results can be quite amazing. Consider Aristotle's summary of what it takes to build a self-sufficient community:

The Right Plan is the one whose ends, means, practical thinking and purposeful action result in a Good Life. A life full of things you need – and not necessarily a life full of everything you want. With a little luck, goods in body and soul, and by making a habit of good choices that reflect moral virtues of temperance, courage, and justice, a Good Life should be sought and found.

Abridged from Politic 322 BC (Messerly, 2013)

To arrive at this summary, I had to consolidate a much longer translation of Politic (Jowett & Aristotle, 2015) - an Ancient-Greek transcript, into our modern version of the language (21st century English in this case) as well. I believe it communicates Aristotle's message accurately without drawing on numerous topics not related to this discussion.

Consider that to arrive at this, Aristotle had to conceptualize what is a meaningful life, one with interesting projects that create a better future, from what he knew of life around him (Hepburn, 2015b). Then he had to bring the idea forward to endure perhaps a thousand years of time and then a year or more of debate, peer review, and team discussion with fellow philosophers. You may realize that this summary does a pretty good job of covering its subject, but you might not also see all of the layers that have been considered in the summary.

Aristotle also noted that it required a man of not less than 40 years of age to comprehend the full meaning of this thinking; a man with the mindset of one who is watching his children transition as adults into society. Once students study and learn how to use academic discussion to move worthwhile projects forward - in both career and personal lives, we all take a notable leap forward.

I had come to the complete first draft of this book, with all high-level steps explained - as needed to build a Good Life. Imagine my surprises when my final research into Aristotle's views on Capitalism and the Code of Hammurabi led me to realize that Aristotle had also come to not only the very same conclusion about a communally self-sufficient "Good Life", but he had also spotted anecdotally - the solution to World Peace directly as well. I will explain more about that in Step Two below.

As a teenager, I had gravitated to engineering and had only been a student of Aristotle's teachings without realizing that our modern University Method of Logic is his. The Puritans and Queen Elizabeth I leveraged his lessons of a good life and the greater good as well, and Protestant Catholicism also added lessons learned during a millennium of poverty-inspired social stagnation in the middle ages leading up to the 1500s.

I imagined myself quite an original thinker until I realized that all of these great thinkers and leaders had worked to build a similar plan in their time. I could not have hoped to find a better validation, as I will explain as we step through this plan and book.

Solving World Peace requires a comprehensive understanding of the high-level needs of interacting societies. It also needs a comprehensive understanding of the building blocks of any society - its individuals, families and communities.

This method of high-level analysis - to view the world at a 100,000-foot level where we can see the forest easily as if from an airplane flying high above the earth. This 100,000-foot vantage point gives a very different view of our lives than a much closer low-level view – perhaps of blades of grass - that we could only hope to describe while standing and even sitting at ground level.

We need to be viewing the planet from space at 100 kilometers up before we can see how similar are the many different forests, deserts and seas that make up our world; and different countries too.

Top-Down vs. Bottom-Up vs. Sustainable Bottom-Up

Planning has a method too. We talk about a process to build solutions in Chapter 17, but it is also important to consider solutions both top-down and bottom-up. For example: To Feed a Man a Fish, is a Top-Down Plan that provides for what he needs today. Next, you want to provide him with the tools and teachings on how to fish for himself, because that is a Bottom-up plan that can sustain him with food for a lifetime. If you next automate or give tools to all of his neighbors - to feed them too, this is a Sustainable Bottom-up Plan because this man's starving neighbors will never need to take what you have given to just one man.

Denying passports to western kids who decide they might prefer the anarchic life of a video game within terror groups in Syria, Iraq and Somalia, is a Top-Down Solution. Showing young people, that they have an important place and can start much more important and interesting lives right here at home, with their families and within a productive society, is a Bottom-Up Plan.

World Peace – The Transition, is a Bottom-Up Plan; also sometimes called a Strategic Plan versus a Tactical Solution. Some Strategic Plans incorporate "Big-Picture Thinking" that includes sustainability, and some do not. Aristotle's "Right Plan" included Big-Picture Thinking and so too does this self-sustaining World Peace Transition Plan.

Intelligent discussion by itself accomplishes nothing, and so I will be taking you next step-by-step through the "*purposeful action*" needed to implement the solutions discussed here in this book as well.

Forward Thinking

Imagine that once upon a time, two hundred or more years ago, that a loud voice in a crowd suggested that a group of people might all travel from New York to Paris for a lunch meeting. That voice would have been shouted over by others in the gathering because this completely ridiculous idea was impossible to imagine at that time.

No reasonable person could suggest lunch in Paris because no human being could swim, sail, nor carriage from New York to Paris. Worse, the suggester might have taken considerable harassment from the crowd for being unthinkably foolish. The crowd would marginalize the man, which - depending on the era, meant he might be labeled a witch, socialist, patent clerk, democrat, locked in irons, shunned, burned, chemically castrated, or he could be marginalized in name-calling or other ways too. Suffice it to say that there is little incentive in human societies through history for thinking, and then speaking or acting beyond the line of site of a crowd – which is why, in progressive societies, we now permit free speech.

Many years later, with jet aircraft in easy line-of-site, we realize that the only foolish voices in this fictional room were the nay-sayers among any group that shouted down a suggestion of easy travel from New York to Paris.

Today, we unanimously and easily accept that any group can arrange and accomplish that meeting if they wish it, thanks to a science fiction presented by Leonardo da Vinci in the year 1500. His initial concept of the Airplane would later be improved to create a British technology and engineering project called the pressurized jet aircraft that first flew between continents in 1949.

Great Process

Let's discuss great process by recalling a very noteworthy example of what it takes to make great strides forward in society.

John F. Kennedy was one of America's very best and perhaps one of its most influential and iconic Presidents. He was an avid reader, and both he and his wife Jackie read a book every day when they could. Like other great U.S. Presidents Washington, Lincoln, and Ted and Theodore Roosevelt, Mr. Kennedy practiced techniques in speed reading, and he was reported to read word groupings at 1200 words per minute; which is roughly five book pages or approximately sixty to ninety minutes per an average book.

Broadly educated, Mr. Kennedy was a student of world economics, a fantastic ambassador, and a storied leader before his assasination at just 46 years of age in November of 1963.

During one of JFK's readings of Jules Verne's book "From the Earth to the Moon", I wonder if he didn't see that Mr. Verne's fictional rocket to the moon was every bit as recreatable as was "Nautilus." Nautilus was the U.S. Navy's second submarine with that name and in 1954 became the world's first nuclear-powered submarine. Nautilus came from another Jules Verne novel "20,000 Leagues Beneath the Sea".

Columbiad was the name of Jules Verne's ballistic cannon and moon-landing ordinance, named after a series of massive cannon from Verne's time which were originally named in tribute to a poem from 1780 called The Vision of Columbus in which "The thundering cannons rock the seas and skies".

Apollo 11's mission teams called their Command Module "Columbia" in tribute and later a Space Shuttle would also be called Columbia in 1981 as well. Their Lunar Lander was called "Eagle" in commemoration of America's national symbol and this explains why Neil Armstrong announced "The Eagle has landed" upon touching down on the moon's surface on July 20, 1969. Note the fig leaf in the mouth of the eagle on the Robbins Medallion commemorating that mission shown below. You see a similar fig leaf in the mouth of the dove on the cover of this book – the fig leaf commemorates Peace.

Like many of the most interesting engineering and technology projects throughout time, scientists and engineers were able to turn the science-fiction of submarines, and a moon landing, into a reality many years after Jules Verne's original concept.

I imagine that JFK might have felt like a kid again when this grand idea entered into his office. Was America to lose the Space Race after Russia had launched Sputnik (the world's first satellite) and then put a man in space in 1961 as well? America had launched Jules Verne's Nautilus, what a thing it would be if America could also create this next invention of his, travel to the Moon, and make that a reality too.

Recall that based on the 50 years of incredible technology innovation leading up to 1961, no-one doubted for one minute that flying a man to the moon and back again was not achievable. America, Russia, and

European engineers had evolved wooden-crate airplanes into supersonic, high-altitude jet bombers, and then built manned rockets. From simple motorized carriages, engineers had incredible mass-produced weapons, stylish automobiles, town-sized ships, submarines that ran on nuclear power, and televisions that broadcast from satellites in space.

As President of the United States in 1962, Mr. Kennedy was in the unique position to be able to call NASA directly. He, with then Vice-President Lynden Johnson, simply asked "can we fly a person to the moon." NASA replied, "It's impossible." JFK asked "Why Not." I can imagine being a fly on the wall and overhearing this incredible and historic conversation.

NASA explained that they didn't have a full explanation of "Why" at hand, but that they would put a credible list together. JFK asked for that list, NASA evaluated everything that was needed to accomplish the task, and then returned with a list of 13 projects. Some of the projects invented new materials; alloys that did not exist at the time. Some of the projects dealt with trajectory, rocket payloads, computer shortcomings, life sciences, on and on.

In late 1962, in the now historic Rice Stadium address before 35,000 people, John F. Kennedy announced to the world that ...

We set sail on this new sea because there is new knowledge to be gained, and new rights to be won, and they must be won and used for the progress of all people. For space science, like nuclear science and all technology, has no conscience of its own. Whether it will become a force for good or ill depends on man, and only if the United States occupies a position of pre-eminence can we help decide whether this new ocean will be a sea of peace or a new terrifying theater of war. I do not say that we should or will go unprotected against the hostile misuse of space any more than we go unprotected against the hostile use of land or sea, but I do say that space can be explored and mastered without feeding the fires of war, without repeating the mistakes that man has made in extending his writ around this globe of ours.

There is no strife, no prejudice, and no national conflict in outer space as yet. Its hazards are hostile to us all. Its conquest deserves the best of all mankind, and its opportunity for peaceful cooperation may never come again. But why, some say, the Moon? Why choose this as our goal? And they may well ask, why climb the highest mountain? Why, thirty-five years ago, fly the Atlantic? ...

We choose to go to the Moon! ... We choose to go to the Moon in this decade and do the other things, not because they are easy, but because they are hard; because that goal will serve to organize and measure the best of our energies and skills, because that challenge is one that we are willing to accept, one we are unwilling to postpone, and one we intend to win...

Kennedy whipped the local community of Houston into a power unto itself with his introduction.

"We meet at a college noted for knowledge, in a city noted for progress, in a state noted for strength. And we stand in need of all three."

For all of the rhetoric, this investment in the space race initiative was rooted in pragmatism. Kennedy justified a $5.4 billion budget as a national security priority – and the program succeeded in a successful lunar landing eight years later in 1969.

To be a person of action, and to be someone who gets things done, is to be someone who takes initiative, wrestles a problem to completion. But - make no mistake – the really important, complex solutions, need inspired goals, great leadership, strong process, great engineers, and the resources sufficient to pay for and provide for all things needed by all participants. Completing big projects is rarely a one-person endeavor, so coordination and communication are key too.

In this way, World Peace will be accomplished as well.

Are there proven processes and procedures that can be leveraged to develop solutions to a worldwide problem with this level of complexity; with this many stakeholders; and with several of the technologies missing as well? Yes; there are proven processes – and you will step through these best-practices as you read on in this book.

The steps are even simple.

In this Plan, I talk about "Steps" because the word infers action; to "Step" is to make forward progress. If you can imagine that some pain of consideration went into choosing the term "4 Steps" over competing other choices like "4 spokes" and "4 wheels", know that by "steps" I simply mean there are four areas that we need to begin to make progress in by beginning immediately. Any notion that Step One must complete before Step Two begins - is unintentional and it may well be that other languages (languages other than English) have a better word to replace "Step" with here.

Step One – A Good Life at Home

The starting point of any plan for World Peace is the construction of a sustainable Good Life for each of us - individually at home – similar to the Knock at the Door above. We could never hope to build sustainably for others without the time and resources afforded by having our lives in order.

Individuals, families and societies have needs of food, shelter, security, healthcare, love, friendship, education, family-friendly societies, and equality of opportunity – to name just a few needs. The challenges that work against providing these needs are - scarcities of resources, territory, food, love, security, greed, labor, time, energy, control, and selfishness – to name just a few obstacles.

So what are all of the needs of a Good Life?

When we are between twenty and twenty-five years old, many young adults are readying to marry and have children. If we have children at twenty, and our children have their children at age twenty as well, we have a very good chance to see our great and even great-great-grandchildren assuming we live on into our 90s.

In today's society, our hiring practices within major companies require that our young adults continue their educations until age twenty-two to twenty-five.

Young adults are also rarely able to manage the expense of all four burdens of home, marriage, children and higher education at the same time. Although there is nothing preventing pure democratic and socialistic countries from creating subsidies to permit this, at the time of this writing I could identify no countries that provided this type of support for a Good Life to its citizens.

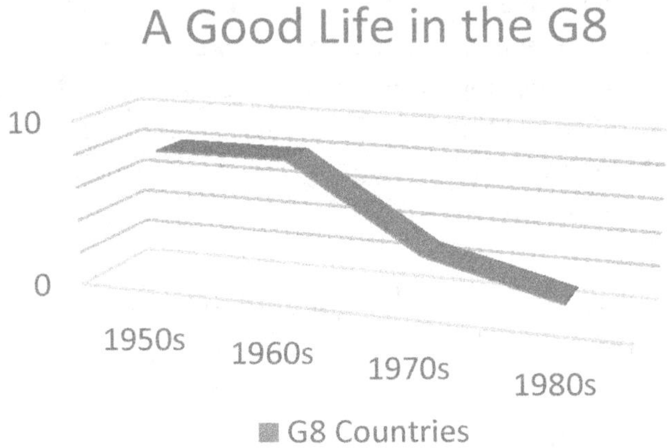

But it wasn't always this way. In many G8 countries, like in the USA of the 1950s up until the 1970s, a high school education was sufficient to find lifelong employment with a full pension, healthcare, benefits, and to buy a home free and clear almost 100% of the time. A young family could begin life at the age of twenty without amassing lifelong debt. So too could Russian families until 1986's Perestroika changes.

Our work and employment provides income, and if income gives us the things we need to provide for our families, then we all need to have access to either work or income, sufficient to give us all of the basic things we need.

If we are lucky or get high marks at school, and find something we love to do, work might also give us both a purpose in life, and a social utility, by building or maintaining meaningful projects. Social Utility, or Merit, is the useful product from our work. Our work might feed others, we might build schools, teach university students to build the future. We might run businesses, provide healthcare, build airplanes or driverless automobiles, and so on.

When we return home from work, we want to find a thriving, educated, well-supported community and secure families. Most but not all can have children and marriage is also important for building families. We want our children to be able to graduate degree studies and start families, homes, lives and careers of their own in the town where they grew up as well if that is their preference, and most of us want to leave this mortal life knowing that the society we built is sustainably improving as well.

So, let's summarize what constitutes a good life from the viewpoint of an individual at every stage of their lives. Rather than rely on Aristotle's definition that reflected the admittedly sexist conventions of his time, I will take another chapter from CSQ 101 and place us on our porch swing in our 100th year, looking back upon a full life lived well.

I suspect that for many, it might be very nice if we had the option to make the following choices.

A Good Life - with all basic needs met

Age	Female	Male
0	A Parent is at Home	"
1-2	Childcare (or Mom)	"
5	Grade School	"
12	High School	"
17	University / College	"
20	Home & Marriage	Home & Marriage
21	First Child (optional) Family Cottage (Dacha)	"
22	Masters Studies (Optional) Or Full-time Mom Or Work	Masters Studies (Optional) Or Work/Take Income
23	Second Child	"
23		Work
24	Masters Studies (optional)	
26	Work (optional) Or Full-time Mom	
40	Work or Grandchild care	
41	Grand Children arrive	"
55	Mini-retirements	Mini-retirements
62	Great Grandchildren arrive	"
75+	Retirement	Retirement
85	Great-Great-Grandkids	"
100	Porch Swings & Community	"

I mentioned before that CSQ defines your society to be your country; this is important for reasons of simplicity as complexities of political and legal differences become too great as we span countries. There is enough complexity in this first step as it is.

Today, young people in Brunei, Qatar, Saudi Arabia, Sweden, the Netherlands, Denmark, Norway and others are able to take a home, marry, continue university studies, and start a family at age twenty within a Parliamentary-Monarchy and Democratic-Socialist System of Government.

The Good Life and American Dream are goals that are important to both the Russians and Americans and so it seems unlikely that they will be comfortable sitting in this particular back seat position for long.

In 1985 Moscow, most twenty-year-old women would have been married with her first child on the way, and this was true for a woman of my mother's generation in Canada in 1963 as well.

Team One is assigned to the six social projects needed to provide support for Step One. See Chapter 6 – Social Projects, for the full list of projects.

Step Two – Technology Projects

There are twelve technology projects running worldwide which are needed to build a sustainable World Peace via an automated production economy.

Projects in Robotics, 3D Printing, Automation, Cold Fusion, Energy-to-Matter conversion, rapid charge batteries - are just a few examples. Hundreds of communicating teams are encouraged to support sub-projects that have been assigned individually and initially to each country on earth.

I am jumping ahead a little, but later in this book I will suggest that an important tool for sustaining World Peace in future is a piece of science fiction called a "self-powered replicator." To complete that technology project, two "hard" technology projects, as Mr. Kennedy called them, and a handful of medium and low complexity technology and social projects are needed. At the same time, other important social and technology projects must continue in parallel until nineteen projects complete over the next three to twenty years.

Are these projects already started? Yes - all of them are started, and world leaders in technology, health sciences, and government are running them as well. Germany announced this past weekend a successful pilot of its $1 billion Nuclear Fusion Plant – endless clean energy. This was a 2023 deliverable running on a production scale in 2015.

In keeping, the status of all Technology Projects is Bright Green and even ahead of schedule.

Technology solutions might come from the least expected places – from businesses, private engineers (tinkerers) and academic institutions - and these sources should all be encouraged - and funded.

In many cases, funding the development of World Peace technology will amount to a small fraction of money given in tax breaks to manufacturers who threaten to pull jobs to other countries.

For this reason, governments should focus on technology leaders directly for the far bigger bang for their buck. Most technology financing costs will be surprisingly small after dispensing with traditional technology incubator agencies. Supercollider time might be the one, and most important, expensive exception.

See Chapter 7 for a full list of technology projects and mitigations.

Step Three – The Transition

As the work to automate our production economies continues, jobs will transition to new income safety-nets at a rate governed by GDP Export Targets, unemployment, and retraining budgets. Without these safety nets, our society risks having our most senior engineers unemployed due to their pension risk and then turned into manual laborers. Nothing undermines a country's wealth faster than imports higher than exports and unproductive hi-tech experts.

Social Safety Nets will enable our best and brightest engineers to be able to build a technology safety net that makes money irrelevant and a Good Life sustainable for all.

At the point where all trading partners have automated their production economies, the transition from monetary trading can begin. At this point, money is of negligible value due to the automation of our production, or due to energy-to-matter replication, so we might also choose to change things at home too; going to work differently or buying homes differently. There is no longer need of going to a store; recreation centers will meet our needs for social interaction; and we will want kids to go to school together and build strong, caring communities.

None of this is fiction; none of these automation technologies are beyond our reach over just the next ten to twenty years. We can ignore this disruptive automation, and it will happen painfully, or we can plan for the change and bypass the painful transition parts altogether – professionally and intelligently.

How will jobs lost to automation transition to other forms of income; how will rent and mortgage payments transition? When do taxes stop? When do accountants have no need to tally financial reports nor trade in financial products any longer? These are just a few of the well-understood systems that need to transition. Managing the change are Team One and Team Two working together in a well-communicated plan to transition every member of society in a safe, peaceful and efficient manner as well.

Step Four – Rollout World Peace Worldwide

As countries around the world begin working on their Step One - Social Projects, and as Step Two - Technology Projects continue to advance, new engineering teams come online to help the automation efforts, sponsored by Step Three Transition Safety Nets and new programs. World Peace Agendas country-by-country will ensure that budgets and status are well tracked and communicated by Team Two Project Managers.

Team Three administers Step Four, the World Wide Coordination, Rollout, as well as the consolidated reporting of all automation efforts running worldwide. World Peace is a Bottom-Up Plan that needs coordinating with Top-down efforts to provide Global Goals at the UN presently. Team #3 administers the World Peace countdown clock and other communications functions as required to optimize the successful and responsible rollout of World Peace.

Within developing nations that do not initially have the resources and wealth needed to afford technology development projects, the UN's Top-down Global Goals continue as they have since 1945, to help meet their basic human needs. With-in a short time, the World

Peace Transition's automated economy will begin to take on the heavy lifting and then make a Good Life sustainable for all.

An automated civilization with infrastructure and production economy that rolls out a Good Life to seven billion people is the sustainable underpinning of World Peace. The TED-Talk in Chapter 9 provides more summary graphic information, and this book contains considerable detail and explanations as needed to satisfy both the cynic and enthusiast that this is Aristotle's "Right Plan" for World Peace as well.

Offending Content and Topics

As we advance into the discussion of situations within example nations, you might get the impression that I am highlighting certain values, countries, or policies, and that might make it look like I am making unfair judgments - this is not intentional. Choosing wordings perfectly is impossible and so I, and editors set out to write this book in a way that conveys an important solution in the simplest, most fully explained terms that we can.

For the sake of readers the world over, I will plainly state that I never want to focus on a country's values nor politics; not American values, nor Russian values, nor United Nations policies, nor any other state policies specifically.

States found to align closely with the plan for World Peace will naturally tend to get a little more discussion – the Netherlands and Norway are examples. Countries with extremes: extreme democracy, extreme socialism, extreme communism, extreme monarchy, etc. – these are countries where clear conclusions can be drawn about that system due to the absence of other influences like socialist health care or land grant policies - for example.

Discussions here will also tend to originate where the most credible statistics can be found easily – like charts pulled from the US Federal Reserve databases during a time of "trickle-down" or other notable economic policy. These datasets serve as a starting point, and then as best can, I try to extrapolate to the G7, G8, G20 and then global discussions.

I frequently make mention of the G20 countries. These are countries joined for reasons of trade and governance. The largest are the G8 – now referred to as G7 since Russia was turned out in 2012. The G7 include the United States, the United Kingdom, Canada, Japan, France, Germany, and Italy. G20 countries include the G8 plus Turkey, Mexico, Saudi Arabia, South Korea, India, Brazil, Indonesia, South Africa, Argentina, Australia, and China; 19 countries in total at present.

Great care is taken to find credible sources for data – but CSQ, DashFlows nor the Author maintain these datasets and for this reason, all data sources are referenced for your follow up and your feedback is invited in our forums at CSQ1.org. I always intend to make accurate representations in this book and, therefore, welcome the chance to correct any data misrepresented here.

Let's get started...

CSQ Certifications & Social Responsibility

CSR – Corporate Social Responsibility is important. For the first time in 2015, corporations are being called into The Hague on trial for Ecocide and Crimes against Humanity, Climate Change controls and changes in Fossil Fuel transition commitments are signed by every country, and Lending Institutions are now requiring due-diligence before CSR considerations in financing.

Protect your organization and investors from socially irresponsible decisions with CSQ Compliance and Certifications. At CSQ you can look to a stalwart and incorruptibly transparent source of guidance within the CSQ Governance and Compliance review and training team offerings that work to ensure focus of your management team is safeguarded by distraction from distraction. You do not have to figure out all of this planning on your own.

We want to remain an engineering organization and we do not want to become a sales organization, so take the initiative to be proactive and get your organization registered at CSQ1.org.

Whether your group wishes to sponsor an important CSQ Social or Technology Project, or you need to confirm how well thought-through are the goals set out by your committees in Strategic Planning, Ethical Investments and Funding Portfolios, or make-or-break Projects, there is a CSQ Compliance program available for your needs.

Benefits include:

- Meet CSR / BSR Corporate Social Responsibility Mandates
- Peace of Mind for Voters & Policy or Law Makers
- Higher staff job satisfaction, productivity, and attendance.
- Higher revenues and sales volumes
- Differentiation among your competitors as a CSQ Certified Socially Responsible Corporate Leadership Team.
- Mature Status Reporting and Project or Process Template Libraries
- CSQ Ethical Business Practices Logo for your website and marketing
- Assurance of Compliance for Ethical Investors
- Strategic Business Planning and Investment

Look for the CSQ Logos "Certified" and "Monitored" on your next project or investment to confirm that social, financial, procedural, and ongoing monitoring for measures of success and performance in Government, Non-Profit, Public and Private Licensed Corporations. **If it's not on our website, the product or organization is not certified.**

Visit CSQ1.org, csq1.org/wpprojects, or #wpprojects to get started.

For Certification and Training Programs, or to fast-track monitoring of your performance targets, visit DashFlows.com or CSQ worldwide offices at info@dashflows.com.

1 (800) 700-1001 in North America
33-184 88 04 60 - London, England
44-20 35 14 03 40 – Paris, France

Our WP Magazine is at csq1.org/mag
Flipboard Magazine App users can search "World Peace" to find us.
#wpprojects – is our Hashtag at Google and Twitter
csq1.org/wpprojects - our Google+ Community
csq1.org/csq-100-year-plan and TEDTalk Slideshows

Check our Online Store for World Peace and CSQ Course Products, Games, and Teaching Products for your office or classroom.

WP-TV Documentary and News Programs

Ask your local broadcaster, NetFlix, Apple and other TV Series providers to make WP-TV Documentary and TV News Series available in your area. WP-TV Certifies and Licenses just one CSQ Compliant Production Group nationally, so be watchful of imitators as this is one incorruptible message that we will all want to get right.

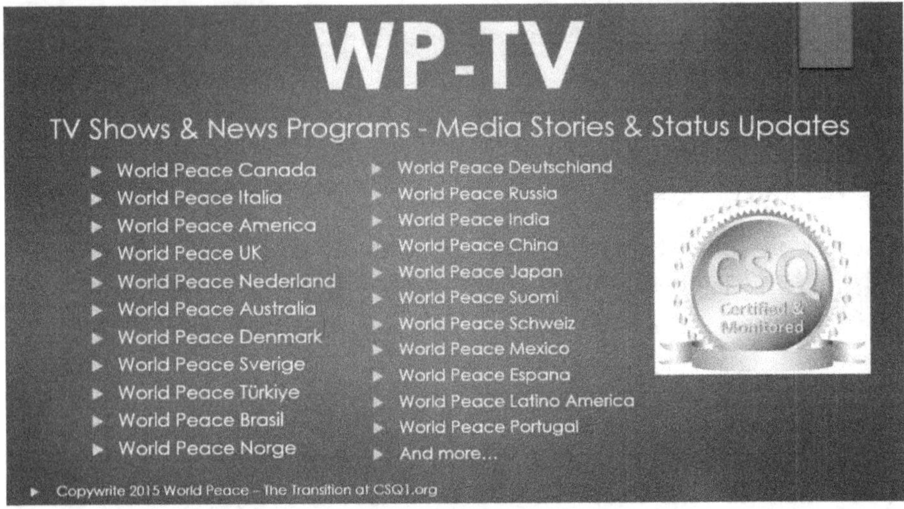

WP-TV tracks the progress of each of our 19 projects on the World Peace Agendas of each country around the world – in Technology, in Society, at the United Nations, and right at home too.

Science, Politics, Economics, War, Technology, The CSQ 100 Year Plan, Human Rights and Development, and of course – our Progress and improvements.

Put World Peace right in your home and community, with WP-TV.

Chapter 2- World Peace

The CSQ Research definition for World Peace is a world in which the human rights of food, clothing, shelter, education and family security are met for everyone, sustainably; where people have the equal opportunity to pursue worthwhile projects and enjoy personal liberty and happiness as well.

Worldwide, a Good Life for everyone seems a tall order, but it is also a very real possibility given technology advances already developed or planned over the next twenty years. World Peace is accomplishable today, where it was not reasonable before because today's technology sits perched on the edge of a breakthrough as big as our change from hunter-gatherers to civilizations ten millennia ago.

10,000 years ago, humanity changed a million years of behavior and moved from hunter-gathers to a community model that we now call civilizations. As hunter-gatherers, we did it all; humans provided for all needs of security, food, clothing, shelter, raising children and building a good life for our families. Life was hard, healthcare primitive, and lifespans of twenty-five to forty years were the norm.

In a civilization, life was much easier because humans could specialize. Some individuals could be farmers; others could be miners, others builders, fishermen, millers, bakers, police officers, on and on.

As our civilizations grew, the natural resources of the land – resources such as farmland, wild game and domesticated livestock, fuel for fires and lumber for shelters, often constrained how large the community could grow and resource availability often determined how easy or hard were our lives.

When resources waned, raiding parties and even wars erupted between neighboring militias or kingdoms. We formed militias when needed for security.

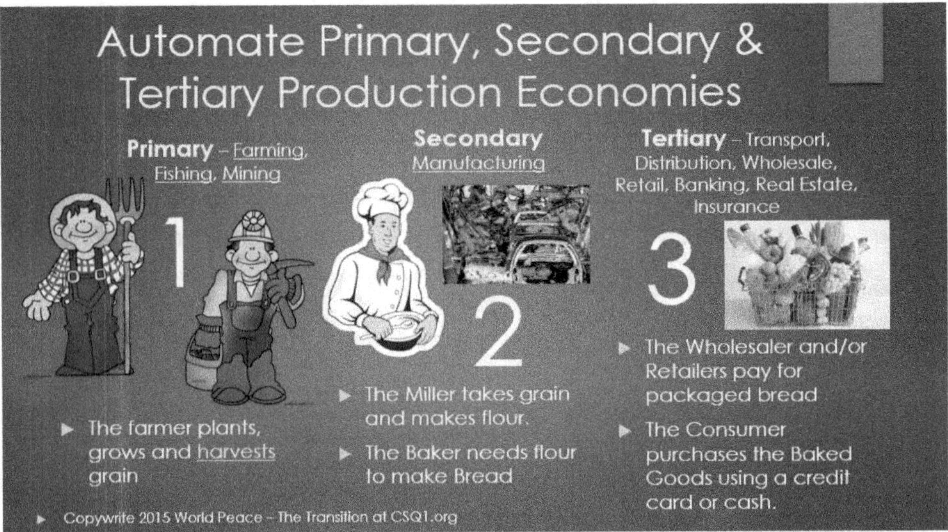

In our next twenty years, however, mankind will develop its automation technology to a point where the manual production economy that has driven our civilization for ten millennia may no longer be required.

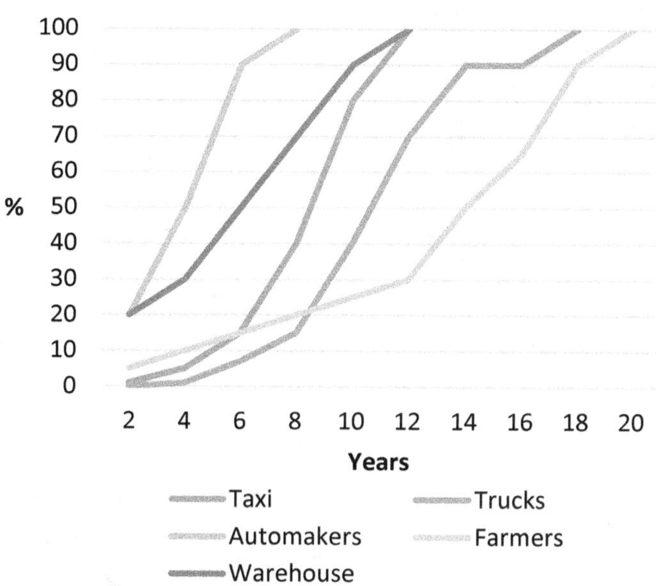

Today, however, a working class is needed as our society could not function without crop pickers, repairers, cleaners, and construction crews. Trucks must make their way to connect goods to markets, and manufacturing; farmers, fishermen, and other actors with roles in our primary, secondary and tertiary production economies must continue.

World Peace through Economic Controls

This World Peace was available to most of the G8 back in the 1950s when there was sufficient wealth to afford all of our basic needs for the first time. Wealth was also distributed well enough in society to sustain an ongoing K-Wave Spring.

What's a "K-Wave Spring"? I will consolidate a large chapter of CSQ Common Sense 101 by summarizing that a well-proven but seldom taught economic theory, called Kondratieff Waves, occur every sixty years in Capitalist Societies (Quigley, 2012). These cycles have been tracked back to 900 AD in eighteen documented sixty-year turns – and recorded in Babylon, Mesopotamia and Egypt 3500 years ago as well (Toussaint, 2012).

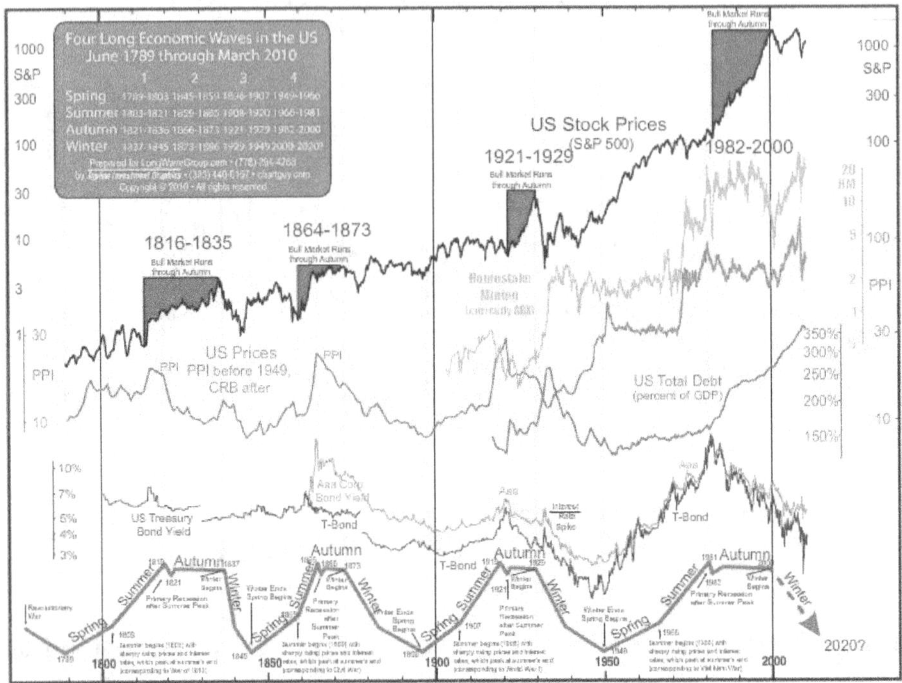

Shown above as a Red line, you've seen a K-Wave Cycle play out with every Monopoly game that you have ever played. In the beginning, the "Spring" - our most recent K-Wave Spring began in 1950, all members of a civilization have enough money and there are properties available to buy; everyone is happy and lives a good life easily.

In K-Wave Summer – 1970 to 1980, some families begin to pull ahead – through luck or strategy; in real-life K-Wave Summers, interest rates rise quickly - creating cash for investors.

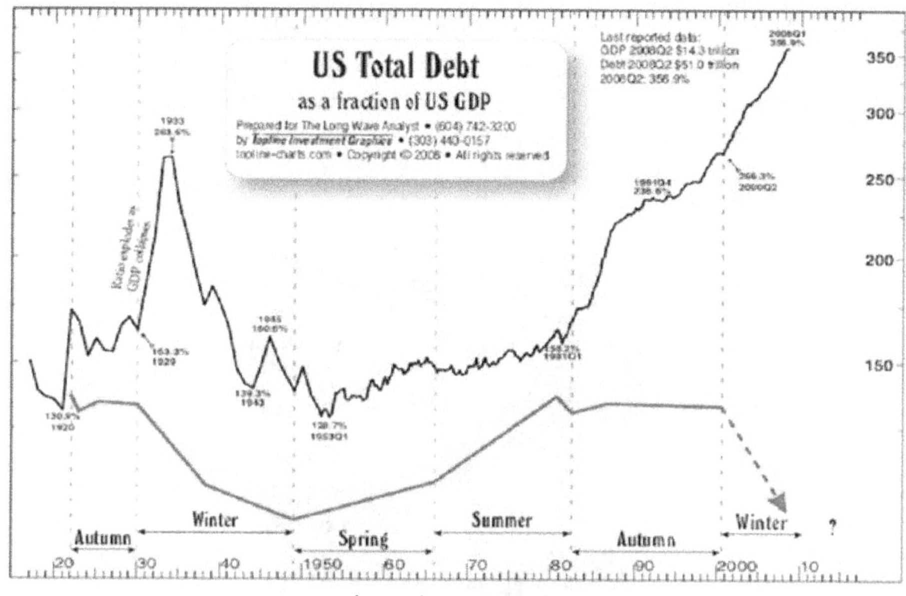

(Gordon, 2009)

By Autumn - 1985-2005, there are no more properties, so home prices begin to soar. The best properties, and businesses, begin generating incomes 7/24 for owners while other individuals are barely affording rents and cost of living month-to-month; Interest rates fall steadily, and lending hits record levels. Autumn K-Waves end in bankruptcies, two incomes are needed, high divorce rates, salaries no longer afford a good life, a few billionaires/hotel owners are winning – and all others are losing as the gap between rich and poor widens sharply.

We call K-Wave Winters “Great Depressions” and they end in a Wealth Distribution that breaks the “air of doom”. 2008 was our most recent chance to redistribute wealth peacefully in our current K-Wave Winter, but the Banking and Automobile debt-forgiveness programs did not go far enough.

In antiquity, in civilizations like Egypt, Mesopotamia, and Babylon 4000 years ago, they lived in a Capitalist society then as we do today. The key difference between now and then being that the system of government was an Absolute Monarchy; they had an Emperor or King.

According to the written law of Mesopotamia from 1750 BC, The Code of Hammurabi - King of Babylonia (a stone copy of this law is in the Louvre in Paris), Emperors imposed country-wide debt forgiveness in Jubilee years either every fifty years; or every seven years, depending on whether debtors could become slaves of the lender.

(Bartz, 2005) Code of Hammurabi - from the Louvre in Paris.

The Torah - The Law - a bible written in approximately 400 BC, insists on debt forgiveness every seven years.

These laws and transcripts developed during a 2000 year period of Capitalism that predated Christ's time. After many repetitions of boom-and-bust occurred again and again, the government of that time developed pro-active Economic Controls to avert the doom and revolutions of K-Wave Winters through Wealth Distribution.

Capitalist Society, in fact, is only thought of as a sustainable system, by K-Wave Economic Theorists, once we describe it as a system that exhausts itself every sixty years – and then restarts a new cycle.

In 2008, Queen Elizabeth II spoke for her nation when she famously asked "Why did no one notice (that an economic disaster was looming)?" during her speech at the London School of Economics.

Despite living 400 years apart, Cambridge scholars from Eton counseled both Elizabeth II and Queen Elizabeth I in the importance of a Monarch's service to the greater good, and also of the correct role of money in a capitalist society. Eton College today, as it was in 1422, is a College based on Aristotle's technologia of physics, metaphysics, and philisophical principles shared by Socrates, Plato, Aristotle and other great thinkers of ancient Greece and Rome - called philosophers.

Aristotle himself called money "sterile - as It cannot enhance in value, because once money enhances in value, it ruins its role as a stable measure of mere exchange, if you will, value."

The Puritan's and Queen Elizabeth I's Protestant Ethic differed from Catholic teachings in that it acknowledged the impact of the previous millennia of human dark ages created by the Christian pursuit of poverty, so instead they monetized their society. From Elizabeth I's father King Henry XIII's time forward, Protestants did not consider poverty a sin nor goal, but debt - specifically, was a concern repeated twelve times in their King James Bible. Shakespeare too references debt in The Merchant of Venice and Hamlet's "Neither a lender nor a borrower be" – both composed at the turn of the 1600s.

Nothing builds a community of voters more terrified to change their oppression like debt. There is a very real genius to it.

"Usury," the immoral burdening of debt forced upon those unable to repay the debt (Zarlenga, 2010), appeared twice in the King James Bible for the first time as well, and many important thinkers of this era wrote about the dangers of both credit and excessive wealth.

If we consider the strong probability that during Elizabeth II's private studies in Constitutional History at the hands of an Eton-Aristotle tutor's advice that "inequity triggers a revolution," both Elizabeths would have very likely been instructed in the cyclic behaviors of capitalist societies throughout history.

K-Wave Economics confirmed capitalist cycles empirically in 1925, but that teaching originated in Russia and wasn't widely available elsewhere in Europe until the 1940s when Joseph Schumpeter named these Longwaves "K-Waves" in Kondratieff's honor at Harvard University.

Was Queen Elizabeth II also aware of the economic control mitigations enforced by past Monarchs of capitalist societies, like King Hammurabi? Hammurabi's father's fathers handed down a code of law that demanded debt forgiveness every seven years (for indentured servitude) and every fifty years, in Jubilee years, for everything else countrywide. I mentioned above that we enacted similar, albeit partial, debt forgiveness in 2008.

Although Queen Elizabeth II has fulfilled a consulting obligation to her twelve Prime Ministers admirably, as a Constitutional Monarch it is not her place to tell administrators how to do things nor how to run the country. She would, however, be more than a little unimpressed to see history repeat when proper management could have permitted capitalism to support a Good Life throughout her kingdom instead.

Our last great depression began in 1929 and ended with World War II, which distributed wealth internationally. Sixty years earlier, the Panic of 1873 was ended by mass new wealth brought by immigration. The California Gold Rush ended 1835's 10-year depression by, multiplying America's gold reserves ten times. War and revolutions ended 1787's Panic as were most other 60 year depressions researched back to 900 AD.

Once a country reaches a GDP (Gross Domestic Product) of approximately $100 billion dollars annually, the importance of Wealth Creation shifts from the number one priority - to a distant second priority. The new priority becomes ongoing Wealth Distribution.

Perfect Wealth Distribution was believed fairly impossible in most capitalist states as there simply wasn't enough wealth to go around sufficient for everyone's needs. In antiquity, 3500 years ago, wealth distribution was needed when money lenders assisted farmers to plant. If when crops failed, money lenders enacted their legal right to take debtors and their families as indentured slaves.

Social Architects like Carl Marx, Henri de Saint-Simon, Charles Fourier, and Robert Owen theorized the development of Utopian societies, but in the end currency was too important to productivity and international trade. First, money was needed to incent individuals to take on important, difficult and menial tasks such as crop picking, garbage collection, mining and so on. Second, as seen in the Russia versus China Socialist and Communist models, it was very important to monetize your productivity so as to be able to afford to trade with other nations.

The reality that capital and labor was not in sufficient supply to afford everyone living in an equal state left everyone to fend for themselves in a Capitalist environment essentially. Businesses employed people, provided healthcare and pensions, and in this way – wealth trickled down through society and was distributed.

Extreme wealth distribution inequity created social problems, militias, and eventually revolutions - until the ruling classes began to ensure that this trickle down redistribution of wealth happened reliably through regular legally mandated "Jubilee."

In the 21st century, both the trickle-downs and the availability of national wealth, of sufficient quantity, changed.

Years of wealth generation and accumulation continued year-over-year until, in the United States, the top 20% of wealthy came to accumulate as much wealth as that of all G20 countries combined (excluding China and Japan).

National Wealth Distribution Policy

The consistent and long-term profitability of Oil allowed some middle-eastern countries to sustain their citizens entirely. Saudi Arabia is an example of a wealthy country that nationalized its biggest companies, and then distributed earnings and wealth to every citizen sufficient to meet all of their basic needs. Saudis are not a welfare nation; rather all citizens share in the proceeds of their profitable oil business. Citizens then go to work and add employment salaries and business proceeds to their home incomes.

The distribution of wealth in Saudi Arabia is class-based, with some 5000 Monarch "Princes" receiving income from the state far in excess of commoners' incomes, but there are sufficient incomes given to all other citizens in society as well.

Economic Control Roadblocks

In ancient times, Emperors implemented Economic Controls according to a rule book. Within the U.S., government economic controls that distribute wealth are almost prevented entirely by a system of Pure Democratic Capitalism unique in all the world.

Pure democracy represents the needs of 51% exclusively, until the needs of the 49% are represented almost not-at-all. It is in this way that the larger wealthy class are protected from income distribution.

Americans are lovely people; a half dozen of my English Puritan ancestors were on the Mayflower in 1620 and an Upper-Canada Loyalist family member made his way from this settlement to become a founding father of the Dominion of Canada as Premier of New Brunswick in 1867 as well.

The original Americans were largely principled and honorable families that provided for the greater good. That seed started to separate from the tree only after generations of good living until, in the richest country in the world, 40% of Americans control just 0.3% of its wealth today.

The pervasive viewpoint in America was that no-one was going to die from their situation; warm spaces abounded; food and jobs could be found. Instead, the poor formed informal militias for their own security and have longevities twelve years shorter than their richer countrymen.

William Penn's Quakers believed in pure democracy but were hung by the British in early days, in part because they gave the poor no vote nor quarter, and also because they had no interest in the greater good of others once their majority had voted.

Even as recently as fifty years ago the problems of the poor, were seen as an expensive burden on the rich. It was an easy concern to ignore as well as no leader could be elected that did not share this viewpoint after all. As the entitled children of the affluent majority of society started to make the decisions, ignoring the poor became socially acceptable and even an important part of the American value system.

America's poor today emulate the strength of the principled people of the 1600s, with values similar to those that built the country. Poor Americans, have no healthcare, pensions, dental care, and live with inferior education systems. They accept that this is their lot in life and are proud to be Americans as well. They join the military; they work harder; keep trying – because that is the honorable thing. Opportunity, however, is not equal as it was for their forefathers in the early days of America.

Historically, and universally in every culture, inequity creates social problems. Richard Wilkinson and his team from the University of Nottingham, plot thirteen indicators of social problems versus wealth distribution at his website (Wilkinson, 2011).

Here is a sample of just one of Richard's analytic summaries that identify a data correlation for Wealth Distribution and Social Problems.

Security shortfalls within a community create militias, and the U.S. supports both the world's largest military and highest incarceration rates in keeping (707 per 100,000 seconds only to Seychelles, Africa); a rate almost five times higher than the next closest G7 nation, the UK.

Country	Incarceration Rate per 100,000	Longevity	Government System
United States	707	75	Pure Democracy Presidential Elections
UK	148	79	Parliamentary Monarchy Public Healthcare Democratic Elections Prime Minister by MP vote
Canada	118	81	Parliamentary Monarchy Public Healthcare Democratic Elections Prime Minister by MP vote
France	103	81	Socialist Parliamentary Republic Public Healthcare Presidential Elections
Italy	100	81	Parliamentary Monarchy Public Healthcare Democratic Elections Prime Minister by MP vote
Germany	78	80	Socialist Parliamentary Republic Public Healthcare Presidential Elections
Japan	49	82	Constitutional Monarchy

US Longevity Stats are provided by Barry Bosworth of the Wall Street Journal and reported by the Atlantic. (Thompson, 2014).

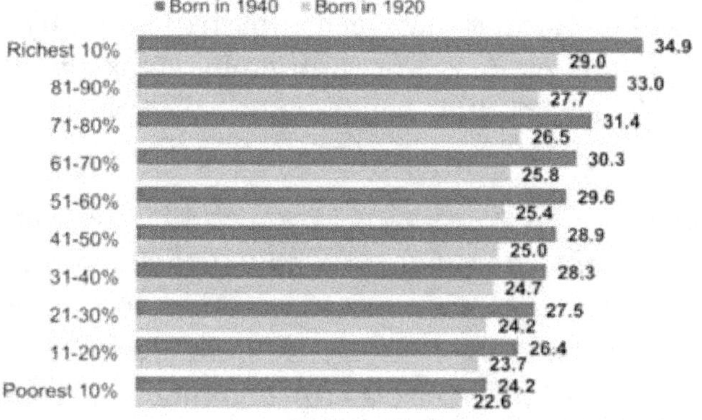

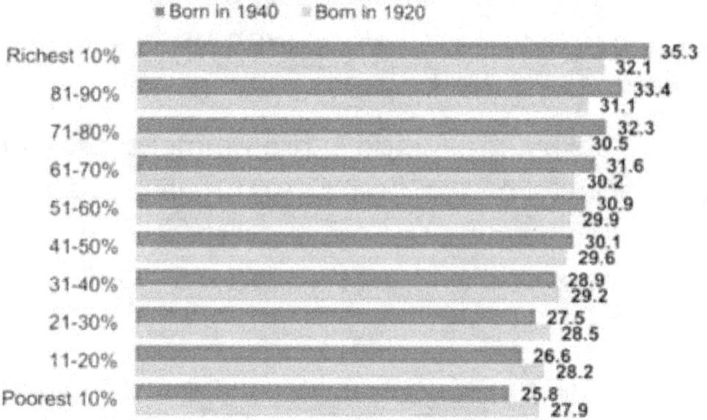

Sustaining a K-Wave Spring

I mentioned above that 1980 was the G7's chance to set in place economic controls to sustain a state of K-Wave Spring. In Spring, major businesses could have been more heavily taxed and assets redistributed to salaries and hiring should also have been reported at year-end to the government.

Housing costs too needed capping at a ratio consistent with salary growth. Instead, housing costs doubled four times and salaries dropped in the same thirty year period.

The Netherlands maintain a GDP per capita equal to exactly that of Canada and Russia despite having just 1000th of the natural resources. The Netherlands accomplished this by requiring a business to be owned in-country; by making offshoring of engineering illegal; and by exporting manufactured goods only as well.

Dutch Engineers and Consultants are world famous, and their high-tech industries are booming. Note how diversified the Netherlands are in the following chart – and notice too how strong are manufactured exports. By comparison, Russia, and Canada to a lesser extent, are greatly exposed to commodity resource pricing fluctuations.

A country generates wealth based on the currency that other countries will pay to it for the products built from its bare resources - and only Japan and Korea do it with as few resources, and as profitably, as does the Netherlands.

China, of course, is the world's pre-eminent example of export wealth generation. For the first time, China surpassed the United States this year to become the world's largest manufacturer – and it was already the world's largest exporter – with over 40% more exports than the U.S.

Naturally, creating more wealth for a country makes tremendous good sense, but statistically wealth doesn't improve the lives of citizens significantly unless you are living within a developing nation. Well before a country reaches G20 status, wealth distribution is the far more important priority that improves quality of life, and that is not to say that both cannot continue side-by-side concurrently.

John McCain made this error in 2010 stating that Obama must prefer wealth generation and not wealth distribution because these are mutually exclusive pursuits. In fact, a one or the other approach is incorrect, you must pursue both simultaneously. Of the two, Wealth Distribution has the far greater impact on a Good Life for citizens. An exclusive Wealth Creation agenda is even irresponsible because of the risk of militias and trough wars (McCain & GONZALEZ, 2008).

McCain lost the 2008 Presidential election by an Electoral Vote of 365 to 173 with a popular vote of 69 million to 60 million; 53% to 46%. Where was this misunderstanding of basic economic management most lacking? (Many, 2015)

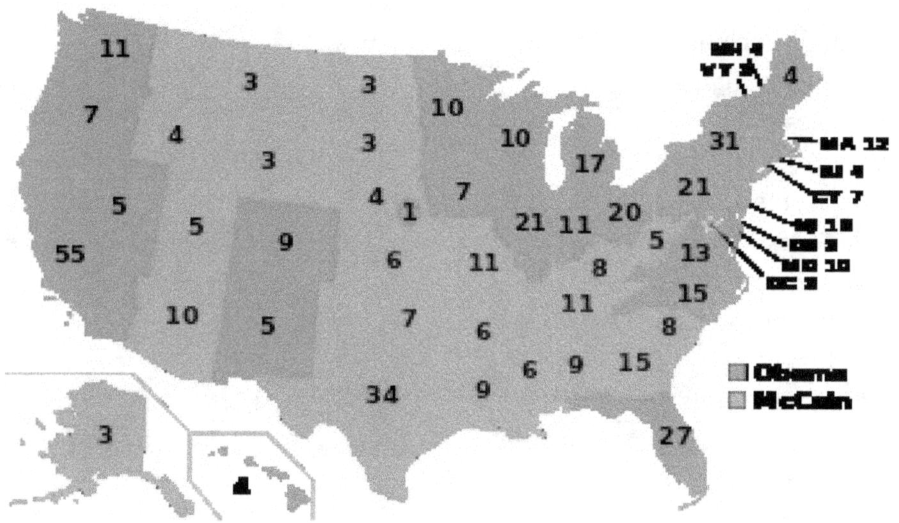

GDP Exports	Netherlands	Canada	Russia
Oil, Natural Gas, etc.	117.2	128.9	288
Machines, Engines, Pumps	93.9	32.6	8.9
Electronic Equipment	74.7	13.6	
Pharmaceuticals	30.2		
Plastics	27.4	13.2	
Medical, Technical Equipment	26.4		
Organic/InOrganic Chemicals	25.1		5
Vehicles	21.9	59.7	
Iron & Steel	15.5		20.2
Live Trees & Plants	11.1		
Agriculture & Fish		27.9	
Other Transportation Equipment		17.3	
Consumer Goods		52.2	
Forestry		12.7	7.6
Metals & Minerals Products		53.9	
Gems, Precious Stones		21.5	11.6
Aluminum		8.9	6.3
Cereals		8.8	7
Aircraft, Spacecraft		12.4	
Fertilizers			8.9
Copper			4.9

In the 1980s the U.S. and the U.K. - under various administrations, most notably President Reagan and Prime Minister Thatcher, enacted Trickle-down Economics - which simply stopped working in 1997 as illustrated here in Federal Reserve Statistics - showing a growing division between corporate profits and household incomes.

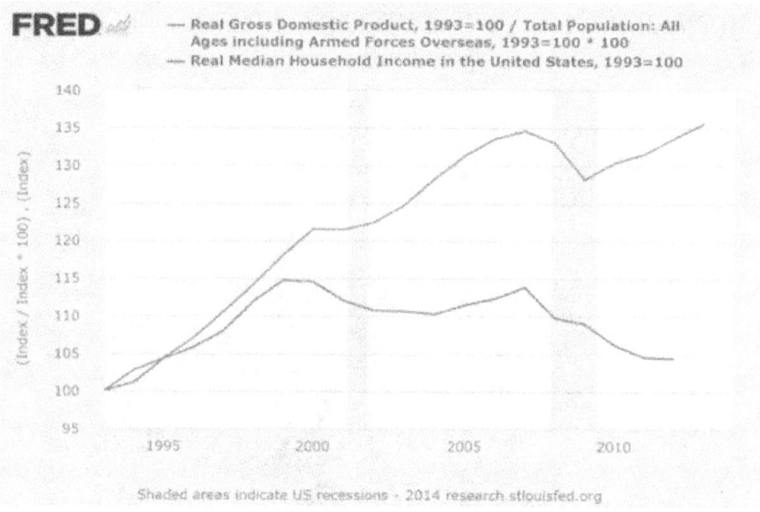

As a result of having no economic controls, and no monitoring, hoarding by big business began and wealth distribution for citizens did not become a priority. Targets for wealth distribution - as this following pie chart might suggest, were never even suggested. Remember that the rich are earning 7/24 and compounding interest annually year over year – and, by "Rich", realize that I am referring to individuals who make $3 million per day for the past 30 years. Most of the rest of us have just enough wealth to keep us voting "conservative".

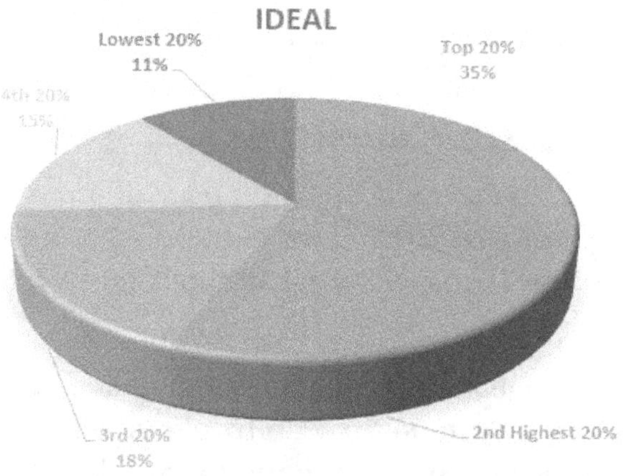

As a result, the actual wealth distribution in the USA came to look as the following pie chart according to Federal Reserve statistics in

2010. In 2010, the top 20% - 80,000 Americans, owned 85% of the wealth.

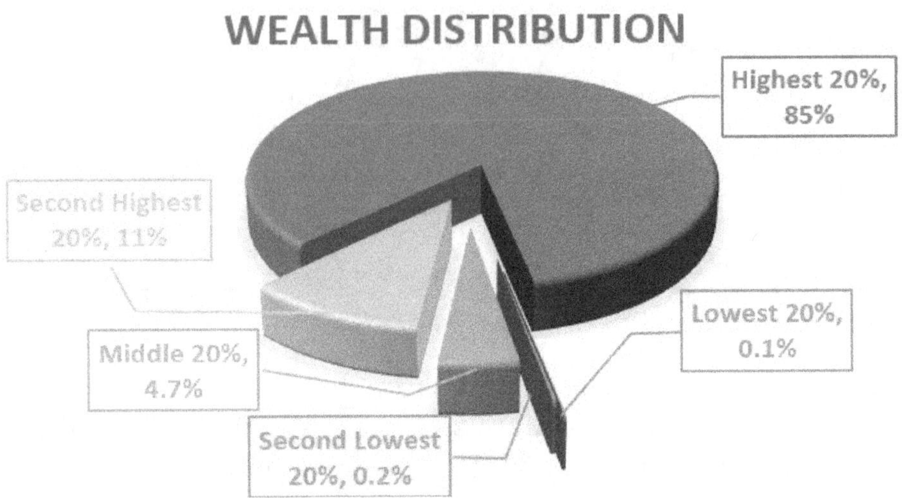

And yes, as discussed in K-Wave Theory explanations above, this is all quite normal as we are well into our latest K-Wave Winter cycle. A Great Depression was delayed only when 2008 wealth redistribution did not go far enough.

The reality today is that the rich could easily write a check that would distribute wealth, and in all likelihood, a new and ***final*** sixty-year cycle of prosperity would begin for all.

The Old War-Distraction Reuse

The economic picture running up to World War II was very similar to the economic picture of 2015. If we extrapolated the stock markets of the 1930s to the 2010s, we are just about three years away from a World War.

Given our nuclear weapons technology capabilities today, the time-honored practice of running a war to either distribute wealth or distract us from our wealth inequity problems, could be an extinction-level event due to the nuclear winter that would almost certainly result.

World War III is avoidable, just by getting to the work of Wealth Distribution. Make this a condition of winning your vote in the next election, and then set to work on the following plan to ensure we never see another great depression again.

Status of World Peace Economic Controls

The Status of Economic Control Projects is Yellow presently – which is a heads-up to say that if these redistributions do not happen voluntarily, and if government is unwilling or unable to take a chapter from history and distribute wealth proactively in law, history has proven that what comes next is doom, despair, depression deep and dark – and then, either:

1) Money and wealth will appear from some other source - such as immigration or the discovery of new wealth such as gold or oil.
2) Revolution will force a distribution of wealth.
3) A World War redistributes Wealth. The last one was a conventional war that claimed 60 million lives; 200,000 by just two nuclear bombs at Hiroshima and Nagasaki.

   Or, for the first time.

4) A technology breakthrough frees us from our dependency on money and wealth - once and for all.

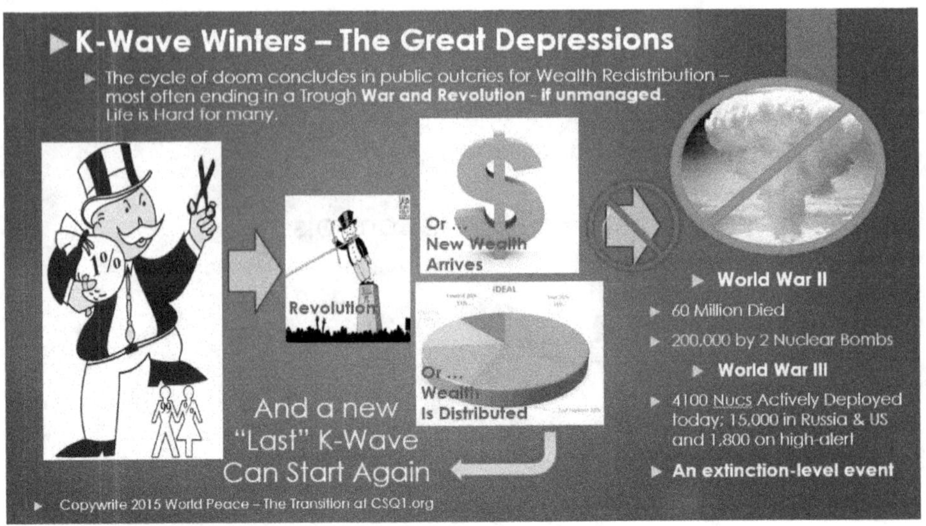

Slide taken from the World Peace TEDTalk in Chapter 9.

A New Option

What if, for the first time in history, we were just a handful of years away from this fourth option above? It has already started in fact. Our production economies are automating. Aristotle observed that if our tools could somehow work themselves; if houses, or cars, could build themselves, "that chief workmen would not want servants nor masters slaves " and this year we see for the first time at Audi, world-class automobiles are built without autoworkers for the most part.

Most of us will not find it easy to imagine the world without labor jobs nor money – but you really should begin thinking about it as a very desirable and inevitable eventuality. Rather than throwing ourselves into despair, throw yourself now into worthwhile projects instead.

World Peace through Technology

Robotic technology will very shortly exist to automate almost all repetitive labors, and our presently manual production economy will continue to automate rapidly too; in the same way that Audi, Tesla, and BMW builds cars on the production floor from beginning to end with almost very few human workers today (Tesla, 2013).

In the same way that Google's driverless cars just hit a million-mile milestone, America's largest employer – the trucking industry, will also automate. There will be robots that pick and harvest crops, mine and smelt sheet metals, and manufacture most of the other finished goods we need.

Too expensive you say; it can never happen because of the cost! It will happen, it is already happening - and it is a good thing too. Consider that costs become progressively lower, and, therefore, less relevant, as our technology advances to the point where a 3D Printer or robot, can print or assemble clothing, blankets, cars, tools, houses, and a table-ready loaf of bread, meats, and other nutritional foods as well.

We'll ask our machines to produce table-ready bread and produce, and those goods will appear – eventually, without a single human being required in their production.

Technology makes money irrelevant in a second way as well.

Our technology will continue progressing until one day, an engineering team will transform energy into hydrogen, then helium, then carbon - and then eventually - Gold. This is an important milestone because Gold is the basis of all of our capitalist monetary systems. Gold was originally created from energy billions of years ago at the time of the big-bang, or in geologic transformations that took place mechanically thereafter – and so the science of producing Gold from energy appears perfectly well proven.

Diamonds too, are being fabricated, with fewer flaws than natural stone, cost effectively today in a similar example of the eroding underpinnings of traditional measures of wealth, as created by technology.

At the moment that we can plug a nuclear reactor into a supercollider and produce unlimited amounts of gold, or produce things we need in an automated way, humanity can record the start of the Transition from our reliance on wealth and money. Scientists at Sandia National Laboratories turned Hydrogen into metal this year and so we could be closer to this science than we think in fact. (Giles, 2015)

Remember that Gold never really had any value to begin with after all; it was a rock – and as such was less important to ancient man than a potato too because he could not eat it. Gold was simply chosen to anchor currency valuations internationally as the universal measure of wealth because it was perceived to be rare and have trading value. Men would go to war for gold – or phrased another way; leaders would raid neighboring countries for their wealth.

Status of World Peace through Technology

If you think that all or any of this is fiction, think again. Most everything discussed above is building and shipping already. Only the cold-fusion and matter-to-energy transformation technology is still in a lab and working through concept and pilot phases.

To my knowledge, the only technology at risk of not arriving within twenty years are the more complex energy-to-matter conversions needed to fabricate the building ingredients, the "goo and glue" needed by 3D Printers and Robotic Assemblers to construct all of the goods that we need.

Those super-collider tests mentioned above are currently scheduled to begin in 2016 by the Physics teams at the Imperial College of London. The mitigation for the risk that complexity will delay these energy-to-matter technologies is to create an automated production economy robotically.

Household and industrial robots, GPS-guided trucks, tractors, planes, and ships that drive themselves will complete repetitive, manual labors with a human at the controls only when safety demands the fail-safe.

As Cold Fusion technology advances, we should have no need of power grids; robotically automated mining and farming will meet our needs for raw materials until we perfect the light-to-matter technology equivalent.

We can generate food too, in theory, from energy within twenty or thirty years, but it is more likely that automated pickers, sorters, and transportation to your door would deliver organic, grown goods instead.

As the work to convert energy to matter matures, we should be able to convert Cold-Fusion-based clean energy into matter sufficient to provide whatever building materials, or cars, or houses that we need.

A wide array of assumptions that govern our municipal planning today will change as well. For example, we have no need of living in close proximity to our place of work at that point. We can spread out and alleviate congestion in major urban centers. See Chapter 8 – The Transition for details.

At the time of this writing, the status of all technology projects required to build this bright future are "bright green" and even ahead of the schedule as laid out in the CSQ 100 Year Plan (Tilley, 2015).

Timeline for World Peace through Technology

The projects needed to build an automated production economy and Good Life sustainably through Technology will require between three and twenty years to develop and implement throughout the G20 and in Pilot Nations too. The initial Pilot countries will test operational readiness and validate that processes can be supported fully and work well.

CSQ 100 YEAR PLAN

PLANNING STRATEGIC LONG TERM GOALS
WHAT WILL BE THE LEGACY OF OUR GENERATION?

DISCUSS THIS PLAN AT CSQ1.ORG/MAG

50 years out

A+
Human Cloning

A+
Reverse Replicators – which convert matter to energy and back again.

A
Consciousness Transfer Technology

75 years

A+
Anti-Gravity

A+
Warp Drive
Negative Ene
Technolog

B
Impulse Engines

A
2055
The Singularity - Computers Programming

A+
2055
End of the Monetary System - in G7 No taxes

Long-Term - 40 Years out

Social Problems Decrease

CSQ
Compliance & Certification

A+
2015
4, 3 and 2 Day work-weeks equal to automation profits
Government

A+
2015
Re-shore to Create Wealth
Government

A+
2035
Start 'The Transition' from Money to Social Utility
Government

A+
2035
In-home Self-Powered Replicators for Food, Gold, Clothing
AutoDesk, StrataSys

A+
2015
Re-shore Engineering
Government

A+
2015
Close Markets to Socially Unproductive Companies
Government

A+
2015
Robotics
Caterpillar, Siemens, ABB, Boston

Wealth Distribution 2015

Bottom 40%

Top 20%

CSQ RESEARCH: WORLD PEACE - THE TRANSITION

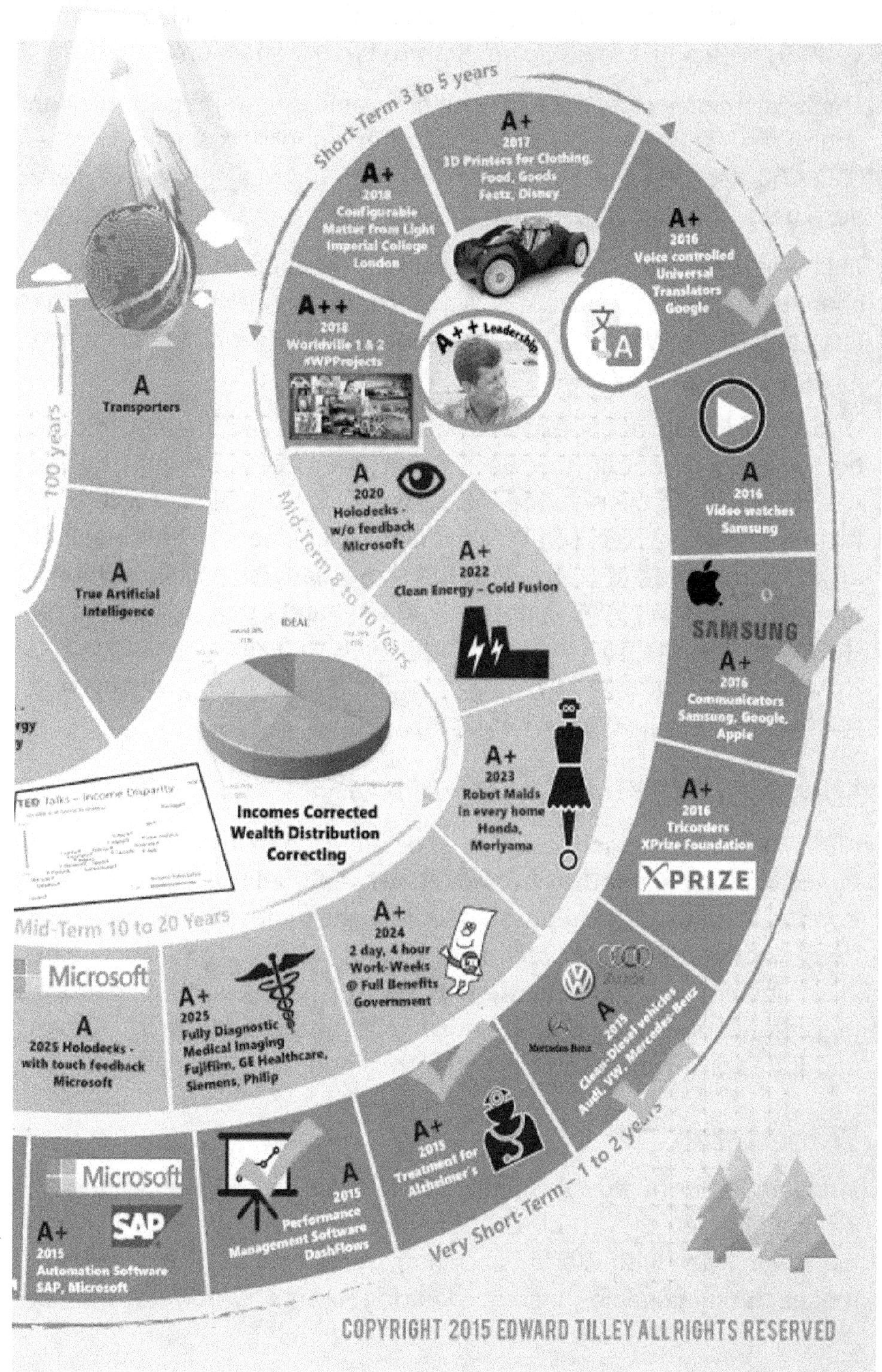
Short-Term 3 to 5 years
A+
2017
3D Printers for Clothing, Food, Goods
Feetz, Disney
A+
2018
Configurable Matter from Light
Imperial College London
A+
2016
Voice controlled Universal Translators
Google
A++
2018
Worldville 1 & 2
#WPProjects
A++ Leadership
A
Transporters
100 years
A
2020
Holodecks - w/o feedback
Microsoft
A
2016
Video watches
Samsung
Mid-Term 8 to 10 Years
A+
2022
Clean Energy – Cold Fusion
A
True Artificial Intelligence
SAMSUNG
A+
2016
Communicators
Samsung, Google, Apple
IDEAL
A+
2023
Robot Maids in every home
Honda, Moriyama
TED Talks – Income Disparity
Incomes Corrected
Wealth Distribution Correcting
A+
2016
Tricorders
XPrize Foundation
XPRIZE
Mid-Term 10 to 20 Years
A+
2024
2 day, 4 hour Work-Weeks
@ Full Benefits
Government
Microsoft
A
2025 Holodecks - with touch feedback
Microsoft
A+
2025
Fully Diagnostic Medical Imaging
Fujifilm, GE Healthcare, Siemens, Philip
A
2015
Clean-Diesel vehicles
Audi, VW, Mercedes-Benz
Mercedes-Benz
A+
2015
Treatment for Alzheimer's
Microsoft
SAP
A+
2015
Automation Software
SAP, Microsoft
A
2015
Performance Management Software
DashFlows
Very Short-Term – 1 to 2 years
COPYRIGHT 2015 EDWARD TILLEY ALL RIGHTS RESERVED

Finally, a full rollout can happen worldwide to seven million people perhaps within just the following ten to twenty years after that.

These Technology steps are laid out alongside long-term priorities for healthcare, space exploration and other important science in the CSQ 100 Year Plan at http://csq1.org/csq-100-year-plan, and seen here below in Chapter 7 – Technology Projects.

This book lays out the detailed planning needed to roll out these changes so that this current K-Wave Winter might very well be the last Capitalist K-Wave cycle that our society will ever have to manage its way through as well.

One would reasonably expect that everything can change for the better for everyone in the world where our needs are met by the tools available in every home. Along the way, acute voices will call it too expensive or too complex, and then engineers will deliver it anyway. Together, in teams with the same leadership skills of John F Kennedy and the NASA scientists and engineers that put a man on the moon; leaders with the technology skills to take complex projects through to predictably successful, stress-free, sustainable conclusions, will just make it happen.

Sustainable Economic World Peace

Again, advances in technology are not required to implement World Peace today. Wealth Distribution all by itself is all that is required to do that. In many G20 nations this could begin almost tomorrow – and then we could build technology that keeps this Good Life "sustainable" and ongoing. Remember, most G8 countries had the ingredients of a Good Life for Individuals until capitalism's K-Waves continued to advance through its normal course.

Three Teams

Sustainable World Peace will take three teams working side-by-side. The technology deliverables will mature over twenty years, but you can begin living with World Peace long before the CSQ 100 Year Plan makes it sustainable, by coordinating teams with two sets of coordinated task lists - and moving forward together.

Team One - Business leads, wealthy residents, and leaders in Government and Economic Policy. You - must sustain our current systems and move forward wealth distribution and wealth creation economic policies with full priority.

Read a few pages of the original US Pilgrims' trials and then be one of those men of integrity; leaders who stepped chest deep into frozen Massachusetts waters while securing the boats of expeditions while building a secure perimeter for a settlement. There is no need to give it all away; rather simply work pragmatically toward the distributed pie chart above.

You are repairing the damage done, and sending a message to the young today that you are going to play a major part in working toward building a start in life that is as easy as the one that you began.

The Team One leadership to-do list:

- Build a Transition Safety Net immediately and fund technology project workers and resources as needed to sustain the work of the technology projects per the Transition Plan in Chapter 8.
- Contact every CEO in every major business in the country and get the new hire paychecks rolling again.
- The minimum wage must quickly reach a number high enough to eliminate homelessness, and for the bottom 20% of earners to take 11% of wealth annually (or some other KPI target).
- Hire quickly; hire fifty-plus year old workers; hire two for every one job; have SMEs hire and scrap HR and MBA (non-SME) hiring manager process layers.
- Put an end to offshore hiring altogether – bring workers onshore once local resources are really exhausted and only if their values are a good match and they can marry within your own society.
- Get your KPIs in order; your business and society have productivity measures to meet.

- The next five years is not the time for record profits; suspend executive bonuses and send those dollars to payroll; cap multi-million dollar salaries and redeploy these incomes to ten or twenty others until employment reaches 100% as you automate your production lines and processes making your company more efficient and profitable.
- Work commutes must be no longer than fifty or sixty minutes which means housing costs must come down or salaries must go up as needed.
- How do you know that you are working with or investing in businesses that have signed up to make this future work as well? Look for their CSQ Certifications.
- Procurement departments need a complete leadership change they touch technology contracts. The popular notion emerging this past five years that "You don't have to technical to lead technology teams" is a very incorrect assertion made by business leads given too much control over engineers. Look to a detail explanation in Chapter 7 – Technology Projects.

Team Two - Technology Engineers and Scientists. You are the leaders of a sustainable World Peace; not business leads, not sales leads, not career politicians. Your contribution will make a difference. The difficult work of continuing to build and advance your family and society rest on your shoulders with government's full support and priority.

On your to-do list are:

- Interesting projects, projects, and more projects. To use an analogy, you are going to be eating an elephant-sized cow all by yourself - and that requires pacing, breaking things down, varying your designs for digesting the next bit, and then moving programmatically to the next challenge with excellent performance and reliability.
- Your List of Projects begins in Chapter 7.

- Be a person of parts and tackle the voice recognition until it works well.
- Next, tackle the robotic picker's 100 hands until they are autonomous, agile, affordable and reliable enough to check that project off the list – and then move on.
- Work out every day, turn it off at six PM, but keep a notepad by your side for ideas, and then spend weekends and evenings with loved ones.
- Ask a woman or a man out on a date and build a family;
- Learn languages
- Take travel breaks.
- Be good at what you do, finish what you start, meet or exceed performance timelines and communicate.

Have fun and get it done!

Team Three - The United Nations: monitor the status of a World Peace Agenda from every country according to the instructions set out in Chapter 10 – The World Peace Agenda.

By mentioning "sustainable" above, I meant to say that the technology projects delivered by our engineers will easily permit this model for World Peace to be rolled out to every country; even into countries where wealth is not sufficient today, and this will permit their ongoing needs to be met without the capitalist system's ups and downs.

Voting and Growing Together

Forty-five years of unplanned and undiagnosed social decline serve as evidence enough that conventional finance and business leads need to lead a change for a Good Life through Wealth Distribution Projects now.

Technologists, Engineers and Scientists have a mountain of Transition Projects ahead. To use the analogy of Alan Turing, administrators will just slow things down. Engineers and Hi-tech Leads must direct the build of an automated production economy from Safety Net funding set aside specifically for automation programs sponsored by Government and Major Businesses.

Fortunately, our Democratic systems permit us to make changes through our votes. Readers might like to pass along the importance of voting for MPs, MPPs, Congressmen and Senators who are experienced engineering leads with the long term planning experience and vision to deliver a sustainably better world for our children over just the next ten years.

Chapter 3 – War

Avoiding war is a highest priority consideration at a time when our civilization literally rides a fence between a step in one direction that leads onto a never before attainable, sustainable Good Life for seven billion people - World Peace – or, in the more travelled second direction; to a possible oblivion via an extinction level Nuclear World War III.

The direction of war is one that mankind has traveled in many K-Wave winters; in many capitalist societies before this one. The message in this book is to Stand Down and bypass that war; there really is no good business case for it – but there is much to be gained by building a Good Life immediately instead.

Understanding how wars are discussed academically, help us navigate through the important elements needed to create wars – so that we can avoid them. This chapter, therefore, is dedicated to the strategies behind sidestepping wars. To understanding why they start, why they continue and then why, and how, they come to an end; so that we might be able to avoid better the middle part altogether in future.

As a member of society, there should never be broad support for aggressive moves against another democratic nation as this is a sign of intolerance and an acknowledged shortcoming in public non-secular education.

Everyone has the basic human right to exist, to education, to basics of life, healthcare, and I submit – to a Good Life as well. Without this, there are social problems, revolutions, and wars that benefit only a few individuals.

In any academic discussion and forensic examination of war, it is possible to upset and unintentionally hurt deeply survivors that lived through conflict first hand; even at the cost of the loss of loved ones.

I warn you now that there is going to be a discussion of everything here; racism, religion, failed leadership, victims, insanity and death tolls. I rely on your consideration and feedback in our forums at csq1.org to help build a better explanation than this beginning; until an explanation that reliably avoids future wars in perpetuity results.

Why do Wars begin?

From a paper on The Reasons for War, Matthew O. Jackson and Massimo Morelli from Stanford University (Jackson & Morelli, 2009).

There are two prerequisites for a war between (rational) actors. One is that the costs of war cannot be overwhelmingly high. By that, we mean that there must be some plausible situations in the eyes of the decision makers such that the anticipated gains from a war in terms of resources, power, glory, territory, and so forth exceed the expected costs of conflict, including expected damages to property and life. Thus, for war to occur with rational actors, at least one of the sides involved has to expect that the gains from the conflict will outweigh the costs incurred. Without this prerequisite, there can be lasting peace.

Second, there has to be a failure in bargaining, so that for some reason there is an inability to reach a mutually advantageous and enforceable agreement. The main tasks in understanding war between rational actors are thus to see why bargaining fails and what incentives or circumstances might lead countries to arm in ways such that the expected benefits from war outweigh the costs for at least one of the sides.

This second point highlights the importance of treaty management in maintaining peace between neighboring interests in the world where there are estimated to exist many more treaties than there are wars.

There were more than 66 known peace treaties signed before 1200 AD. The Treaty of Kadesh in the Temple of Amun a Temple erected in 3200 BC, in Karnak, Egypt is the earliest Peace Treaty to survive to date (Bryce & Klengel, 2015).

The late middle ages of Europe (1200 to 1800 AD) created an interwoven tapestry of complex and ever-changing peace treaties and trade agreements. These treaties evolved quickly as weapons technology and war tactics evolved from bayonetted muskets with Napoleonic tactics, to trench warfare, to motorized artillery, armored navies, automatic weapons, tanks and finally air and nuclear weapons.

A treaty even exists to prevent weapons from being deployed in space.

Non-Combatants

In 1933, Germany voted Adolf Hitler's National Socialist German Workers Party (NAZIs) into power with a popular vote of seventeen million of the country's sixty-five million population (26%).

Within just months, Hitler exercised a loophole in Germany's Constitution. This loophole allowed him to bully the passing of an "Enablement Act" which made his party the sole party of parliament. He next suspended the role of President when Charles Lindberg passed away in 1934, and this propelled Hitler into the role of Head of the Army and Parliament, a legal dictator of Germany. When the dominos finished falling, a Fascist Authoritarian Government resulted from a vote of just 26% of Germans.

Germany has long been one of the most populous and immigrant-reliant countries in Europe; just as they are today. Hitler's notion of socialism included a strong racial bias in deference to the meritorious class bias suggested by Marx's Socialism. Different from Marx also was Hitler's preference for land ownership, and the expansion of territories - which required a larger military presence.

In his early writings, Hitler stated both that "Germany's future has to lie toward the east at the expense of Russia", and he also expressed his loathing for democratic process as "those who reach power are inherent opportunists." Class by Race, land ownership, and German expansion appealed to the German Elites, which is why they supported Hitler's original election to Chancellor, Head of Parliament, in 1930.

It was for these reasons that Hitler rejected communism - he believed in the ownership of land; Communists did not, and communism's brand of socialism was quite parliamentarian, and Hitler loathed parliamentary and democratic process.

On one storefront closed by the NAZIs in 1938, a sign of public shaming was placed, "The Jewish owners of the five P.S. shops are the parasites and gravediggers of German trade! They pay German workers starvation wages! The principal owner is ..." and this sentiment was very appealing to financially oppressed depression-era German workers and citizens who were also suffering under the burdens of the Versailles Treaty's First World War debt.

North Americans were born of racial diversity after four hundred years in a "melting pot". Immigration practices encouraged intermarriage, and this proved a healthier society-building approach to immigrants too, who often sought freedom from racially motivated oppression that was common in Europe. For this reason, North Americans do not easily accept the differentiation of class by race as did European Elite ninety years ago.

The Jewish community had traditionally kept a strong business focus and had also kept the proceeds within their religion, race, and community as well. As this elevated their community financially, it also made them envied and an easy target for the racist NAZIs.

Messages which portrayed a rich Jewish community enjoying wealth while uncaring of their starving workers' needs, compelled Germans to support Hitler's police who soon took instructions to stop and reverse these inequities by an inferior race as an example to others of the penalty for not complying with a German socialist ideal.

Here in North America, a good-natured old Irish joke showcases the unimportance of race. "There are two kinds of people – those that are Canadian/ Polish/ American/ Israeli / anyone - and those that want to be."

Ironically, the Israeli notion of Chosen Ones echoes a similar racial bias, and this will seldom impress a North American. We look for these sentiments in someone who lacks a sense of humor regarding one's opinions or self-importance. The British call it "taking the pee out." The process is a good one as it tends to make us more interesting, considerate and thoughtful people as well generally.

By preferring immigrants with similar values, who also "marry into" the community of the host society, racial tensions can often be eliminated in a world that permits easy transport to anywhere we might like to travel.

A similar case where a lack of diversity drove civil war occurred in Ireland where historical countrymen drew lines around religion. Three hundred years of English Protestant occupation oppressed Irish liberty to the extreme point of mass "prisoner" population deportations (to Australia and eastern Canada), child slavery, guerilla warfare. Irish, who lived in Protestant areas, had little choice but to submit to a new similar religion; accused of sympathizing with tyranny and oppression these people became hated by other Irishmen until conflicts changed from an English versus Irish conflict, into a Protestant versus Catholic conflict.

Between these two derivatives of the same religion, there are very few differences, basically who is the head of the church and what is the role of poverty to be in society – and that's about it. This case study is extreme, so drawing valid conclusions is possible like, Religion and Racism are not a reasonable justification for formal "war". In this case, an "eye for an eye" and "measured response" with reparation for damages and loss, are a far more appropriate response to wide scale violence and attack.

In most examples of wars, the ratio of combatants and supporters of a war are often quite a low percentage of the public. Vikings, who participated in raids throughout Europe at the turn of the first millennium, represented just 3% of an otherwise peaceful agriculture-based civilization.

Support for militarization in Germany peaked at an all-time high of just 26% of the population in 1933, and yet that small amount of support was sufficient to nearly destroy the country and to kill 7.4 million Germans from an original 69 million people in total.

Death Rates in War

The highest death rate per capita in World War II was inflicted upon the Poles due to their early occupation (6 to 11 million casualties) and next was the Soviet Union with 24 million dead. Higher German death rates were averted by the suicide of Adolf Hitler on April 30th, 1945, which effectively called an abrupt end to the war in Europe on May 9th.

Another factor that may have saved German lives by the tens of millions was the positioning of Eisenhower's front-line American Troops at the time when Germany signed its surrender. In the last campaign of the war, the Soviet Army was positioned behind the relatively fresh U.S. and British front-line troops.

U.S. Troop positions had the effect of protecting Germans from USSR's army somewhat and, in the end, a relatively short period of the war was fought on German soil.

As it was, the victorious Soviet Army's pull back across Poland was one to the most dangerous periods of World War II for Poles; just as was the defeated Japanese Army's pull-back through China, for the Chinese, in August of 1945.

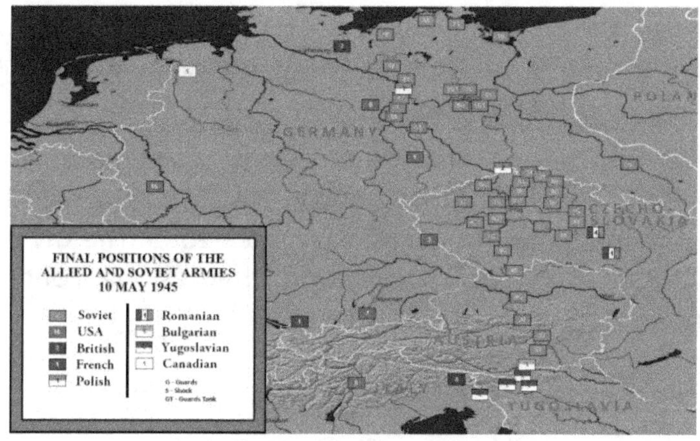

(Allied Army Positions)

Refusing to submit in the closing months of World War II, Japan's leadership endured the immensely destructive firebombing of its major cities until two nuclear bombs were deployed at Hiroshima and Nagasaki on August 6th and 9th - killing 200,000.

In any society with an active army at war, whether in Viking times of old; German Citizens in 1933 or at many other times in history, non-combatants represent the very great majority of persons affected by any war. Non-combatants are especially affected by wars that occur on their soil.

Americans lost 400,000 in WWII, composed almost entirely of servicemen stationed abroad. But 620,000 Americans died in the American Civil War of 1861 to 1865 on American soil. By adjusting these numbers to World War II population levels, a Civil War on American soil would have killed six million Americans, and the 1942-styled bombings of major cities in WWII would have killed at least as many civilians again as well.

In antiquity, the outcomes of battles were decided between champions, from time to time, rather than to destroy all of the men and armaments of two opposing kingdoms. Battles could be delayed for months, allowing Champions to arrive at the battlefield for just this purpose. When looked at in this way, one cannot help wonder would it not be ideal to take the armies and leadership of warring groups – off to a remote location to conduct the fighting of a war, or simply work through the issues that prevent a solution to any impasse. This approach would not put non-combatants and their families and homes in harm's way as well.

Protecting non-combatants in just this way, and through armed peacekeeping campaigns, has been the R2P (Responsibility to Protect) Agenda of the United Nations since 2005. The UN's key role in World Peace will be explained further as we continue this discussion in more detail in Chapter 14.

Democracies don't war with Democracies

With few exceptions, democratically elected leaders have seldom been able to garner sufficient support to launch and maintain a war against another democratically elected nation. 99% of Wars have begun between nations where at least one nation was led by a strong President, Monarch, Authoritative Regime, and Dictator or similar.

War of Needs

For centuries, war has been viewed by world leaders as a somewhat normal tool, in a toolkit of options available as required to maintain or improve their country's economic cycles, basic needs, power, and standing with neighbors. If a country needed living space, food or resources to build stronger arms, they looked to available supply nearby and then sought out those situations where the cost of the war could be lower than the value of the resources won.

These are referred to as Rational Wars, even though this term is also given to Religious Wars as well in most cases.

The rationale for wars is born of basic needs like food, farmland, clean water, gold, security, living space in cities, or some similar need of a better life.

The American Revolution

The 1776 War of American Revolution is thought to have been born of needs financial, but was also presented to the public as a concern for Taxation without Representation; and as a war of principal. It became, therefore, in the eyes of the colonists, a war over "values".

Consider that Britain had split from the Catholic Church in 1500 because King Henry VIII sought to protect his subjects from Inquisition, Catholic Church taxation, indentured poverty and the interference of the clergy in secular matters of state.

English Protestants focused on wealth creation for the next 100 years. Britain's caste class-based society was one in which nobles were lifted to positions of personal wealth and leisure - that hard-working families could never advance to, and this was a struggle to make work well within even England's mature social order.

After Queen Elizabeth had died and Mary Queen of Scots' son James assumed the throne, many British sought a return to Catholicism and these voices were greatly concerned by the growing new wealth and importance of the well-educated and growing Puritan Merchant Class. One such Puritan group fled to Protestant Holland from Bedfordshire, an hour's drive northwest of London, but then set out again to Massachusetts on the Dutch charter ship the Mayflower, in 1620, after they realized that their children would not be able to continue as Puritans within their new host country.

In the colonies of New Virginia, Pennsylvania, and others, survival depended on limitless labor. With abundant land and natural resources, the colonists began to realize that they had no need of a social pecking order nor taxes. Many had fled the oppression and taxes of distant Monarchs in Europe and were fearful.

In 1681, King Charles II handed over a large piece of his American land holdings (Pennsylvania) to William Penn to satisfy a debt the king owed to Penn's father. Penn was an early advocate of pure democracy and religious freedom, and his followers, called the "Quakers," taught that all men and women were created equal in the eyes of God.

Proud Englishmen saw the behavior as self-centered and not in keeping with the greater good of the realm based back in England. With abundant wealth, all English colonists enjoyed more social equality than did people in England as well. There were Gentry (the

Rich), Middle Class (75%), and even Indentured Servants were given fifty acres of land upon five years of service to whoever paid their ocean passage. The land was plentiful and easy to buy. By the 1700s, half the populations of the southern states were African Slaves working on rice and other commodity plantations.

During the ten years of French and Indian Wars ending in 1763, King George III had spent a great deal of money financing his army and the colonies, and so to pay the debt, he enacted taxes that the colonists had never before had to pay. King George, like other "blue blood" Royals of this time, was afflicted with Porphyria, which caused him to suffer temporary lapses in judgment and even intermittent insanity. The King's condition and learning disability resulted in cognitive impairments that meant that he was not a clever man - who learned to read only at the age of eleven.

In defiance of a tax, Colonists began a boycott of British goods that culminated in the Boston Tea Party. At the time, tea was a very expensive and King George welcomed a 100,000-pound-sterling act of impertinence enacted by a group that had just spent ten years defending as you might expect. At a time when budget problems drained English coffers elsewhere in Europe as well, this added frustration drove the King "half-mad" by historical accounts.

It was in this state of mind that King George enacted the "Intolerable Acts" of 1774; a series of four seemingly vengeful and tyrannical laws with the goal of bringing the colonists in back into submission. The most reasonable "Act" was the Boston Port Act, which shut down the Port of Boston until $1 million (in today's money) in tea was repaid. Fearful colonists rebranded the Administration of Justice Act, the "Murder Act" because a British governor could move any trial to Great Britain and so Colonists feared the worst; that whole crowds could even be whisked away and murdered.

The colonist's First Continental Congress created a "Declaration of Rights and Grievances" and awaited Britain's reply which culminated in Paul Revere's ride through Lexington and Concord calling "The British are coming" in April 1775. Lexington produced just eighty-eight American minutemen so the British simply invited them to go home, until an accidental discharge by an American soldier resulted in eight deaths. A few miles further, the second battle - of Concord showed colonists were well prepared and the battle and pursuit in retreat resulted in 300 British casualties to 100 American casualties.

The use of Guerilla warfare tactics in killing British soldiers and countrymen as they retreated after the battle were tactics considered ungentlemanly at the time.

This convinced King George to ignore the Continental Congress's follow-up "Olive Branch Petition", an assurance of loyalty to Great Britain, and forced the American Congress to draft its Declaration of Independence bearing America to a War of Independence. From 1775 for eight years, war waged until British Treaties with France, Spain, and the Dutch Republic (who had supported an American defense in exchange for new lands), was signed collectively as the Treaty of Paris in 1783.

Wars in Europe had continued in some form for 4000 years, and at this time leaders were well trained in the big picture causes, costs and benefits of war. Wars became very like a game of chess played by Monarchs, where smarter "moves" resulted in bigger financial gains. Nowhere was this truer than in the American Republics where no social values changed as a result of this war, rather only land and trade rights were supplied to sponsoring countries, the Netherlands, France and Spain. In the Treaty of Paris, England gave America very generous concessions in return for it's continuing to be a major trading partner with England. A French Ambassador remarked that "The English pay for peace; where others create it."

Unlike Revolutions in France and Russia which forged new societies, the American Revolution retained its current values and merely created the Independent Republic; the new autonomous state was very like the previous. England sought to uphold its duty and defend its colonists against lawless murderers, and Americans sought freedom from taxation without representation and social equality, both had causes. In the end, it can be said that only money and territory changed ownership.

Thirteen years later, in 1796, a deep depression came throughout the Americas and Western Europe as a result of the spending on wars and security in the colonies. King George died in 1820, quite mad after thirty years of being considered unfit for his position by his country and most of Europe's monarchy.

The Iraq War

September 11, 2001's devastating terrorist attack of the World Trade Center Twin Towers in New York City was reported initially to have been led or sponsored by Iraq's leader Saddam Hussein; later reports would prove this incorrect. The Iraqis agreed to an extensive number of disarmament steps before the initial U.S., British, Australian and Polish troops attacked in March of 2003.

Military step-downs resulted in the then fifth largest military in the world, and one of the largest air forces, being fairly incapable of mounting a defense when President George Bush Jr. authorized "Shock and Awe", a Blitzkrieg-like rapid air assault on the capital city. Iraq's Air Force was eliminated within the first night or two.

The US-led coalition forces occupied Iraq until 2011 with more than one million Iraq combatants and non-combatants casualties in all over the next eight years. External military withdrawals completed in 2011.

In 2014, 17,049 civilians - people with family values similar to those in the west, were murdered annually in Iraq. That's a murder rate 10 times that of the U.S. and 40 times Japan's murder rate. Iraqis live without the security of knowing if they, or their families, won't be killed in their homes on any given night and this lack of security has historically given rise to militias and revolution. This was the human toll.

The financial benefits realized in this war by Coalition countries are substantial owing to a direct control of Iraq's extensive oil fields. America's War on Terror turned next to Afghanistan in 2004 and to the Taliban sponsored terrorist group Al-Qaeda who were responsible for the terror attacks of 911.

War in the Congo

Stanley and Livingston's royally funded adventures to find the source of the Congo River in Africa reported a country rich in slaves and natural resources such as rubber and ivory. In 1884, Emperor Leopold of Belgium – a "truly evil man" by the admission of the King of Austria - marched troops into the region and murdered, by various reports, upwards of ten to thirteen million indigenous Congolese.

A common and brutal discipline of his Belgian Administrators was to cut off the hands of people who didn't obey the smallest instruction immediately, and then victims' children and wives were also disciplined the same.

To keep the population in balance, and hard at their assigned labors, Leopold sent Missionaries to the Congo with instructions in 1883, on how to interpret gospel so as to promise reward after this life. Throughout his new colony, Colonial Missionaries taught "religious studies" – and not Mathematics, not Language; not science nor agriculture, and they were instructed not to teach "Thou shalt not kill" because "they already know that". (Jaide, 2013)

Administrators were given full reign to control the people, and large production quotas to fill with no monitoring nor regard to the conditions of the people. Religious studies evolved in time to teach children how to defend themselves against the "infidels" or "heathens" on the other side of the country, and how to seek poverty, and it taught children of the Congo these fundamentals from a very early age. The indigenous students set about their learnings and followed instructions per this plan beginning almost immediately.

Back in Europe, Leopold's response to the criticism of polite society over his human rights violations was to rationalize that he was "bringing heathens to Christianity". Even the most balanced accounts of Leopold's legacy acknowledge his willingness to deploy religion to control both population and labor.

His approach to population control was both reprehensible and effective. It worked so well, in fact, that other colonizing nations followed suit; the Portuguese, Spanish, and others adopted similar strategies, and - in the Congo - 130 years of civil war has endured and controlled the indigenous population - through Religious teachings.

Modern examples of religious control are seen and felt today all over the world. An important value and lesson here, is that violence is never tolerable and war is seldom an appropriate remedy. In the discussions of level-setting and Human Rights to come, we generally want to treat problems like we treat projects in Chapter 17. We level-set with language that all understand the same, we gather facts sufficient to permit supportable next steps, and then we plan changes responsibly so that families, and even the right to a Good Life, is safeguarded.

Wars over Oil

Just as wars started for new territory in the 18th and 19th centuries, wars were started over Oil in the late 20th and very early 21st centuries. As oil prices and demand soared in this time, the value of this easily monetized resource was too great a lure for most powerful businesses and country leaders to ignore.

Oil prices are set high in times of plenty and sent low as a tool to stimulate target countries and businesses. The price of Oil has fluctuated between $40 and $135 per barrel between the years 2008 to 2015.

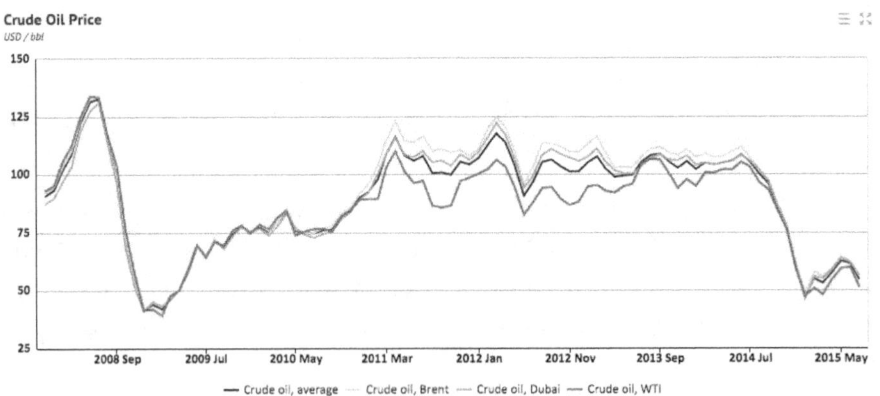

Good news heralds from our technology teams here again as we now realize that wars for oil could shortly be a thing of the past as well. Two technologies will probably lead the way initially – Blue Crude and Rapid Charge Car Batteries. Over time, I assume that cold fusion charging systems will keep vehicles running after that.

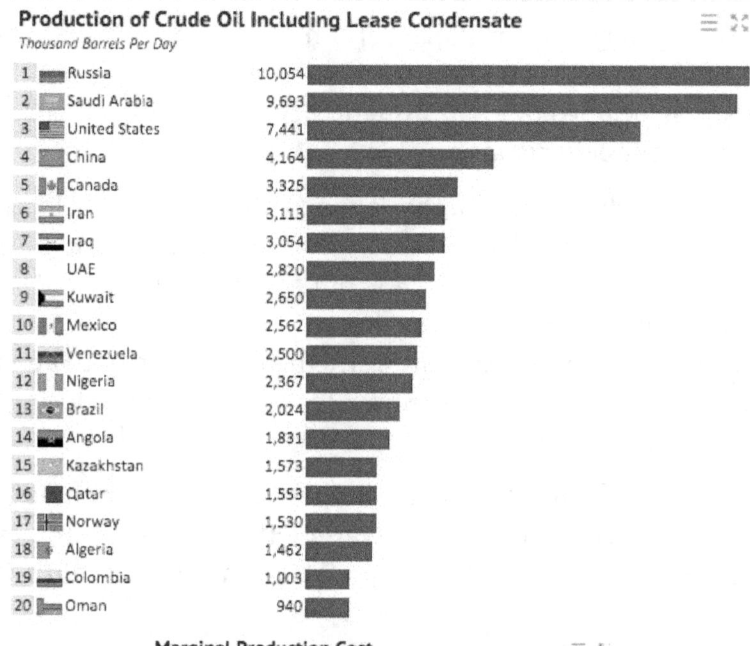

Marginal Production Cost

Marginal cost of producing one new barrel of oil (USD / bbl)

		Marginal Production Cost 2014
Russia	Arctic	120.00
	Onshore	18.00
Europe	Biodiesel	110.00
	Ethanol	103.00
Canada	Sand	90.00
Brazil	Ethanol	66.00
	Offshore	80.00
United States	Deep-water	57.00
	Shale	73.00
Angola	Offshore	40.00
Ecuador	Total	20.00
Venezuela	Total	20.00
Kazakhstan	Total	16.00
Nigeria	Deep-water	30.00
	Onshore	15.00
Oman	Total	15.00
Qatar	Total	15.00
Iran	Total	15.00
Algeria	Total	15.00
United Arab Emirates	Total	7.00
Iraq	Total	6.00
Saudi Arabia	Onshore	3.00

Replacing Oil

The reason that our cars run on oil-based fuels in the first place is that fossil fuel oil, like alcohols, contain hydrocarbons. Hydrocarbons are composed of, as one would expect, Hydrogen and Carbon.

Alcohols that have five or more carbon atoms per molecule are no longer soluble in water due to their hydrocarbon chain's dominance and this is an important environmental consideration because most alcohols are deadly poisons.

When a fuel spill occurs, Methanol and Ethanol dissolve in ground water and can create a dangerous environmental problem. Methanol is cheaper to produce than ethanol, but a half-teaspoon of Methanol can cause blindness in humans. Ethanol in 3% to 40% mixtures is the only alcohol that we can consume, and the difference between these two is just a few degrees of temperature difference during their distillation process.

Today, Audi and parent company Volkswagen, are manufacturing crude oil made from electrolyzed water (H^2O) and recovered carbon dioxide (CO^2) in a 70% efficient process that creates Blue Crude. (Gray, 2015)

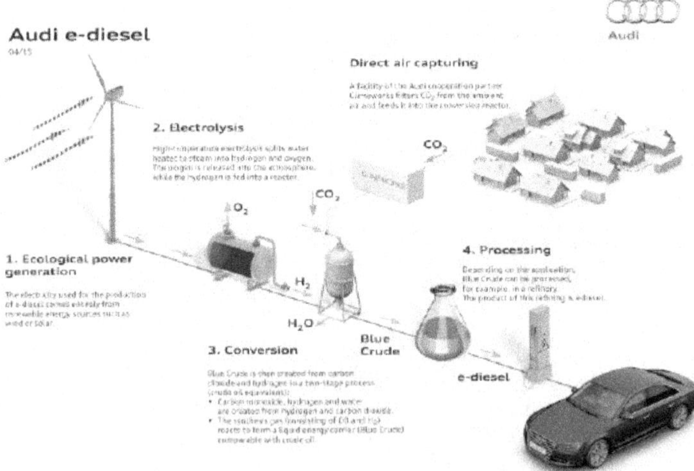

Audi's clean hydrocarbon crude oil uses the same Fischer-Tropsch process that has been around since the 1920s (Davis, 2015). Its zero-carbon-footprint process makes use of wind energy. Carbon Dioxide sources include simple, safe home recipes like baking soda and vinegar (Calkins, n.d.), Air Recapturing, Natural Gas and others. This Oil needs only refining further to create component diesel fuel, gasoline, kerosene, and aviation fuel as needed, just like . You could choose to refine gasoline from this crude, but fuel in diesel form takes less energy to refine, has better combustion for a more fun ride, and is almost twice as efficient at highway speeds. Audi's President runs this fuel in his A8 and reports that this fuel makes a car run quieter too.

The reasons to adopt Blue Crude are compelling. The combustion power of Audi's e-Diesel fuel is greater than that of fossil fuel diesel, it makes cars run quieter as well, and it is expected to be available to consumers for approximately $1 per liter at retail pumps – equal in cost, or less than, fossil fuel diesel. It can also be used in our current fuel distribution system.

The only waste by-product from manufacturing and burning this pure Blue Crude is oxygen – which might lead to bigger insects and smarter people after much time - but the problem of what to do about all that clean air is a problem for another day.

Assuming that clean fuel additives can be found to guard against solidification in cold temperatures, and other practical storage considerations, I imagine that it might even be possible to create a food-grade version of this product – although, why would you choose to drink it?

Cost is of little consequence when a combination nuclear reactor and nearby Blue Crude refinery could provide a limitless supply of clean burning mobile fuels in the same way that the USS Nimitz Aircraft Carriers desalinate water through electrolysis for a crew of 6,000 every day for the past forty years.

Clean Fuel Puzzle

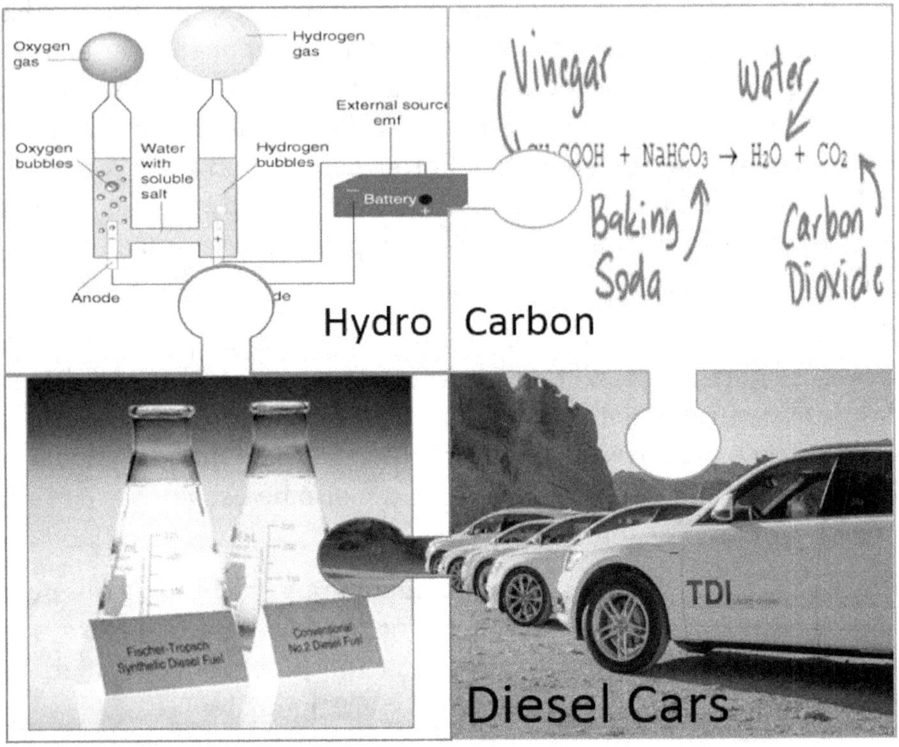

Oxygen gas
Hydrogen gas
External source emf
Oxygen bubbles
Water with soluble salt
Hydrogen bubbles
Battery
Anode

Hydro Carbon

Vinegar
Water
$CH_3COOH + NaHCO_3 \rightarrow H_2O + CO_2$
Baking Soda
Carbon Dioxide

Fischer-Tropsch Synthetic Diesel Fuel
Conventional No.2 Diesel Fuel
TDI

Diesel Cars

Russia's brilliant new floating nuclear reactor power plant ships are another option based on proven 50-year icebreaker design (Diggs, 2015) (OKBM, n.d.).

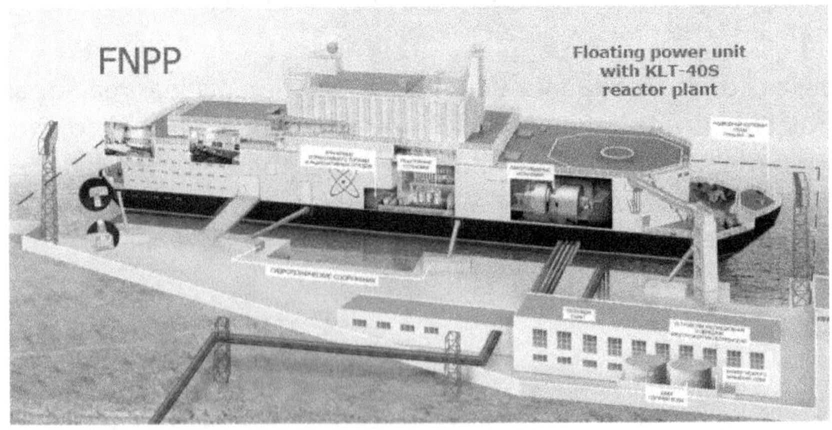

Germany's Fusion Reactor creates limitless clean energy. When it began operations in December 2015, this project was fully eight years ahead of the CSQ 100 Year Plan.

Volkswagen, Audi's parent company, is one of the few companies to bring an affordable diesel car to North America which is surprising considering Europe's predominant preference for more efficient Diesel vehicles. Diesel fuel, and diesel vehicles, are important because:

1) Diesel made from water and Carbon Dioxide, can be fabricated cost effectively in Zero-Carbon Footprint processes, and burns with Zero Emissions. The waste byproduct of oxygen makes humans smarter.
2) Gasoline-based cars prevent us from developing cleaner, cost-effective alternatives to fossil-based Crude Oil.
3) Diesel was originally developed to burn Paraffin Oil made from coal and natural gas. This hydrocarbon burns with 50% fewer emissions and Crude Oil was a dirtier alternative that Oil Companies made to work in Diesel engines.

4) Diesel takes less energy to refine, and its higher combustion efficiency takes the same car almost twice as far on the highway over gasoline.
5) Diesel does not dissolve in water making spills and environmental cleanups easier.
6) Diesel is a lubricant that ensures long engine life – where gasoline is "a corrosive" that wears out engines more quickly.
7) Diesel cars last longer, have fewer maintainable parts, lower running costs and set the high-bar for reliability, often continuing to operate for 500,000 to 1,000,000 kilometers and beyond during their service lives (double, and more, than gasoline engines).
8) Diesel fuel detonates through high compression only and is not flammable.
9) One can run a diesel passenger car for 10 hours without stopping as a tank of fuel will often sustain almost 1000 km of highway driving.
10) Replacing all cars with Diesel equivalents would reduce total energy needs by 15% overall (depending upon the country).

In the early autumn of 2015, Volkswagen was centered out for adjusting their in-car programming to turn off emission controls on their diesel passenger cars. EPA officials claimed that these changes resulted in diesel cars emitting nitrogen as high as ten to forty normal gasoline cars in operation.

As a CTO (Chief Technology Lead), I can assure that a quick reprogramming effort is all that is needed to change auto settings so that emissions controls resume upon "idle" - instead of just running during "sensor-plugin" testing. A little more work with adjustments would deliver a balanced performance and emission formula that makes overall diesel emissions equal to those of gasoline engine cars given gasoline's much higher fuel burden upon the environment; Gas versus Diesel is not an apples to apples comparison - as mentioned above.

I am assuming that the EPA is credible in their assertions here too.

The EPA has come under criticism in numerous cases in recent years, including if I recall California's removal of 1997 GM EV1 electric cars in 2001 - upon the advice and direction of committee hearings filled with Oil-backed special interest panel members. (EPA, 2001)

The buying and burning of trolley cars in the 1930s and 1940s ended North America's love affair with a Europe-like public transportation network - and set an 80-year trend by businesses to prefer the "greater good" realized through profits.

During research work for this book, I was surprised to find well-documented cases of extensive conspiracy and absolutely illegal practices forwarded for investigation by U.S. Senator Ted Kennedy in 1972. (Kassel, 2012)

"Mass transit didn't just die, it was murdered" (Kwitny, 1981)

If a computer reprogramming effort brings the overall auto emissions back equal to that of gasoline-powered cars, without forsaking the benefits of diesel, I would see that the greater good is served by supporting diesel, and the more important clean "e-diesel" fuels, as energetically as possible.

Rapid Charge Vehicle Battery Systems

Today's best electric cars, the Tesla, requires the installation of a special charging bay in your garage and then it can take up to eight hours of charge time to give your vehicle a 245 mile (394 km) range of travel. The Tesla's 'S' model has a 300-mile range but often 150 miles (241 km) is the max range if you do not want to be careful to drive it for optimal battery length.

Tesla Superchargers can recharge up to 50% of the car within twenty minutes and are designed to move you along from a gas station during a long trip. On a 240 volt system (North America runs on a 120v standard), it takes one hour to charge for each 30 miles – therefore 9.5 hours for a full 300-mile charge; and on a standard 110v system, one hour of charge is required for every five miles or fifty-two hours to charge fully.

These battery charging overheads keep electric cars constrained to daily city use primarily, and then gas and diesel vehicles meet our long-haul transportation needs.

Battery powered cars are much more expensive than gas cars too. The Tesla is an $80,000 to $120,000 CAN purchase, and this does not include the additional cost of installing a non-standard 240v charging station in your garage. Lithium-Ion Batteries too, have a lifespan of just six years and an environmental cost. The cost of a new Tesla battery will be between twelve and $29,000 to replace.

Once cost effective, longer life span, faster charge time batteries can replace current technology, then, the battery powered car, will become a superior alternative to gas and perhaps even diesel vehicles too.

Oil is not even needed really

In 2012, researchers at Princeton University (Elia, Baliban, & Floudas, 2012) confirmed that a combination of coal, natural gas, and non-food crops to make synthetic fuel, could replace all of America's need for crude oil altogether. This plan would also reduce emissions by 50%.

Their work went on to estimate that it would cost $1.1 trillion to refit all pieces of the system as the need to roll out the change over the next twenty to thirty years when current facilities needed replacing anyway.

This planning is detailed enough that researchers were able to determine the locations of each of the large, medium and small processing centers where Paraffin Oil, and other manufacturing plants of substitute fuels, would need to be deployed to replace all oil. The map on this page shows the locations needed for a 50% replacement of Oil throughout the United States.

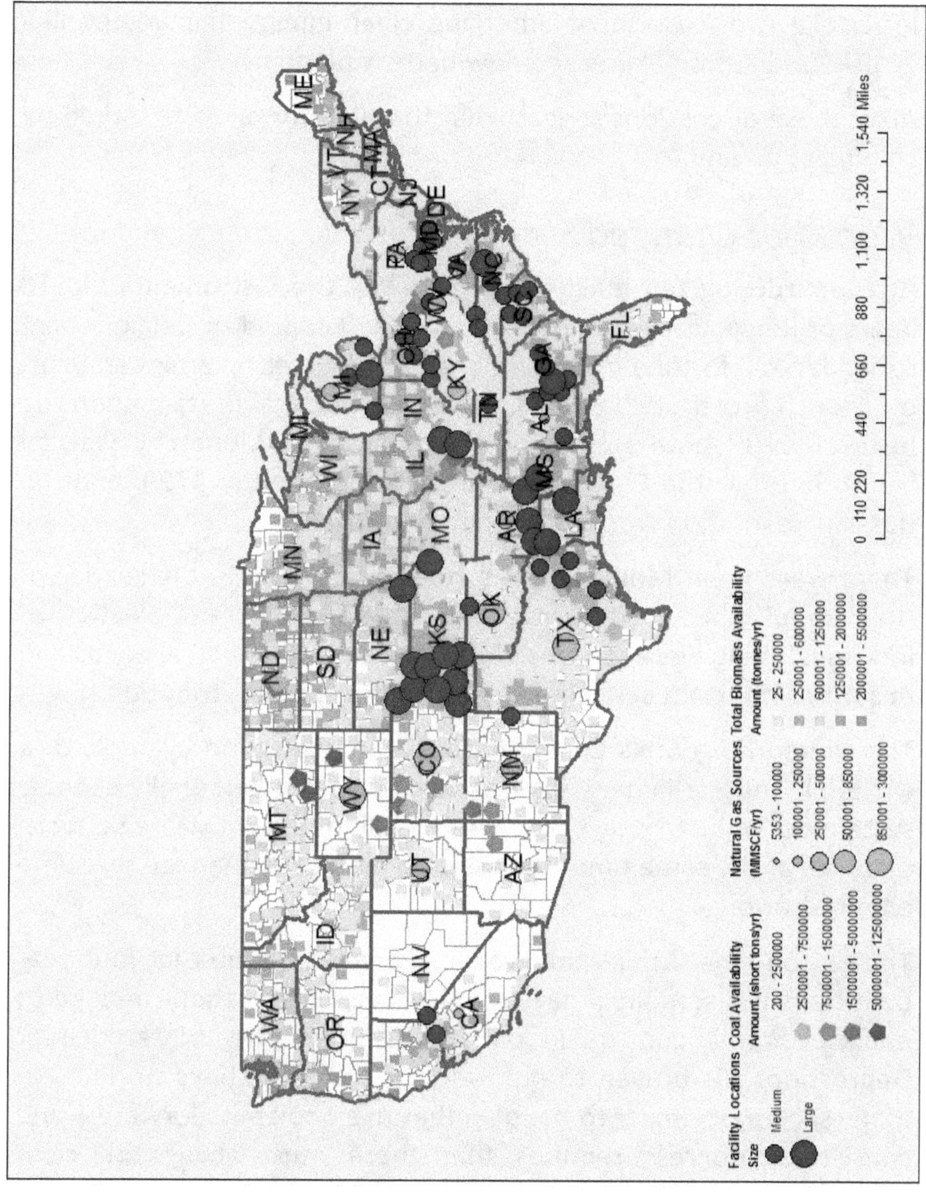

We know from discussions of Blue Crude and Ted Kennedy above, that i) powerful forces opposed to change certainly do exist within our society, ii) we also know that we control these powers with our votes, and iii) we learned that the cost of alternatives to oil will be equivalent with at least 50% and lower emissions as well.

So, we have a few energy alternatives that replace oil, even before I introduce a discussion of emerging clean-energy technology like 2017's Cold Fusion Plants and new battery planning.

All of these discussions suggest that the time for oil wars should be coming to an end.

Wars that Changed Society

After murdering ten million and installing the instruments of 130 Years of Religious War in the Congo, King Leopold of Belgium next squandered a fortune from the Congo's resources on a sixteen-year-old French prostitute. Leopold placed a considerable stain upon the institutions of both Humanity and Monarchy. Monarchy was an institution that the French had already abolished in 1799, and the Russian Revolution would abolish it in the 1910s.

There were great Monarchs who ruled wisely and in the service of their people, but citizens also wanted the democratic election of individuals who were responsible for managing their local needs and ensuring important services like food and income distribution.

The United Kingdom's Constitutional Monarchy is an example of a Monarch and Parliament separation of duties. I personally like this system potential to deliver the benefits of long-term strategic social planning at the same time that MPs are elected to manage real-time administration.

The reasons that I think this way are, first; the hope that four-year politicians will somehow navigate the course corrections needed to enact economic controls in K-Wave Summers and Winters (Great Depressions), is proven to not work. Second; we have no Plan. In many societies, similar to the plan that this book puts forward – and consistent progress requires that there must always be clear next-steps expressed in projects and process.

Building a sustainable society is one project where you need to have specific "technical" training to do the job well. I have seen again and again, with my own "Consultant Eyes," dreadful results when unqualified program leads sit on top of large projects. Within major corporations, I see what we see in society exactly: Panics, poor planning, denial of the reality of depressions, and no idea as to what are next steps. Anxiety and drama are daily norms instead of enjoyable professional progress. It's not quite an anarchy, but sometimes the result is the same, and progress is often two steps backward for every one step forward.

The Chinese Emperor system ended in 1934 when Japan seized control of China. The U.S. tried to broker a peaceful succession back to Chinese rule at the end of World War II. Instead, a full civil war began between China's Nationalist and Communist parties which each controlled troops in the millions. Nationalist armies held an upper hand in capability, but their most experienced troops had been laid low by the Japanese. Rampant corruption, economic chaos, and suppression of dissent by Nationalist Party leaders persuaded even Chinese elites to regard the Communist Party more favorably until Mao ZeDong came to power in 1949.

All four of these example "wars for social change" replaced one system of government for another. In France, a Monarchy was replaced by the Socialist Republic. In Russia, Monarchy was replaced by Communism – a classless form of socialism where the state controls land. In China, a classless socialism was chosen. Land grants were given to citizens originally and then later individuals were required to purchase their properties.

The Two Faces of Every War

Invariably, the pragmatic math-driven business value of war has nearly always been prefaced with a morale-building message of deserved retaliation, religious higher purpose, birthright or some other similar explanation to an uneducated army and public who were then forced to live at the front lines of that war as well.

"Canon-fodder" was a derogatory term used for combatants who were treated as expendable or who were forced to deliberately fight against hopeless odds.

Historically, the initiators and Generals of war traveled with their armies. Today, leaders very seldom experience the danger that the soldier and non-combatants on the field see; nor pay as high a cost. This can give rise to a decrease in the value given to life; reducing deaths in the field to an inconsequential necessity instead of a deep and personal loss.

Historians use the term "irrational" wars, to describe those wars that were fought without a pragmatic goal.

A forensic review of all wars throughout time show that 99% of high death toll wars are considered rational in this way by historians; containing a solid business reasoning hidden behind a façade of rhetoric (also called propaganda) presented to troops and citizenry. Propaganda has included the need for living space, retaliation for criminal acts, religious retribution, racial superiority, and many more.

Rational Wars

This central theme of financial or territory gains could be seen in a majority of wars, and this led war historians to categorize these as Rational Wars.

World War I and II also, were both born of a series of failed treaty violations and, in 1939, a coordinated plan between allies Germany and Russia, set out to gain new territory from Poland. New Territory and manipulations of contracts and treaties constitute a "rational" war, in the eyes of historians.

Non-Rational Wars

Just so that you can regale your dinner guests, as I have, with stories of some of the silliest wars of all time:

Lijar vs. France – a 98-year war with no casualties "erupted" when 300 citizens of a small southern village in Spain, heard reports that their King Alfonso XII had been insulted and even attacked by Parisian mobs during an official visit in 1883. In 1981, town council elected to cease hostilities after reports that King Juan-Carlos' visit of 1976 to Paris was welcomed with great respect.

The War of the Oaken Bucket – 1325 saw a twelve-year war declared when a group of Modena soldiers raided Bologna and stole a large wooden bucket. Yes - a bucket. War raged for 12 years, and Bologna never got their bucket back; it remains in Modena's bell tower today. Casualties are unknown.

The War of the Stray Dog – in 1925, a Greek soldier chased his dog over the Bulgarian border and was shot dead by a sentry. Greece invaded Petrich the next day and halted when the League of Nations ordered Greece to desist and to pay 45,000 pounds in reparations. 52 died in the ten-day war.

The War of Pork and Beans – Aroostook War. 550 died of disease, winter hardship, and accidental injuries at the end of the War of 1812 when American woodcutters began clearing land in Maine held by the British Army. As the American Army moved to defend, war seemed imminent. A supplies department slip-up sent American troops an enormous amount of Pork and Beans – which led to the name of the conflict.

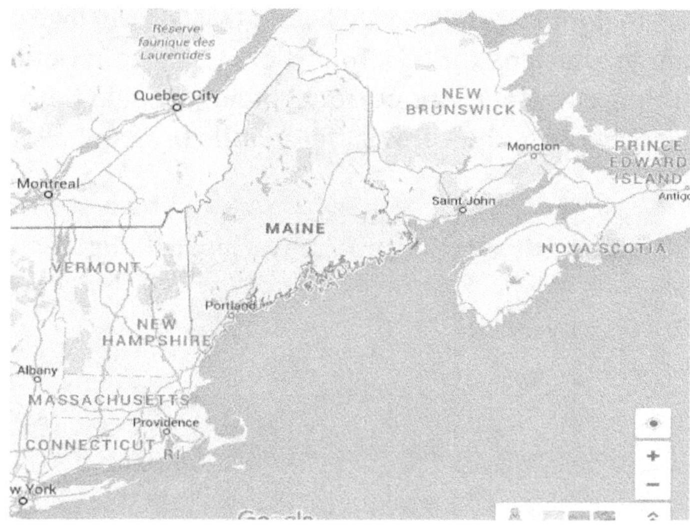

The Pig War – June to October 1859. British infantry shot a pig on American soil. A group of American Militia moved to the border and waited for war. The British eventually apologized and the four-month war ended.

The 375 Year War – the Isle of Scilly off the Southwest Coast of Britain declared war on the Netherlands in 1651. Many wars of this time were not taken seriously and was forgotten before the two countries finally agreed to a peace treaty in 1986. No casualties.

The Football War – the El Salvadoran Army launched an attack on Honduras in 1969 which started a four-day war that killed 3,000 people. The cause? El Salvador lost a football game (North Americans call this game Soccer).

Effect of Democracy on War

Democratic nations have rarely declared and sustained war upon another democratic nation. In practical terms, the number of people needed to support that war simply become too difficult as casualties and costs mount and time passes sufficient to devise a satisfactory workaround or compromise.

A very great strength of Democracy is that the needs of the non-combatant majority are felt in this way. In Monarchies, situations could emerge that created disproportionate personal benefits for the elite ruling classes. These personal benefits could both incite or prolong wars because the elites suffered little or not at all.

Democratic elections make war unsustainable.

Chapter 4- A Government Systems Primer

Permit me a little literary license at this time in the interest of academic discussion, and let me ask the reader to forget a lifelong collection of definitions and labels placed upon government systems for just a bit.

My request in the opening pages of this book asked that you part ways with past definitions and now level-set anew with a definition of democracy and socialism in keeping with both international norms and simpler categories.

Most definitions of Capitalism, Socialism, Communism, Monarchy and even Democracy overlap, are broad, and can easily be understood differently by individuals within even the same workplace, school, and community. Country to country, these definitions vary tremendously.

My categorizations and definitions will, therefore, probably differ from every definition in every country in the world. They are accurate for use in this book and important to understand if you want to master a shared solution for seven billion Good Lives and a sustainable World Peace.

Government Leadership Categories

The following definitions are suggested for government systems discussed within this book. The last section of this chapter provides a list of specific government types alphabetically, but there are so many overlapping and similar definitions, I have chosen to simplify many government types into categories here.

Note also, that there is little correlation between Government Type and its policies, as you will see in Chapter 5. Governments of the same Type, United States, and Finland for example are both Republics so both elect their Presidents. However, Finland's strong Socialist Policies inspire excellent income distribution that are not seen in all other Republics such as the United States and Brazil - where more Capitalist policies are prevalent.

Policies, therefore, really tell the nature of a government and these categorizations try to capture this.

Absolute Monarchy – A single individual inherits power to rule and then bequeaths it upon their death back to their bloodline. Often regents, ministers, and courtiers, are allocated power to administer specific policy areas of government as needed to administer the kingdom.

Constitutional Monarchy – a Monarch presides over affairs of state within the limits of a constitution. Some Monarchs employ a parliamentary system in which the Monarchy is a ceremonial role only, and an elected Prime Minister has full political power.

Democracy – Government by the majority vote.

Democracy is not a left nor right leaning but rather a system of election. Whatever the majority states, becomes the rule. In extreme examples, such as William Penn's 1647 Quakers, 49% could, in theory, be permitted to starve if 51% were unknowing or uncaring of the plight of their fellows.

A country tends to be measured nondescriptly as a "Democracy"; I suggest that a country's policies are more telling of its category or leanings - whether Capitalist, Socialist, Communist, Monarchist. All of these categories may use democratic elections, but Capitalist-leaning states more often refer to themselves as a democracy.

In an extreme democracy, there is no obligation to others; no accountability to the "greater good" – there is only the majority's decision. For this reason, Constitutions are often set in democratic nations to guarantee that important human rights and values must be upheld. See "Republic".

Rarely is a Good Life defined among protected Human Rights – which is why I suggest that it should be in Chapter 20 of this book.

Watch an episode of a game show called Survivor to see how simple democracy works. In this contest, the basic strategy needed to win is to ensure strong leaders are voted out quickly so that a more average group of individuals can leverage their collective bargaining power by daily election until eighteen contestants dwindle to a final winner.

Election democracy is more complex in that each Party must put forward a leader strong enough to win an election.

Republic – A democracy with a constitution and order of rule. France's implementation of a republic is notably socialistic as well with a larger government and pro-active wealth distributing economic controls.

Socialism without Classes – a state where everyone has equality and equal opportunity without regard to class, caste, or social status. Everyone progresses together as a society in all things. Property ownership is permitted.

Socialism with Classes – a state where everyone has equality and equal opportunity within classes. By in large, everyone progresses together as a society in all things including property ownership. Classes might include:

i. **High-Income Earners or Elites** – Successful business owners, Ministers, Royals

ii. **Utility or Merit Class** – Engineers, Scientists, Doctors; this class rewards high performers, skills and contribution to a higher class or social position.

iii. **Race** – Master Race, Chosen Ones, etc.

Meritocracy – a class defined by the merit and contribution of an individual. Business leads may be seen as more important to reward in the world where profit is of primary importance; Engineers are seen as more valuable in a rapidly advancing society; doctors would be most important in a longevity crisis, and so on.

The Transition suggests that engineers who drive social benefit will drive financial benefit at the same time and are deserving of higher merit; medical practitioners, such as nurses, palliative care workers, and teachers are deserving of higher merit, and these folks must begin building automatable solutions as well.

Communism – a classless social structure with government control of property ownership. Within communist states, democratic elections vote members of Parliament who then vote for a President. Presidents are elected every five years and may be reelected only once consecutively.

Seeking to reward and not discourage performance, Communist states often implement classes that reward officials, scholars, builders, top engineers, and even good students - with larger salaries, homes, and other perks. See Meritocracy.

Wealth Distributed Capitalism – citizens are given money sufficient to afford all of the basics of life including a home, food, travel, healthcare, clothing and money for furnishings, cars, etc., at a level in keeping with the wealth of the entire country. Today, this is often referred to as a Socialist State. Saudi Arabia might be considered an example within an Absolute Monarchy.

Capitalism – the business class governs. Play a game of Monopoly to understand how this system of government works. The system serves the majority, which will tend to become the wealthy members of a capitalist society over time as well. Conservative or extreme right work to maintain status quo and protect their economic advantage with the highest priority.

Authoritarian regimes centrally hold all powers to plan for the economy and all production means.

Common Misconceptions

Socialist Governments

All political parties take policy positions on dozens of topic areas including Natural Resources, Immigration, Human Rights, Education, Unemployment, Business Development, Environment, Housing, Economy, Banking, and so on.

"Socialistic" is a measure given to each policy on an individual basis. A Conservative right-leaning government, for example, might have several or many socialistic or social policies and even communal or communistic policies as well.

Canada is an example of a Government with a Conservative Government this past eight years and, like eighteen other G20 nations, it ensures universal health care – a decidedly socialistic policy. Californians too, enjoy similar socialistic policies with their state-funded healthcare plan.

Ownership of land is the norm in Canada, but in its national parks 100-year leases are granted only. Denial of land ownership is a communal or communistic policy.

The Term "Big Government" is often associated with Socialism. Certainly it will be true to say that a left-leaning party, one that is sponsoring a larger number of social programs, may require more government employees and resources to administer these policies. The following is proven false as well:

i. Left-leaning Political Parties have less interest in good financial management.

   Right and Left governments are equally responsible for managing economies well. Look to failings in leadership bench strength (unqualified candidates) if you see that candidates left or right are missing these targets.

ii. The more social policies a Government supports, the more inefficient it becomes.

    Socialist Countries like the Netherlands, Norway, Sweden, Denmark, Switzerland, and others offer universal health care, low unemployment, high minimum wage, most citizens have a good life, most can afford a home, a car, all citizens get pensions and enjoy some of the highest longevity statistics in the world. Their governments are some of the most efficient, best-management teams in the world as well.

iii. Asking the rich to pay higher taxes and having more social programs will result in less business and lower standard of living. (Blodget, 2011)

    The Netherlands, Norway, Sweden, Denmark, Switzerland all have higher Wealth Generation (GDP Export per Capita) than any of the G8 nations and all boast better income equality, home ownership rates, they retire comfortably, longevity is higher, and so on.

Communist Government Misconceptions

True or False? Communist Governments are Totalitarian Dictatorships without free speech.

False. Communists elect officials to lead with the mandate to provide a Good Life to all citizens; Free Speech is a component of Human Rights. Countries with all Systems of Government struggle to provide Free Speech wherever a Good Life cannot be afforded.

There are few communist countries functioning as such in the world today. The former Union of Soviet States changed to Democratic Republics in 1986 and China monetized in the 1980s successfully, but China today is much more capitalistic in its policies these past fifteen years. A Communist Russia was the last of the G8 countries to stop supporting the American Dream for its citizens in 1986.

Wealth Distribution becomes a challenge in counties with strong Capitalist influences. Much of China has transitioned from the Stone Age to have two or three cities, like Shanghai, which are among the most sophisticated in the world.

Ensuring a minimum standard of living for 1.3 billion people with basic human rights is a herculean task - and free speech was seen to be an acceptable casualty given the size and importance of the primary task at hand – rightly or wrongly.

Communist country leaders are democratically elected in a five-year election cycle that is little different from other democratic nations. Presidents are elected by members of Parliament similar to the election of a Prime Minister in Constitutional Monarchies like Canada, the UK, and Norway.

China's Communism succeeded very well this past 30 years, and Russian Communism ran out of money before 1986. The two states differed in just one way predominantly: China monetized its production economy successfully - and Russia did not. Success for most nations have come after monetizing their a manual production economy through exports.

Since 2003, China has changed its housing policy to an ownership model that has created ghost cities and a hard life for tens of millions of citizens and families.

Freedom of Speech and Human Rights in all societies has been a challenge.

In Capitalist countries and in Communist countries too, free speech is limited to varying degrees in keeping with that country's ability to provide a good life. America does not actively broadcast a 25% unemployment rate among veterans and high homeless rates nor high-income inequality, and China does not permit the broadcast of equally embarrassing statistics that describe a hard life among its citizens either.

I would hope the reader appreciates that public safety and social transparency is a balancing act when a good life cannot be made available to all. This World Peace Transition Plan's automated economy is intended to correct that shortcoming.

True or False? Communism is Very Very Destructive to the individual and to the society.

True and False – All of Marx's 10 Planks of the Communist Manifesto are adopted to some degree in policy within the United States - as makes best sense. (Various, n.d.)

Marx's socialistic and communistic theories worked when that country monetized. This model would also be the more obvious fit for a system of government once an Automated Production Economy-based society made money irrelevant and a Good Life was guaranteed for all as a part of our basic human rights.

Today, 38 million people in the socialistic Scandinavian countries and the Netherlands live the American Dream that all G8 countries lost in the 1970s and 1980s.

True or False - Communism is UnAmerican.

False – The MayFlower Compact of 1620 describes a democratic society committed to the greater good of the community. The American Constitution supported taxation for the General Welfare in education and other planks of Marx Communist Model. William Penn's Quakers believed all men were created equal and that 51% vote governed all things. Colonists are recorded in history to have hung Quakers for their immoral social disregard.

Choosing a System

Citizens vote for both their leaders and for a system of government that most closely matches their values and beliefs. Voting is the important responsibility of all citizens once they reach an age of majority – usually eighteen years of age.

Poor performance by parties over time can often lead to voter disillusionment and quite low voting turnouts. In many democratic nations, federal elections can see just 60% of its eligible citizens' vote.

Here is a quick chart to help you understand the system of government that most closely matches your values and wishes.

Do your needs of a society include ...

1. **All people in your society should move forward together:**
   a. **Equally** - in Pure Socialism and Communism; where everyone is equal
   b. According to Class by:
      i. **Income/Elites**
      ii. **Utility/Merit**
      iii. **Race**

      If yes, then policies which are Socialist, Communal, and even a Constitutional Monarchy are a fit for you.
2. **Don't care about the other guy?**
   i. Pure Capitalist Democracy – 51% dictate everything without regard for the 49%.
   ii. Authoritarian, Non-Benevolent or Fascist Dictator Models – although a great majority of dictators are not remembered as acting benevolently to bring a Good Life to their population.

   The smaller list of beloved Benevolent Dictators, leaders who had the best intentions of their people in mind, might surprise you. Fidel Castro was such an individual, United States President and my cousin Franklin Roosevelt, Ferdinand Marcos - Philippines.
3. **Do you want citizens to own Land?**
   If "Yes" – Socialism, Democracy, Constitutional Monarchy, Chinese Socialism (China's brand of communism asks citizens to buy or rent their homes; this is more typically called socialism.)
   a. Pros: In theory, everyone owns their home and can profit from its sale. The statistics, however, show that only 30% of Americans own homes free-and-clear; and

that all others are financed 50% as the U.S. national average.

b. Cons:
   i. Usury – debtors lured into loan agreements they can never exit illegal in most countries and on the rise. Just 30% of Americans own their homes, and the stats indicate this number is getting worse and not better; given the move rate, this implies that **65%+ may never be able to pay off their mortgage debt**.
   ii. Unmanaged Housing market bubbles and speculation force up to 10+ years of saving and then a 25-year reliance on mortgage lending. If savings are too difficult to accumulate, family starts are delayed. In China, families are even separated from young children when the slums that working parents must live in are unsuitable for kids.
   iii. These are traits often associated with a "hard life" and are seldom considered attributes ideal per the definition of World Peace above.

If "No" – Monarchy and Communism (as in pre-Perestroika Russia), are a good fit.

See Chapter 13 – Land Ownership for more discussion on sustainable housing solutions.

Right vs. Left

Right versus Left viewpoints are terms originating from the French Revolution when countrymen aligned with the King were seated to his right. Voices aligned with the revolutionary forces were seated at his left. Right-wing ideologies are status quo thinkers; aligning with traditional points of view be they capitalist, conservative, fascists, monarchists, nationalists, or traditionalists. The right is usually the party sector associated with the interests of the upper or dominant classes. Left are more generally the sector expressive of the lower economic or social classes and include communists, democratic socialists, feminists, green, and social democrats. The "Centre" intermediate stance moderates the interests of the middle classes.

Political Parties are not bound by any particular ideology and it is not unusual for predominantly left or right parties to support both right and left policies at the same time.

The values that define a Good Life above prefer humanistic targets and goals that permit all young people to choose higher education, marriage, and parenthood at twenty years of age as well – if they choose. In our description of a Good Life, young families are provided a home, and even a family cottage for vacations if they wish it.

From our wealth distribution discussions above, we do not accept special interest arguments that society cannot afford to provide for its poor and we also realize that technology will provide all of the basics of life for us soon enough. The systems of government that support these perspectives are often historically called Wealth Distributed Social Democracy, and then Communism only differs in its management of land ownership.

Systems within Systems

Election of a President or Prime Minister

Most government systems provide citizens the right to vote for their leadership. Canada, Great Britain, and China are examples where democratic elections do not elect their President and Prime Minister, however; instead, these countries elect a Member of Parliament, and then that MP-elects a Prime Minister.

America, France, and Germany are examples of Republics or Democracies whose voters elect their President directly.

Types of Government

From Theodora.com, here are a list of common Government Types that fall into categories mentioned above. This section is for reference only and readers can skip and return whenever needed.

Anarchy - a condition of lawlessness or political disorder brought about by the absence of governmental authority.

Commonwealth - a nation, state, or other political entity founded on law and united by a compact of the people for the common good.

Communism - a Carl Marx system of government in which all means of production are owned in common rather than by individuals. Government controls communal property or capital while making progress toward a social order in which all goods are shared equally by the people (i.e., a classless society). Communism is at the opposite another end of the spectrum to Capitalism.

Confederacy (Confederation) - a union by compact or treaty between states, provinces, or territories, that creates a central government with limited powers; the constituent entities retain supreme authority over all matters except those delegated to the central government.

Constitutional - a government by or operating under an authoritative document (constitution) that sets forth the system of fundamental laws and principles that determines the nature, functions, and limits of that government.

Constitutional democracy - a form of government in which the sovereign power of the people is spelled out in a governing constitution.

Constitutional monarchy - a system of government in which a monarch is guided by a constitution whereby his/her rights, duties, and responsibilities are spelled out in written law or by custom.

Democracy - a form of government in which the supreme power is retained by the people, but which is usually exercised indirectly through a system of representation and delegated authority periodically renewed.

Democratic Republic - a state in which the supreme power rests in the body of citizens entitled to vote for officers and representatives responsible to them.

Dictatorship - a form of government in which a ruler or small clique wield absolute power (not restricted by a constitution or laws).

Ecclesiastical - a government administrated by a church.

Federal (Federative) - a form of government in which sovereign power is formally divided - usually by means of a constitution - between a central authority and some constituent regions (states, colonies, or provinces) so that each region retains some management of its internal affairs. This differs from a confederacy in that the central government exerts influence directly upon both individuals as well as upon the regional units.

Federal Republic - a state in which the powers of the central government are restricted and in which the parts (states, colonies, or provinces) retain a degree of self-government; ultimate sovereign power rests with the voters who chose their governmental representatives.

Islamic republic - a particular form of government adopted by some Muslim states; although such a state is, in theory, a theocracy, it remains a republic, but its laws are required to be compatible with the laws of Islam.

Maoism - the theory and practice of Marxism-Leninism developed in China by Mao Zedong (Mao Tse-tung), which states that a continuous revolution is necessary if the leaders of a communist state are to keep in touch with the people.

Marxism - the political, economic, and social principles espoused by 19th century economist Karl Marx. Marx viewed the struggle of workers as a progression of historical forces that would proceed from a class struggle of the proletariat (workers) exploited by capitalists (business owners) to a socialist "dictatorship of the proletariat," to, finally, a classless society - communism.

Marxism worked in a free market manual production economy when monetized such as in China. His models appear better suited to an automated production economy – monetized or not.

Marxism-Leninism - an expanded form of communism developed by Lenin from doctrines of Karl Marx; Lenin saw imperialism as the final stage of capitalism and shifted the focus of workers' struggle from developed to underdeveloped countries.

Monarchy - a government in which the supreme power is lodged in the hands of a monarch who reigns over a state or territory, usually for life and by hereditary right. The monarch may be either a sole absolute ruler or a sovereign - such as a king, queen, or prince - with constitutionally limited authority.

Oligarchy - a government in which control is exercised by a small group of individuals whose authority is based on wealth or power.

Parliamentary democracy - a political system in which the legislature (parliament) selects the government - a prime minister, premier, or chancellor along with the cabinet ministers - according to party strength as expressed in elections. By this system, the government acquires a dual responsibility: to the people as well as to the parliament.

Parliamentary government (Cabinet-Parliamentary government) - a government in which members of an executive branch (the cabinet and its leader - a prime minister, premier, or chancellor) are nominated to their positions by a legislature or parliament, and are directly responsible to it. This type of government can be dissolved at will by the parliament (legislature) by means of a no confidence vote, or the leader of the cabinet may dissolve the parliament if it can no longer function.

Parliamentary monarchy - a state headed by a monarch who is not actively involved in policy formation or implementation (i.e., the exercise of sovereign powers by a monarch in a ceremonial capacity). True governmental leadership is carried out by a cabinet and its head - a prime minister, premier, or chancellor - who are drawn from a legislature (parliament).

Republic - a representative democracy in which the people elected deputies (representatives), not the people themselves, vote on legislation.

These Socialism - a government in which the means of planning, producing, and distributing goods is controlled by a central government that seeks a more just and equitable distribution of property and labor.

Sultanate - similar to a monarchy, but a government in which the supreme power is in the hands of a sultan (the head of a Muslim state); the sultan may be an absolute ruler or a sovereign with constitutionally limited authority.

Theocracy - a form of government in which a Deity is recognized as the supreme civil ruler, but the Deity's laws are interpreted by ecclesiastical authorities (bishops, mullahs, etc.). This is a government that is subject to religious authority.

Totalitarian - a government that seeks to subordinate the individual to the state by controlling not only all political and economic matters, but also the attitudes, values, and beliefs of its population.

Chapter 5 – Government Economic Policy

The role of Government in Economic Policy is straightforward – keep the economy's wealth distributed and growing. Unfortunately, this responsibility of governments both left and right, is not well understood by democratically elected politicians, nor voters, who need to make correct course corrections at specific turning points in a Capitalist K-Wave cycle.

Capitalism troughs every 60 years due to compounding Interest, the year-over-year expectation of higher profits; the rich begin earning day and night; after enough time passes there are only rich and poor – just like any game of Monopoly that you will ever play. These cycles have repeated within our capitalist civilizations for at least 4000 years, and we can only change this inherent instability when we manage wealth distribution proactively.

The start of a new economic cycle, the one right after World War II for one example, is a time in which the distribution of wealth within a society is relatively even. Everyone has what they need, a single family income provides all the basics, and a good life is available to everyone in society. This beginning is referred to by economists as the Spring of a Capitalist K-Wave.

Spring carries on like a monopoly game would - players buy houses easily, begin to collect preferred, high earning property and businesses and then start generating revenue by renting them out and earning income from these investments. Incomes were distributed across society somewhat evenly – similar to this pie-chart below where there are rich people, but that those top 20% of high-income earners control 35% of the wealth; the next 20% had about 25%, and the Lowest had 11 % of the wealth.

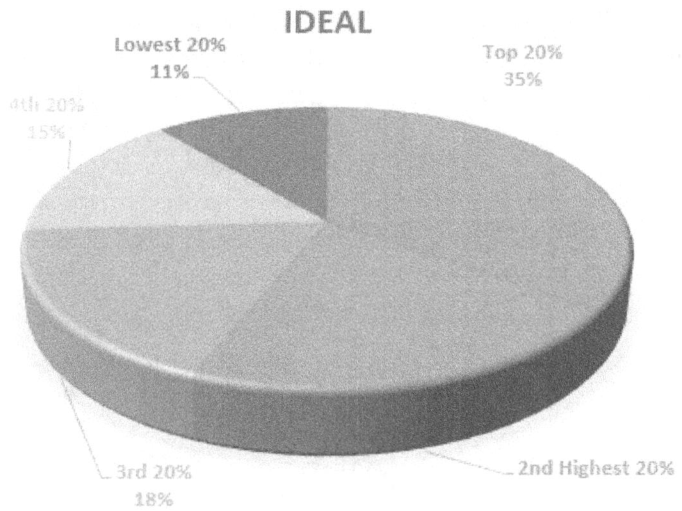

Michael I. Norton and Dan Ariely suggested these arbitrary targets in their report "Building a better America – One Wealth Quintile at a time" (Norton & Ariely, 2011). The actual values assigned to each quintile are less important; the important task here is to create targets and manage wealth distribution to those targets.

In K-Wave Summer, some business owners start to make investment money available – initially at a higher interest rate. There are relatively few takers but still a few members of society start amassing wealth. Speculating and rental investment continues until all houses are gone and competition for homes drives up prices quickly. In the Toronto suburbs in 1982, a home could be bought for $30,000 at 22% interest annually. By 1986, four years later, the same house cost $90,000, and rates were 10%. By 2012, the house could be worth $500,000 with an interest rate at 3%. The notion of reality left real estate prices somewhere within this window of time.

As K-Wave summer ends, the Good Life too is ending now for many, and will be available for fewer and fewer families going forward now - until young adults do not have the same easy start in life that their parents did. Many families feel they need two salaries to afford a living at this time.

The end of a summer K-Wave is the time, in economic terms, to begin reducing tax to low-income individuals and increase the tax to successful businesses and high-income earners. You can see in this next chart "Federal Receipts by Source" from Wikipedia that the opposite happened in U.S. economic policy.

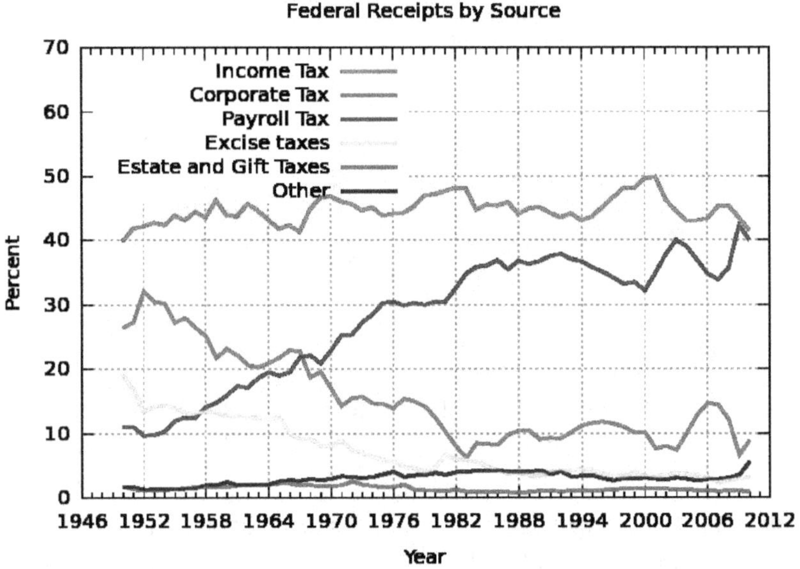

In K-Wave Autumn, interest rates can be expected to fall and fall, houses should soar in cost, and bankruptcies should rise in response to higher mortgage needs. Higher rates of mortgages will default as people lose jobs or cannot keep up with their bills.

The correct economic controls would have taxed the rich more heavily, and the poor to a lesser degree – but in the following chart "Historical Marginal Tax Rate for Highest and Lowest Income Earners", the opposite happened throughout the 1980s.

Tax rates of 91% and more were levied on High-income earners in the 1940s through 1960s. By 1985, high-income tax rates were reduced to 28% and 35%. In 2012, the highest tax rate was lifted to 40%.

Taxes on the lowest income level hovered around 15% throughout the boom years of the spring and summer K-Wave, until the tax rate was reduced to zero for the poor in 1977. In the late 1980s, the tax rate for the lowest income earners returned to 15% and then has remained at 10% since 2003. (Blodget, 2011)

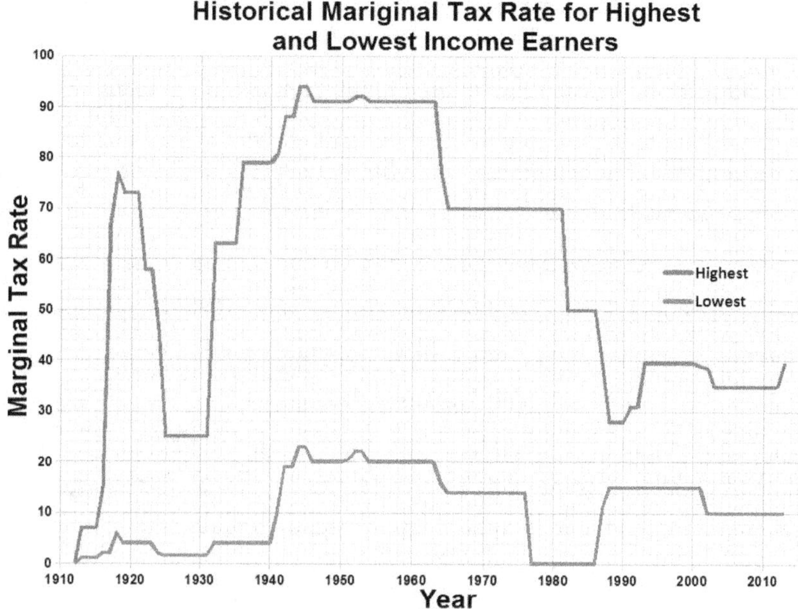

You have seen the Federal Reserve Pie Chart in Chapter 2 that describes 2010's income distributions – where today the bottom 40% of American's share 0.3% of the wealth, and the top 20%, share more wealth than all of the GDP of the G20 combined.

The point here is, that we all should have been able to see inequity of this magnitude coming - and then our government and voters should have prevented it from becoming as extreme as this was obviously going to become as well.

Today our income distributions are extreme for two groups; the rich, who should have stood up to insist on doing the right thing by society; and second, to every other one of us who voted for low taxes in every election since the 1970s.

The Pitfalls of Managing Proactively

What if our country had a President or Prime Minister who knew enough about Capitalist Economies and K-Waves to realize that a sixty-year cycle of wealth inequity was headed our way? As a good and responsible leader, he or she would have decided to prevent a calamitous outcome by implementing appropriate economic controls when wealth was still quite distributed in the Spring or early Summer - back in the 1950s through 1970s.

Finding a leader with a good grasp of K-Wave Economic Controls would have been a tall order. Although documented in 1925 - and in many economic thesis since, my five High School and University economics courses made no mention of it.

John F Kennedy acknowledged the need to manage society's revolutions proactively in 1962 with his famous quote:

"Those who make peaceful revolution impossible will make violent revolution inevitable."

Without fail, every politician that ran for election in the United States with a message of the importance of economic controls that tax the rich and proactively ensure wealth distribution – was voted down and never had a chance of making it into office.

At this crucial time, Americans voted instead for a wonderful, charming, untrained leader with a heart of gold - in Ronald Reagan. Reaganomics was an economic policy that embraced the Trickle-Down Theory and next it was also adopted into the U.K. as Thatchernomics. Trickle-down said that "What was good for the rich and good for business, was good for society." Reaganomics gave money to the rich, and the rich would administer its trickling down into society with virtually no monitoring of the policy's success or failure by the government after that.

The campaign message of "Low Taxes" got Reagan nominated; "Low Taxes" got him voted President in 1981, and it reelected him again in 1985 – echoing a too long period of previous campaigns promising "low taxes" at a time when roads and highways were crumbling. Citizens began to abandon Major city centers due to neglect and security concerns. I can remember that administrators decided to close sidewalks in Buffalo rather than to repair collapsing buildings in their city center.

To not campaign based on "Low Taxes" was not to get elected in the United States and soon "Low Taxes" encouraged irresponsible administration too.

Outside the G8, Norway enacted strong economic controls proactively 25 years ago. Today they have very low unemployment (3%), universal healthcare, a $20 minimum wage, the seventh best income equity in the world, and the highest GDP exports per capita (wealth generation) annually for the past 20 years. Anyone looking for clues as to where the "American Dream" went, can look to Norway to find it protected and thriving here.

Norway's economy looks like a very smart Government Management Team's Performance Report. I suspect that it is no mere coincidence that the Nobel Peace Prize, and the Nobel Prize for Economics in neighboring Sweden, were founded here within Norway's very sustainable society.

Economic Controls - Right and Left

When wealth distributions can become this extreme, with all the negative social problems and embarrassment associated, neither Right nor Left parties are enforcing responsible economic controls.

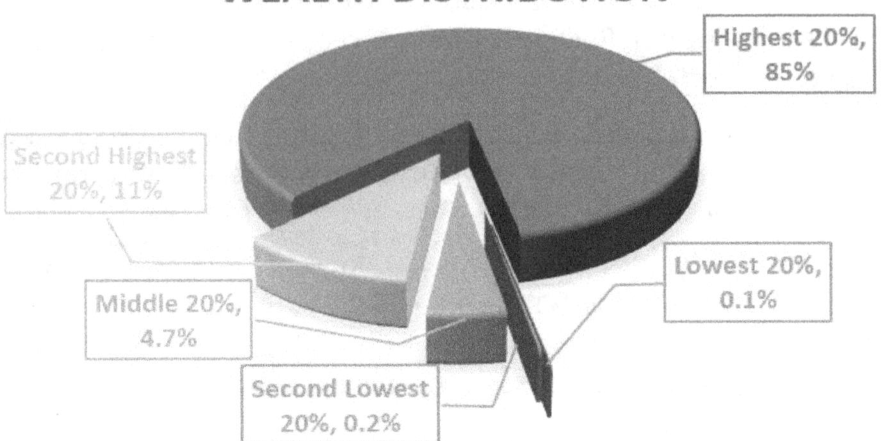

Essential services like hospitals must remain open, the ambulance must come when called, school children have to take a good education within secure communities – and we must avoid trough wars that have gained support from destitute citizens in past economic depressions.

For the first time, our technology is within just twenty years of ending our reliance on money altogether as well. Perhaps now is finally the time to stop voting for "Low Taxes" and look at our Wealth Distribution challenge as one that just needs resolving now.

The government's "Right" tend to be the more conservative managers between the two sides. The Left will support more social and communal services at higher cost and risk to the government – but both have the mandate to present an affordable suite of policy offerings to put to the decision-making hands of their voters.

For either Right or Left parties to ignore its responsibility to Economic Control is not Left nor Right behavior specifically – rather it is poor management, as permitted by poor voter education.

In 1980 America, and other G7 countries followed suit, political campaigns exploited the public's low education level to win election and further business interests - rather than to educate the public in how to judge what responsible financial controls for society are. Political campaigning appealed to our emotions rather than to logic because they could. Voters were told that if they didn't support "Low Tax", they were "a socialist" or "Un-American" - basically. The American Dream ended for most of the poor minority who's voting strength was smaller than the richer majority.

Democracy is an important building block of World Peace, so sound Financial Controls are important to a Good Life as well. Every democratic voter must be educated well enough to stand on guard against this, or Democracy itself becomes unsustainable over time.

The responsibility of the right and left should always extend to the "greater good" as a minimum precept, general rule, and guiding principle. If there are not enough resources to move everyone forward, then work on revenue growth until there is enough wealth – but once there is enough wealth, as is the case for the G20, keep it distributed by graduating tax.

Economic Sustainability Goals

The goal here is always to reward performance without creating gross inequities that cause suffering, militias, and social problems. Had this been done in the G8 in 1980, the world would have avoided today's K-Wave Winter (our current economic problems) altogether.

In summary, John Kennedy's "Peaceful Revolution" by Economic Controls, could have accomplished this by:

1) Enforcing "High Tax" and wealth distribution targets - on big business and high-income earners in Fall and Winter K-Waves.
2) By supporting a minimum wage ($15 to $20 per hour) that also supported sustainable wealth distribution targets, see Chapter 6.
3) Insisting on near 100% employment through controls on offshoring; especially in engineering, would save billions spent on jails and other social problems that create administrative overheads today as well.

4) Control Access to Markets - Always remember that the G8 has populations between 350 million and 30 million consumers and no CEO could continue to manage in a scenario where his products were pulled off of every shelf worldwide. Aligning in this way with allies just gives more balanced business and social benefit.
5) Housing – ensure that housing does not become an investment vehicle that leads to unsustainable bubbles and the inability of children to start families in the same town as their parents.
6) Correct wealth inequity once it occurs in efforts to avoid poverty, social problems, militia, and trough wars.

These and other financial controls are important to democracy and economic sustainability, as is the knowledge of when it is important to implement them as well.

Chapter 6 – Social Projects

The Social Projects in this plan are sufficient by themselves to build a Good Life for seventeen of the G20 countries. There is enough wealth here to distribute, and we could vote it in and build a Good Life for all tomorrow – in theory. Finland took on a Pilot Project it called Universal Basic Income in December 2015 to see if the program would help build a Good Life within its population of 5.4 million. What could explain why the country that is host to the Nobel Prize for Economics is looking at this new program?

Over a long period, islands of civilization have always fallen because of shortcomings or interruptions in either their manual production economies or that of their neighbors. If this can happen to the thousand year Roman Empire, then it would be naïve to believe that it could not happen here again to us - at the hands of a World War III or other series of events.

If we leverage our technology, however, to build an Automated Civilization and Infrastructure, the Transition's Plan for both Social and Technology Projects work together to build World Peace; a world in which many countries, languages, values, and cultures share a common bond of sustainable human rights and a Good Life.

There are few valid reasons that the Transition's "Right Plan" cannot be built quickly as well. In any recession, infrastructure spending is recommended - and all that this plan is altering is that now we are building automated road builders - instead of simply roads.

A Team One - Social Project Leader's role is to secure the finances and laws needed to redefine our Human Rights, to include a sustainable Good Life, and to support the needs of Technology Teams that automate our production economy. Where past government administrations have failed to build simple gun registries, entrepreneurial engineers will be building projects with local resources that just work.

Many of us aspire to find work that sets us for life; one among thousands of us finds incomes that set up our children for life as well. This work builds just that while we spend our days doing all the right things for all the right reasons – which doesn't sound so bad either.

Should we resolutely stand on guard against voices in leadership and society that insist this is not important nor urgent? The answer is yes. When World Peace Agenda projects do not meet targets within your country, vote defensively. A Democratic election is an underpinning, a foundation, of a Good Life, just as does our support through our votes give that plan the traction and forward momentum that it needs.

Technology projects target food delivery, clean water manufacturing, shelter, education, healthcare, security, and other basics are laid out in Chapter 7 – Technology Projects.

Social Projects target Wealth Distribution, like Graduated Tax Programs, 100% Employment, Pension and Benefit Plans, Safety Nets or Guaranteed Incomes, Affordable Housing, Hi-tech Funding; and Wealth Creation projects like the automation of profitable exports and re-shoring of Manufacturing and Engineering.

In Chapter 5, I discussed how "Low Taxes" - for 35 years, mean that our wealth distribution projects now have a steep uphill climb ahead. The leading cause of wage stagnation between 1995 and 2013 included technological changes that supported globalization, and the decline of labor unions and other forms of group bargaining power. We will address these issues in this chapter as well.

Chart 2: Distribution of wealth in the US by quintile, 2010

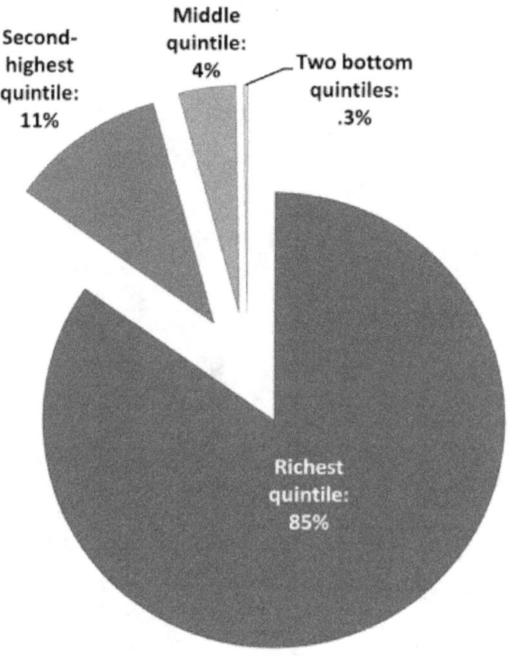

Source: Adapted from Norton & Airely, 2012, http://ppd.sagepub.com/

Wealth Distribution

By 2010, the poor had just 0.3% of the wealth in the United States. That's 160 million people struggling for a 0.3% stake in the wealth. Much worse than this, according to "Robert Reich: Income inequality the defining issue for U.S.", **95% of all gains since 2009's supposed recovery have gone to the top 1%** as well. *The Denver Post*, January 26, 2014; see also (Wolf, 2015).

Remember that wealth and income are two very different measures of economic prosperity, and that high income does not necessarily correlate to mean high wealth or "worth". Net Worth is the sum of all assets, including the market value of real-estate, like a home, minus all liabilities.

How do other nations stack up in measures of Income Distribution? The GINI Coefficient Index, named after its statistician inventor Corrado Gini, is one measure of income inequality depicted here globally.

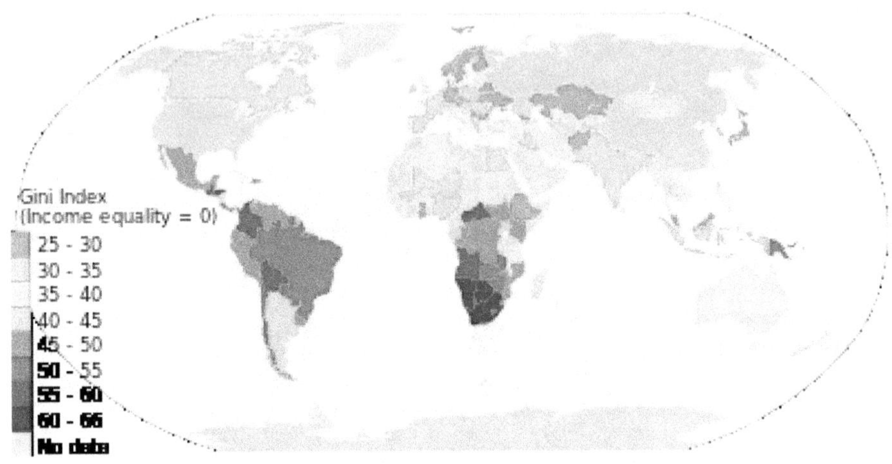

Online at https://en.wikipedia.org/wiki/Gini_coefficient.

#	Country	GINI	Source	Cat	Type
1	Sweden	23	2005	SOC	CM
2	Slovenia	23.7	2012	SOC	REP
3	Montenegro	24.3	2010	SOC	REP
4	Hungary	24.7	2009	SOC	DEM
5	Denmark	24.8	2011 est.	SOC	CM
6	Czech	24.9	2012	SOC	DEM
7	Norway	25	2008	SOC	CM
8	Slovakia	26	2005	SOC	DEM
9	Luxembourg	26	2005	SOC	CM
10	Austria	26.3	2007	SOC	REP
11	Finland	26.8	2008	SOC	REP
12	Germany	27	2006	CAP	REP
14	Belarus	27.2	2008	SOC	REP
18	Ukraine	28.2	2009	SOC	REP
19	Switzerland	28.7	2012 est.	SOC	REP
20	Kazakhstan	28.9	2011	AUT	REP

21	Kosovo	30	FY05/06	SOC	REP
22	Australia	30.3	2008	CAP	DEM
23	Pakistan	30.6	FY07/08	CAP	REP
24	European Union	30.6	2012 est.	CAP	HYB
25	France	30.6	2011	SOC	REP
27	Netherlands	30.9	2007	SOC	CM
28	Armenia	30.9	2008	SOC	REP
29	Cyprus	31	2012 est.	CAP	REP
30	Korea, South	31.1	2011 est.	CAP	REP
31	Estonia	31.3	2010	SOC	REP
33	Italy	31.9	2012 est.	CAP	REP
34	Spain	32	2005	CAP	PM
35	Croatia	32	2010	CAP	DEM
36	Canada	32.1	2005	SOC	CM
37	Bangladesh	32.1	2010	CAP	DEM
38	United Kingdom	32.3	2012	CAP	CM
45	Ireland	33.9	2010	CAP	REP
47	Poland	34.1	2009	CAP	REP
48	Taiwan	34.2	2011	CAP	DEM
49	Greece	34.3	2012 est.	CAP	REP
55	New Zealand	36.2	1997	CAP	DEM
61	Indonesia	36.8	2009	CAP	REP
62	India	36.8	2004	CAP	REP
65	Israel	37.6	2012	CAP	DEM
66	Japan	37.6	2008	CAP	CM
73	Venezuela	39	2011	SOC	REP
92	Russia	42	2012	SOC	REP
100	Philippines	44.8	2009	SOC	REP
101	United States	45	2007	CAP	REP
109	Malaysia	46.2	2009	CAP	CM
110	Singapore	46.3	2013	CAP	REP
115	China	47.3	2013	CAP	COM
118	Mexico	48.3	2008	CAP	REP
130	Hong Kong	53.7	2011	CAP	DEM

Legend: From Most Equal to Least Equal

Categories: SOC=Socialist, CAP=Capitalist, AUT=Authoritarian

Types: CM=Constitutional Monarchy, PM=Parliamentary Monarchy, REP=Republican, DEM=Democracy, HYB=Hybrid, COM=Communist

Patterns in Income Equality

Looking at the Category of Government that is most prevalent in the GINI indexed countries above, the top 10 are all acknowledged to favor socialist policies regardless of their Government types be they Constitutional Monarchy, Republic, or Democracy.

Germany, at 12th position, is the first G8 country to appear and then France at 25th. Italy 33rd, Canada is 36th, UK 38th, Japan 66th, Russia 92nd, and the U.S. is 101. All G8 countries shrink from being called Socialist, not wanting to defend a derogatory agenda, and most G8s have predominantly Capitalist policies with Republic or Constitutional Monarchy Government Types.

Germany and France own to a great number of socialistic policies. Germany, for example, has high taxes, universal health care, universal college education, great infrastructure and government control of the banks. Germans do it all; they do not buy from China; Dollar stores are non-existent; German workers elect boards of companies. It has the largest ownership of its production economy in Europe giving it a competitive advantage because profit and new investment can be wrapped back into the company rather than going to shareholders and it stays in Germany. Germany owns Deutsche Bahn, Hapag-Lloyd, Airbus, Landesbank, Sparkassen, 20% of Volkswagen, 32% of T-Mobile, KfW Bank, 25% Deutsche Post, Hypo Real Estate, Federal Print Office, and many States own businesses within Germany as well.

Theirs is a socialistic capitalist model open to the free market and non-government ownership within the production economy. Government monitors cell phones, TVs, just as does Canada's CRTC. Germany is growing strongly where other G8s are not.

The GINI Top-twelve here boast at least seven of the most successful countries, on a per capita basis, in the world today.

Setting Targets – Minimum Incomes & Tax

I promised in my opening remarks of Chapter 1 that I would not hover too closely on any one country, and then I qualified that I would be constrained by the availability of statistics from time to time, to prefer to discuss a country based on ease of data collection. In this regard, America has every statistic needed to explain the process of Wealth Distribution within easy reach - and so I will use them as an example here.

Plug your own country's numbers into the formulas here to come up with Quintile income and wealth distribution targets for your country.

Wealth

Available Assets – per page 199 of the 2010 U.S. Budget containing the following wealth and income statistics for 2008.

$ 10.2 trillion - in publicly owned assets
$ 54.2 trillion – in privately owned assets
$ 57.2 trillion - in education capital
$ 3.9 trillion – R&D capital
$ 118.1 trillion in total after assets claimed by foreign interests

1995 Assets were about half these numbers at $ 54.1 trillion.

The Reader's Country Total Assets __________

Debts

$ 18.15 Trillion – up from 16.1 in 2012 when it was 108% of GDP

Debt was presented by the Federal Government Debt Clock at usgovernmentdebt.us on October 15, 2015

Debt to Asset Ratio

18/118 = 15%

Although this makes one wonder why to carry debt at all, the current debt load appears to be within any mortgage lenders serviceable range that is often 34% of assets.

Income

$12.95 trillion annual (2012 – U.S. Bureau of Economic Analysis)

The reader's Country Total Income __________

The Lowest Quintile (20%)

To adjust all citizens to wealth and income levels suggested in Chapter 5's IDEAL Income Distribution Chart (11% wealth for the bottom 20% of wage earners), wealth within the bottom group of 80 million U.S. income earners, would have to rise to approximately $74,387.

That's 11% of $54.1 trillion of assets (wealth in 2003) divided among 80 million. Also, these 80 million people would then need to take home incomes of $30,000. These incomes appear to be more than double current U.S. household income levels. See Census.gov at the link here (US Census Bureau, 2015) in Bibliography.

To achieve a **zero-tax income** of $30,000 per year would require 100% employment, or equivalent income, with a minimum wage of approximately $18 per hour plus health coverage benefits (allowing for 4% vacation pay).

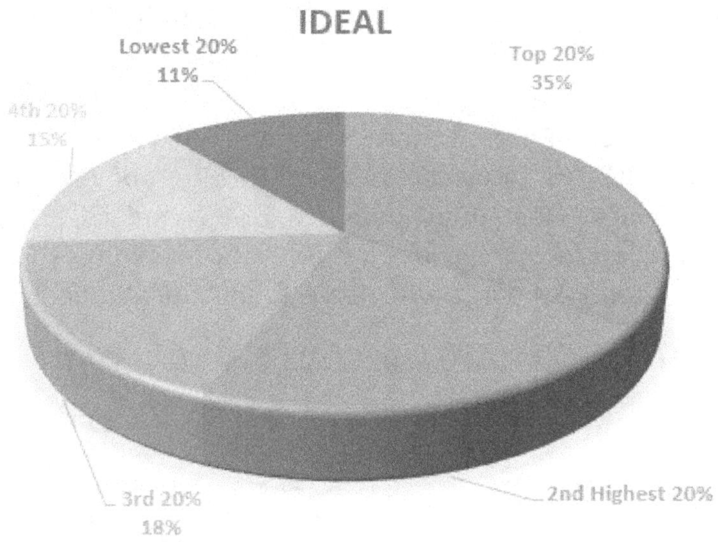

Next Lowest Quintile

By the "Ideal" income distribution chart above, 15% of total wealth is assumed for the second lowest Quintile. 15% of $54.1 Trillion divided among 80 million calculates to wealth per individual of $101,438 and family incomes of $60,000 annually or a minimum hourly wage of $34 per hour with benefits after tax.

Middle Quintile

The Middle 20 Percentile targets wealth of 18%. 18% equates to wealth per individual of $121,725 and family incomes of $90,000 annually or a minimum hourly wage of $49 per hour with benefits after tax.

Second Highest Quintile

21% amounts to a wealth of $142,012 and family income of $120,000, or a minimum wage of $65 per hour with benefits after tax.

The Highest Quintile

35% average wealth of $236,687 with an income of $150,000

annually, or a minimum wage of $82 per hour plus benefits after tax.

Wage Stagnation

In this next Federal Reserve chart, household income is the red line beneath GDP (Gross Domestic Product) in blue above it. Leading causes of wage stagnation between 1995 and 2013 include technological change, the decline of labor unions and more specifically, their joint bargaining powers, and globalization.

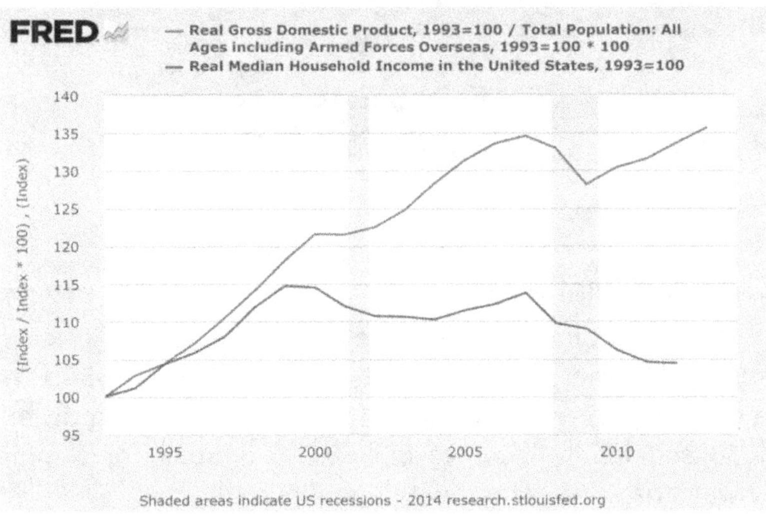

Minimum Wage

By taking this opportunity to target national goals of wealth distribution and employment, governments put an end to "Starvation Wages" and reduce the many and varied costs of Social Problems created by Wealth Inequity.

Creating a governing principle within the laws or guidelines of your government's economic policy, should ensure that these minimums are regularly reviewed and kept current. Over time, this will not guarantee perfect distribution, but it will go a long way to protect the Good Lives of everyone while permitting movement between the Quintiles for High Performers.

Graduated Tax

Implementing Graduated Tax structures ensures that wealth distribution meets national targets for equity sustainability. In a graduated system, both large businesses and rich individuals pay a larger percentage of the total costs of running the country and keeping all employed and salaried.

As discussed above, there are often five quintiles used to represent the population of a country. This model might be too simple for very large countries, but it serves as a starting point that can be used to monitor the success of graduated taxes actively over time.

As wealth creation remains strong and wealth distributes to targets, Governments can be said to have managed the country's finances very well.

Wealth Creation

The way to generate wealth for a country is to take the raw earth and things that grow on it, and sell it as productized resources or manufactured goods in exchange, or in trade, for currency from other nations.

The challenge for economic planners is that you are creating wealth, and you will become wealthy only, based on how much product you can sell versus how much you have to import to create that export. Being self-sufficient is also important.

For this reason, the GDP of a country is not enough information to tell you about its Wealth Creation efficiencies by itself. You have to ensure that GDP Imports are less than Exports first, and then I find that normalizing exports per population tell much about the productivity and use of human capital within a country. A country with highly skilled and educated engineers, who are enabled to make high-value manufactured goods that are sought the world over, like Germany with its cars and America with its software, tells me that the government gives well-considered priority to wealth creation.

The United States leads wealth per capita for the G8 – which is tremendous of course, but it is also the worst among that group in measures of their income distribution. Norway's exports, by comparison, are 40% higher per citizen than the United States, and income equality in Norway is 7th best in the world. Norway also ranks first on the UN Human Development Index as they have low unemployment, universal pensions, healthcare, education, state day-care, affordable housing, $20 minimum wage, and generally – a Good Life for all.

In fairness, there are just 5 million people in Norway, 10 million in Sweden, 16 million in the Netherlands, and another 6 million in Denmark – but these 37 million (the same population as California) combined produce 19% higher exports per capita than the top G8 exporter.

As an aside, Hong Kong's 7 million population has twice the Wealth Generation of Norway per capita with the highest income inequality of all countries mentioned. Hong Kong is unique in that it does not have a democratic election of their government, 45% live in state housing where living area is 47.8 square feet per capita – essentially a box with a window only if you are lucky. Rent will cost 40% of household income that is $120,000 HK or $14,000 US, but there is universal healthcare and low unemployment (3.5%).

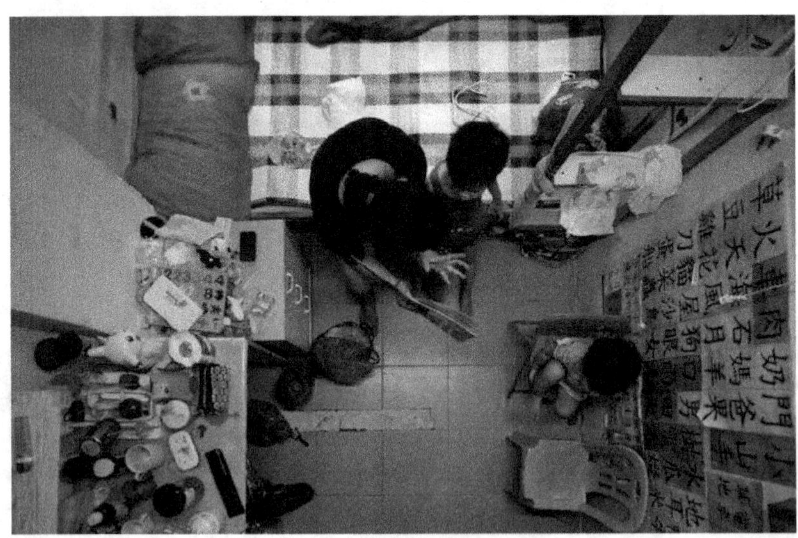

(Ho, 2015)

I mention these considerations because I would not have discussions of economic policy turned in knots over Hong Kong. It's a country purpose-built for international business with an HDI on par with the UK – but clearly, the UN's HDI does not reflect GINI Wealth Distribution influences very well, as seen in this photo of a typically subdivided apartment in Hong Kong. Hong Kong ranked 130th in GINI, 15th HDI, and Top 3 in Export per Capita.

China is the leading wealth generating nation in the world with almost twice the Export GDP of the United States. China also surpassed the U.S. as the world's largest manufacturer in just this past year. See the darkest colors below for highest GDP countries worldwide.

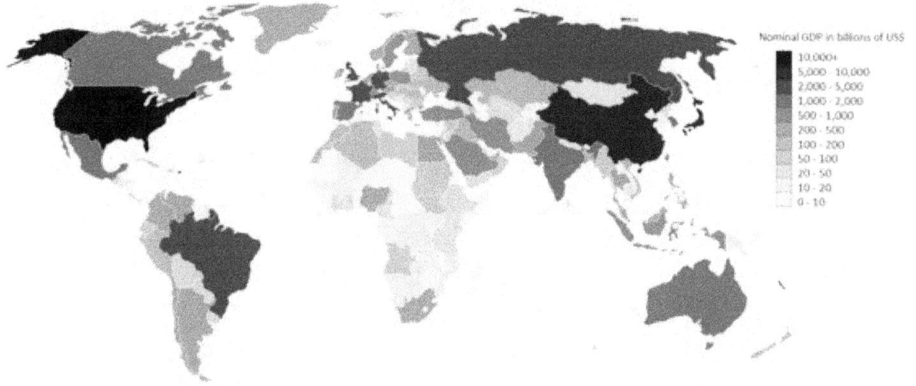

If the GDP Exports per Capita are high for your country, you are probably exporting high-automation, high-value manufactured goods to other nations and customers.

Socialistic Policies Mean Business

I decided to test the concerns expressed by many fellow Canadian Conservatives, that socialist and anti-offshoring policies would drive away wealth from the country. In the chart below, Wealth Creation for each country is sorted in descending order by Export GDP per Capita. G8 countries and strongly socialist nations (shown here in bold) are represented here with exceptional GDP Export countries for comparison.

Country	Pop (mill)	Export %	GDP Export	Export /Cap	Area	UN Index
Luxembourg	0.5	203 %	122,258	244,516	811	0.881
Singapore	6	190%	563,451	93,909	459	0.901
Hong Kong	7	229%	629,139	89,877	316	0.891
Switzerland	8	72%	494,541	61,818	41,284	0.917
Netherlands	17	82%	707,925	41,643	41,850	0.915
Norway	5	38%	203,089	40,618	323,802	0.944
Denmark	6	54%	182,728	30,455	43,094	0.9
United States	323	55%	9,344,834	28,931	9,526,468	0.914
Sweden	10	43%	253,842	25,384	450,295	0.898
Germany	83	50%	1,890,123	22,773	357,114	0.911
Finland	5	38%	102,066	20,413	338,424	0.879
United Kingdom	63	42%	1,150,129	18,256	242,495	0.892
Canada	36	30%	553,160	15,366	9,984,670	0.902
Australia	24	19%	304,419	12,684	7,692,024	0.933
France	65	28%	793,659	12,210	640,679	0.884
Italy	61	28%	613,893	10,064	301,336	0.872
Spain	47	31%	439,643	9,354	505,992	0.869
Japan	127	14%	721,554	5,682	377,930	0.89
Russia	142	28%	594,855	4,189	17,098242	0.778
China	1,393	26%	2,423,838	1,740	9,572,900	0.719

This chart shows clearly that socialist nations are consistently at the very top of Export GDP per Capita ranks worldwide.

World Bank Statistics above are at data.worldbank.org for 2014-estimated. Every effort is made to be accurate and transparent despite resource constraints, so look for a copy of this spreadsheet at csq1.org.

Strongly socialistic-policy nations consistently exported (created wealth) at a higher rate per capita than G8 nations. Export per capita is three times higher for the Netherlands than Canada for example. There appears to be a direct correlation between the United Nation's HDI (Human Development Index) and Wealth Generation but HDI seems less telling of Wealth Distribution.

The next chart and charts above speak to the value of managing human capital **Efficiency**. I concluded that having citizens with better wealth equality, is shown statistically here to result in higher Wealth Creation and therefore, is very good for business.

Stepping back for a minute, consider that this makes sense intuitively for per-capita calculations too, because when 100% of people have the means to contribute to Export Sales, they will probably outpace a population where only 51% have the opportunity open to them to participate - as in an extreme Capitalist Democracy example.

GDP Export Detail	Netherlands	Canada	Russia
Oil, Natural Gas, etc.	$ 117.2	$ 128.9	$ 288
Machines, Engines, Pumps	$ 93.9	$ 32.6	$ 8.9
Electronic Equipment	$ 74.7	$ 13.6	
Pharmaceuticals	$ 30.2		
Plastics	$ 27.4	$ 13.2	
Medical, Technical Equipment	$ 26.4		
Organic/InOrganic Chemicals	$ 25.1		$ 5
Vehicles	$ 21.9	$ 59.7	
Iron & Steel	$ 15.5		$ 20.2
Live Trees & Plants	$ 11.1		
Agriculture & Fish		$ 27.9	
Other Transportation Equipment		$ 17.3	
Consumer Goods		$ 52.2	
Forestry		$ 12.7	$ 7.6
Metals & Minerals Products		$ 53.9	
Gems, Precious Stones		$ 21.5	$ 11.6
Aluminum		$ 8.9	$ 6.3
Cereals		$ 8.8	$ 7
Aircraft, Spacecraft		$ 12.4	
Fertilizers			$ 8.9
Copper			$ 4.9

If your exports are simple commodities and resources, the country's export income tends to be relatively low-profit, high-volume, and sales by the ton. When you look at large and resource-rich countries like Russia and Canada, you may be very surprised to see that the Netherlands has a 20% higher GDP export despite $1/1000^{th}$ the resource base. Norway too is very small by comparison, and yet it has a larger export GDP than any three of these countries.

So, improve your exports and create Wealth, and also study the export **Quality** of other countries to improve your export quality as well.

Lastly, consider **Risk**. When a country's GDP Exports are heavily resource based when commodity markets tumble, as they did for oil in mid-2015, these countries will struggle to maintain their committed level of spending and will have a difficult time averting bankruptcy. Russia is a major oil exporter, and almost half of their export incomes are oil-based. That is one example of a country that will be hit hard by oil prices dropping to $40 per barrel.

By sifting through the Export GDPs of other countries, we can begin to see that there is much to learn by studying the difference between Russia, Canada, Mexico, Netherlands, and Norway - which appear to have similar GDPs per capita, but which are quite different in **Risk, Quality, and Efficiency**.

There are many websites that catalog and present Export GDP and this can be considered required reading for government leaders asked to make policy and investment decisions concerning exports.

Reshoring Production & Engineering

For 100 years or more, toward the end of the industrial revolution, offshoring production to countries where products could be built more inexpensively became a normal practice of business in North America. Pollution controls were less onerous abroad, and the damage to the environment was not felt here at home either.

Governments permitted or ignored this practice in countries where jobs were thought to be readily available in other industries or by retraining workers. Politicians believed that healthy businesses that made healthy profits would make a healthy and wealthy country as well.

This trickle-down was the model that worked throughout the early part of the 1920s when Henry Ford rebuffed the expressed opinions of peers, and insisted that his factory laborers receive high wages by the standard of the day. Mr. Ford realized that good wages and employment built good communities and that if he could be smart in his production, the highly incented workers would also ensure high quality and reliability of his products for a lifetime.

The strategy worked very well for Ford as every worker bought his cars and then generations of children did too. Truly, the interests of business were aligned with the interests of the country. The Great Depression and World War II trained a generation to understand very well that building strong societies was the only goal that mattered – and business was to be the tool for society's construction.

Since the year 1995 or so, the effect of causes automation and offshoring, was to sustain high levels of unemployment, and as workgroup collaboration tools were improved, the engineering and programming work also went offshore. I noticed that the practice of offshoring engineering became flagrant by 2014 or so here in Canada.

The Netherlands had none of these problems. Today the GDP in one of the smallest nations in the G20 is roughly the same as both Canada and Russia, and their technology industry is world renowned and thriving.

How did the Netherlands achieve this? They made offshoring their engineering illegal twenty years ago, and only permitted companies access to this market of 30 million when those companies produced good and jobs in the country.

The Netherlands was too small to sustain the job losses caused by offshoring early on and became an early adopter of reshoring twenty years ago.

Recently, I read an article that interviewed the CEO of GoodYear tire. He had just returned from a visit to his plant in France and announce to the reporter that "I'm not stupid," I watched workers labor 4 hours a day and sit around unproductively for the remainder. I will be closing that plant and moving it offshore.

What the Netherlands did was to refuse access by GoodYear, to sell their products, if not manufactured in the Netherlands. The CEO of GoodYear Tires would remain CEO for about a month if France denied his company access to sell tires to 66 million French car owners – not counting trade allies. This is the power of on-shoring.

Multinational businesses will externalize social costs and play a game of benefit and profit to their advantage until governments collaborate and take back their markets. Since the Bush administration, the U.S. has played Mexico and Canada off one another in this way quite deliberately; pulling jobs from one country to give to the other country unless each country contributes sizable ransoms to sustain its jobs.

Mitigating Multinational Extortion

Extortion is a strong word and one that should not have to appear in a book architecting World Peace. Especially at a time when it is in the interests of both business and consumers to share economic benefit until a redistribution of wealth can sustain strong capitalist momentum again in a new K-Wave Spring.

No politician wants jobs lost within their electoral boundaries, and so the Canadian government has made generous payments to a brand of extortion that could have benefited society elsewhere many times – most recently with GM in Oshawa in early 2015. Lexus was given a generous subsidy immediately following this move without even asking, to pre-empt the possibility of an announcement that they too could leave.

To mitigate extortion – a problem that costs both Canada and Mexico money and jobs lost back and forth over time, all that these two NAFTA Trading Partners have to do is to work collaboratively to ensure that companies cannot operate in socially irresponsible ways. GM in this example would be asked to set up production in both countries – or have their market access closed in both or all three countries, as well. A win-win-win for Mexico, Canada, and the car maker who continues to have access to sales markets and skilled labor.

If the Netherlands' advice holds merit in the eyes of U.S. lawmakers, a production plant could be set up in the United States as well.

Philanthropy

Many of the biggest names in technology and business earn upwards of $3 million per day and more. Fans, employees, and dependents in politics surround most billionaires in all other areas of their lives for more years than not. Even the very few billionaires who have had tough years; have now been successful and powerful for most of their adult lives as well.

It's interesting to see how individuals from different cultures react to constant attention. American Rock Stars seem to lose themselves in their popularity where the Brits – like the Rolling Stones and even Paul McCartney, never really have lost their roots and grounding quite as much.

To be fair, some individuals lack the basic ability to empathize with others as well. Whenever someone simply does not have to interact with others due to their station, or when they have been programmed not to – that is an obstacle. The Queen of England has taken a lifetime of instruction on how to handle this burden professionally and still see to her obligations and duties; most other billionaires must figure it out for themselves and will get it right as often as not.

The reality is that there are just too many voices asking for help, and too many different directions offered to apply that help to best effect. It is seldom clear to the giver that the organization administering the distribution are directing funds by a worthwhile plan. The Right Plan, again, is a recurring reference within this book to Aristotle's insistence that most plans are not comprehensive enough nor targeted enough to reach the all-important goal of a Good Life and Self-sustaining community.

I hate to single out best intentions, but just to give a frustrating example of how difficult the choices can be for philanthropists. The Gates Foundation and many others followed the advice of professors and PhDs at the UN's World Health Organization (WHO) and spent millions of dollars to purchase and ship condoms to Africa.

Adoption was comedically abysmal, and millions of unopened condoms created an environmental problem that filled landfills by the truck-load. To make matters even worse, recent reports revealed that condoms are making AIDS worse and not better in Africa – apparently condoms are only effective among high-density groups like Bangkok prostitutes and San Francisco's gay community.

Were it me in the shoes of the giver, I would be rethinking philanthropy because it can be frustrating. But philanthropy is not difficult really, you just have to resign to a little stock taking and cut back on the gala events. Gala operators are high achievers and wonderful individual contributors, but might not be terrific builders and engineers - as evidenced by their career choices.

Instead, hire engineers quickly, and then assign as many as possible to work on World Peace Transition bottom-up projects in coordination with top-down efforts by the United Nations. Supplement the safety net incomes of the unemployed in the countries that your businesses take revenue and then train workers of all ages that want to help with this project work too. It costs relatively little and opens up new worlds of opportunity and possibility for everyone.

The plan is self-directed, simple and given time and resources, will care for all regions of the world over just the next twenty years while we automate our manual production economies at the same time.

Make sure that you take a few minutes to review the Founding Father's discussion in Technologia below in Chapter 11.

Billionaire Bequeathals

Some years ago, Warren Buffet made news headlines by pledging to the world that he would give away his fortune leading up to, and at the time of, his death. Other billionaires followed suit by pledging to give away various percentages, half, etc. of their wealth upon their death, and I have called this selfless notion the "Billionaire's

Bequeathal" here.

Most people in our society will not earn $3 million in their lifetimes, and it comes as quite a surprise to me that our society is so dependent on the good graces of these individuals that no-one has called them out to say – what's wrong with tomorrow? Recalling language that I learned from the days when I used to play penny-ante poker with the guys in my basement: Throw in your chips!

Charity begins at home, so ensure you have made provisions for all others in your country to have a good life. In this way, they are then able to help support efforts abroad as well. It's very like the old "give a man a fish" story; instead of a meal, give him the tools to fish for a lifetime.

Examples are everywhere: Jim Simons, of Renaissance Technologies, supplements the salaries of 800 Math Teachers in New York State, the Gates Foundation, and Google philanthropy allocate tens of millions of dollars to World Health Federation projects, on and on.

Billionaires can be rich; there is no need for this to change. Wealth distribution simply works to ensure that 40% of society do not have to struggle needlessly and unproductively; living with no plan nor hope for a Good Life, competing for just 0.3% of all of the available wealth that there is.

Here is an interesting statistic. U.S. Federal Reserve Statistics supports that the Billionaires of the United States could get together and write a check tomorrow, still be rich, and enable every individual in the USA, and even seventeen of the lesser wealthy G20 nations, to begin living a Good Life. A wealth distributed life that would then permit perhaps a billion people, to begin also to support the rest of the planet in attaining their Good Life as well.

The Case for Distributing Wealth Quickly

Assuming that this redistribution of wealth was to happen quickly enough, this would probably have us all busy and productive enough to allow us to skip right past any fear of World War III's obliterating the planet as well.

The WWIII discussion is thought provoking and worth bringing up now because any bequeathals planned after that extinction-level event, matter little. In this context, Wealth Distribution becomes the ounce of prevention that creates a pound of cure.

Income and Wealth Distribution is especially important at a time when one great technology push could make money irrelevant once and for all within twenty years. Isolationists love over-population scares, but there is no foundation for it just like there is no foundation for terrorist fear mongering in our media. There is plenty of energy, lebensraum (living space), and resources. World Peace is going to be a terrific thing, and calmer heads will prevail.

By way of a higher validation, all of the great thinkers – Plato, Aristotle, Socrates, Elizabeth I, Buddha, etc. without exception spent lifetimes confirming and articulating that selfishness is unsustainable. That only the Greater Good, with the guidance of a Right Plan and action, could attain a self-sustaining community - and World Peace. By Good's grace if you will.

Team One - Next Steps and Targets

I think that I have covered all of the important discussion points and targets that need to be understood before Team One can begin their project work.

Look to Chapter Two above for Team One's list of projects and get online at CSQ1.org to ask questions in our forums and to discuss your progress with fellow Teams in other parts of the world.

Step One has much to do with getting project team resources funded, and so you will want to get your Project Charters kicked off based on the examples in Chapter 17.

Voters will be looking for results that include:

Wealth Distribution

- Graduated Tax
- Transition Safety Nets or Guaranteed Income – Afforded by both Business and Government

        - Safety Net One – Engineering Safety-Net used resourcing Transition and CSQ 100 Year Plan projects
        - Safety Net Two – Social Safety-Net a guaranteed income that ensures a minimum Good Life for all.
    - Low Unemployment, Good Jobs for New and 50+ year workers, with pensions and benefits.
    - Minimum Wage adjusted to transparent income distribution targets
    - On-shoring – Engineering and then Robotically automated other export and import products
    - Leverage Record Profit sectors for new jobs – Banking will tend to peak during recession for example.
    - KPIs for business to meet social responsibility CSR measures
    - Affordable housing and an end to Mortgage Usury
    - A solution to the Divorce and Housing Projects discussion in Chapters 12 & 13
    - Leadership by Engineers & SMEs
        - Procurement departments need a complete leadership change when they touch technology contracts. The popular notion is emerging this past five years that "You don't have to technical to lead technology teams" is a very incorrect assertion made by business leads given too much control over engineers.

Chapter 7 – Technology Projects

The Technology Projects will automate our production economy until we have a Technology-based Social Safety Net. World Peace Technology Projects target those projects that are required **to automate the production of the basics of life and a good society.**

> "If every instrument could accomplish its own work … chief workmen would not want servants, nor masters slaves."
>
> Aristotle 322 BC

Using the example from the opening chapter of this book – the knock on the door is a delight as you have been waiting for this day for a lifetime.

This person is living in a home with the basics of life; like potable water and sanitation. If this person's home already exists – as in an existing city, then there is little to build and maintenance bots will keep the building up to code throughout its useful service life.

If the home had to be built new, then a robotic building system would have to grade and service the land - according to a subdivision plan. A House Plan would then either 3D Print the house - or robotically assemble it from a bill of material of parts that arrived by automated transport just-in-time as needed.

Automated digging and pouring of the foundation was followed by framing, roofing, windows and doors, and then wire and plumbing. Drywall, flooring, and baseboards were installed, painted and then cabinets and finished fixtures plugged themselves together. Finishing touches included lighting, furniture, and entertainment and computer equipment as needed.

By adhering to conventions for rapid development, the building systems have taken about 18 months to develop and then another six months in integration tests to get the bugs out. The finished house goes up in about a week and shows pretty well for a version 2.0.

Version 2.0 also includes the automated construction tools and processes needed to build this robotic tool so; now we can roll out both tools and homes automatically. All designs including how to improve the next release of the system are conveyed to the World Peace Charter as needed to let engineers get the remaining bugs out of the system by version 3.0. By version 2 or 3, there should also be at least three versions of the solution for the choosing. For the example project of a robotic house builder, there should be three styles such as a townhouse, bungalow, and two-story home; or perhaps a variety of townhouse designs to choose from in the same way that cars are built to order.

Assembling building materials takes place in a variety of automated cement plants and transports; forestry fabrication robots create wood; and automated mining and smelt solutions create metal framing. Window and door glass was created in a central plant in large sheets and then cut to size as needed in local distribution centers. Automatically manufactured door plastics and locking hardware transfer to the site just in time to install as needed. Adjustments to all plans for weather delays and material shortages were also factored into the project software and adjust lead-time plans automatically.

Food systems planted, cared for, harvested, cleaned and assembled vegetables, dairy, meats, canned goods and other grocery items into bags and delivered them automatically as orchestrated by ERP software once requested by members of each home. Groceries and other ordering takes place via a web-based grocery app on the homeowner's TV, smartphone and computer systems and is supported by the Help Desk as needed.

The "Worldville" Competitions

"Worldville" 1 & 2, whose final naming will be determined by a local competition, are the first Pilot Towns with an automated production economy towns.

"Worldville 1" is an existing town with new automation as needed to automate 90% of the production economy. With 100% automation worked toward within five years.

Worldville 2 is new construction with all systems tested to work together to deliver an automated town. This production economy can sustain 50 to 1000 residents in modern, comfortable, constantly improving, sustainable accommodation - complete with a park, school, marina, hospital, old age facilities, good jobs and a recreation center with growth plans as needed to ensure a terrific future for all members of the community.

Worldville Towns and cities are no island communes, nor amusement park attractions, Worldville is a plugged-in vital part of the planet's economy. A place where real people want to live.

A Knock at the Door in "Worldville"
pepper
WORKING
WORKING
WORKING
WORKING
Copywrite 2015 World Peace – The Transition at CSQ1.org

Read "Getting the Word Out" in Chapter 24 and assist the development of World Peace projects in your country as quickly as possible. By late in 2016, we should have a release 1.0 of many of these technologies, and hopefully enough of these systems should be under development to continue interoperability work as needed get these systems working together well enough to build one or two towns by 2017 - with 2018 as a rain date.

If that does not sound like an easy test, I do not know what does. Everything is up for discussion during design here so get working and stay tuned as the work continues.

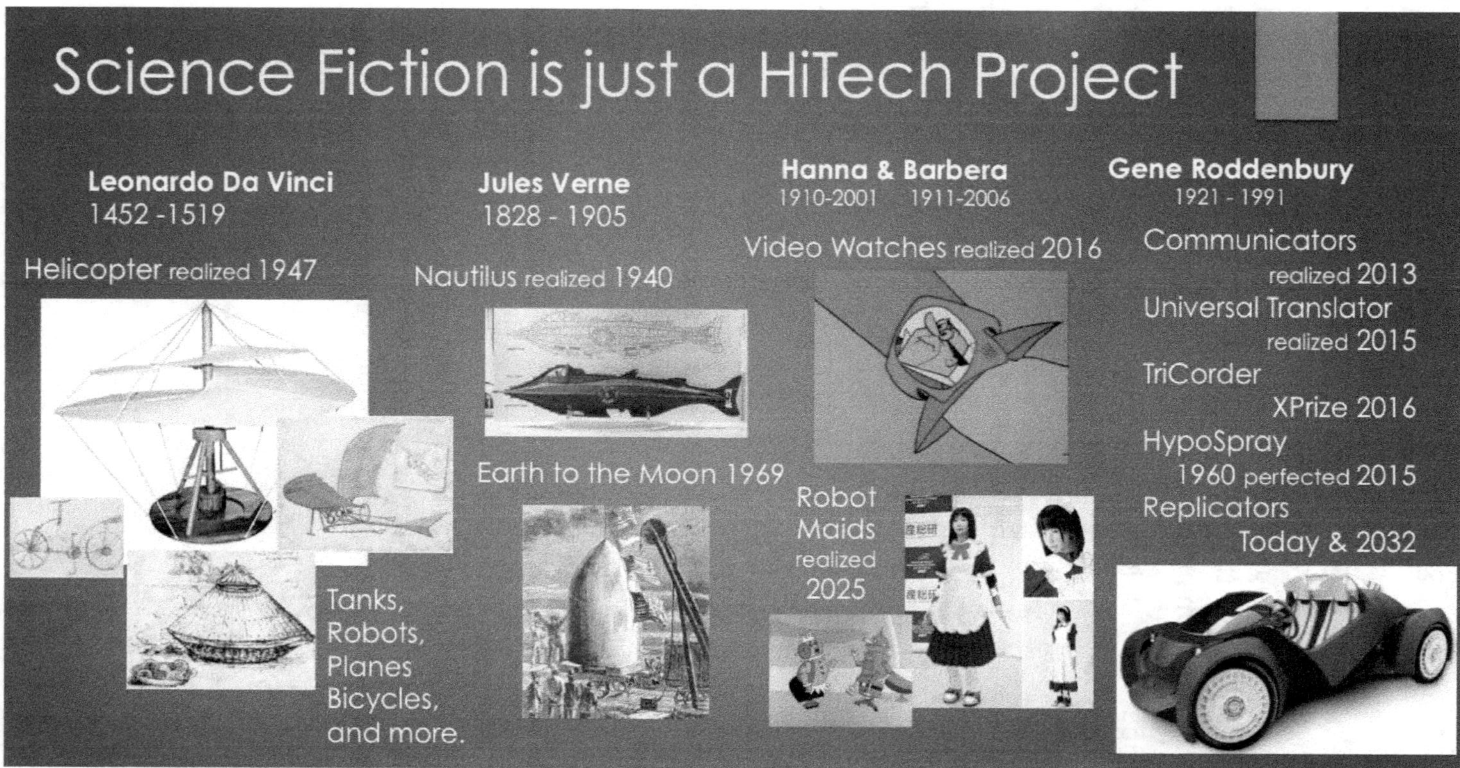
Science Fiction is just a HiTech Project
Leonardo Da Vinci
1452 -1519
Helicopter realized 1947
Tanks,
Robots,
Planes
Bicycles,
and more.
Jules Verne
1828 - 1905
Nautilus realized 1940
Earth to the Moon 1969
Hanna & Barbera
1910-2001 1911-2006
Video Watches realized 2016
Robot
Maids
realized
2025
Gene Roddenbury
1921 - 1991
Communicators
realized 2013
Universal Translator
realized 2015
TriCorder
XPrize 2016
HypoSpray
1960 perfected 2015
Replicators
Today & 2032

Science Fiction is just a Technology Project

We have already made a lot of forward progress – as seen in the complete automation of car manufacturing at Audi and Tesla, and much of the auto industry, and in the Driverless Cars Program at Google. As it turns out, four of the projects identified in the plan are already completed this year as well – Communicators, Universal Translator, Performance Software and Clean Diesel.

Throughout time, visionary Engineers have imagined great works of technology that we come to accept as quite commonplace and not extraordinary at all today.

Leonardo da Vinci designed Helicopters, Airplanes, simple animatronic robots, bicycles, tanks and other technology that we accept as fact and commonplace today. It took 400 years for da Vinci's designs to take flight.

Jules Verne imagined sophisticated technology including Submarines, Time Machines and a Trip to the Moon. It would 100 years to build his Nautilus and Flight to the Moon.

More contemporary engineers employed contemporary mediums too. Hanna and Barbera envisioned Robot Maids, Video Watches, Anti-Gravity, and a broad array of social improvements, like two day work weeks, that would most probably need to be added to the fabric of society in the future as well.

Gene Roddenberry hasn't been gone long enough to be fully appreciated for the contributions that his visual creations gave us in support of space travel and a sustainable Good Life for all. Gene conceptualized Replicators of food, air, clothing, vehicles and everything else that we see today in our 3D Printing Industries.

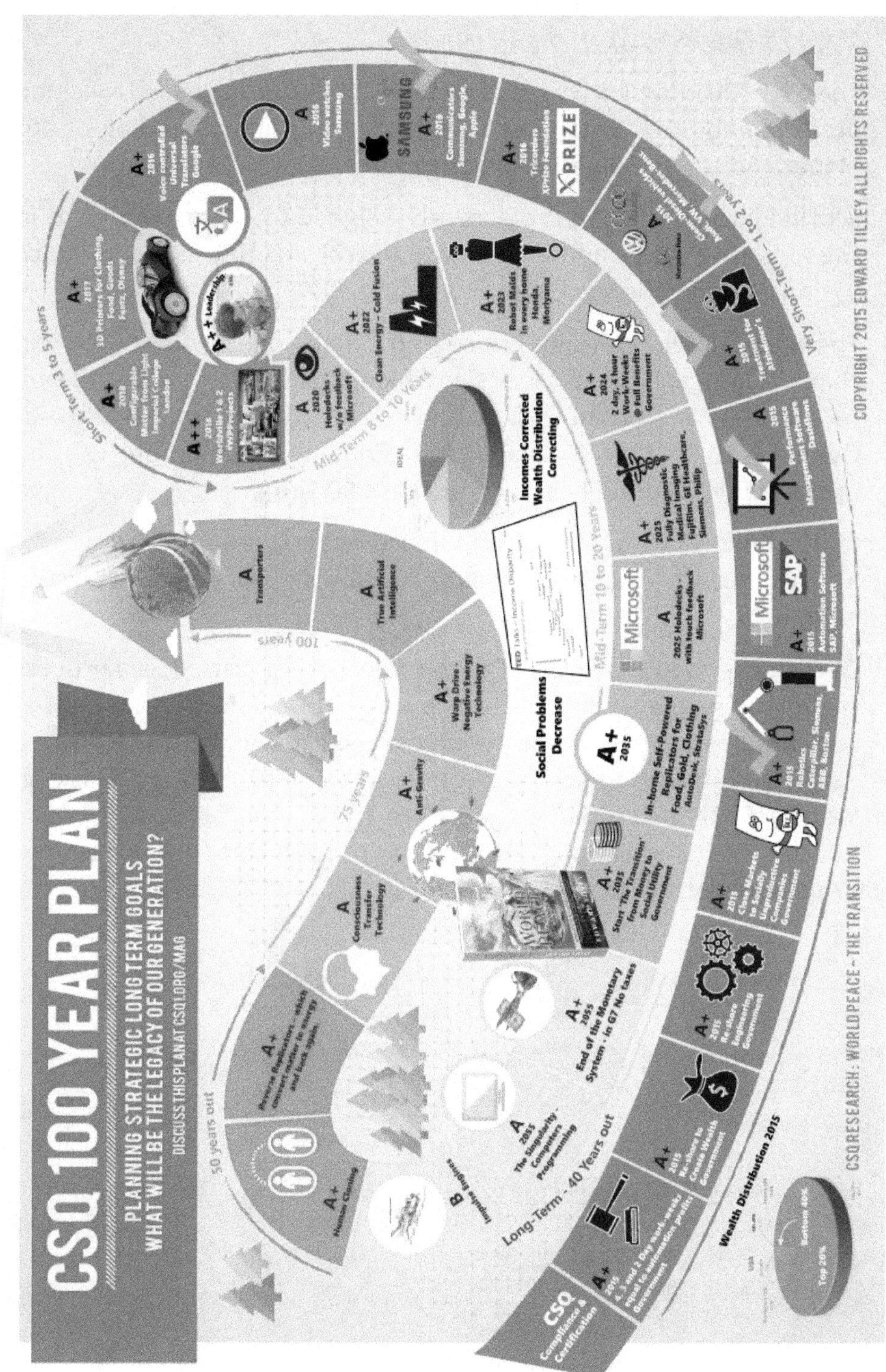
CSQ 100 YEAR PLAN
PLANNING STRATEGIC LONG TERM GOALS
WHAT WILL BE THE LEGACY OF OUR GENERATION?
DISCUSSTHISPLANAT CSQ.ORG/MAG
Short-Term 3 to 5 years
Mid-Term 8 to 10 Years
Mid-Term 10 to 20 Years
Very Short-Term – 1 to 2 years
Long-Term - 40 Years out
50 years out
75 years
100 years
Transporters
True Artificial Intelligence
Warp Drive - Negative Energy Technology
Anti-Gravity
Consciousness Transfer Technology
Human Cloning
Social Problems Decrease
Incomes Corrected
Wealth Distribution Correcting
In-home Self-Powered Replicators for Food, Gold, Clothing
AutoDesk, Stratasys
End of the Monetary System : in G7 No taxes
Start 'The Transition' from Money to Social Utility
The Singularity - Computers Programming
Robot Maids in every home
Honda, Moriyama
Clean Energy – Cold Fusion
SAMSUNG
XPRIZE
Microsoft
SAP
Wealth Distribution 2015
Top 20%
Bottom 40%
CSQ Compliance & Certification
CSQRESEARCH: WORLDPEACE - THE TRANSITION
COPYRIGHT 2015 EDWARD TILLEY ALL RIGHTS RESERVED

World Peace Agenda Projects

The CSQ 100 Infographic above contains both World Peace Transition projects and other initiatives that extend our understanding of our planet and surroundings and make the World a better Place.

World Peace Transition Projects are assigned to Team Two Leads in the following tables and documented according to the World Peace Agenda specification in Chapter 10.

The Team Two list contains Priority, an initial Project Number, Project Names, and each project is assigned leadership at a very high level. As countries and businesses take ownership of the projects, Project Owners, Stakeholders, Leads and Contact information will be updated in the World Peace Agenda at CSQ1.org.

Status Reports within the World Peace Agenda are updated weekly in real time by project leads to alert stakeholders when project teams need help.

Getting assigned to a World Peace project, and salaried, is a process that will change with adoption by the United Nations, member countries and sponsoring companies. Look to our forums for more detail instructions on how to get onto a team as this plan releases in late 2015. CSQ Certified Projects are those who align with World Peace Transition Planning, and you can request certification for your project at cert@CSQ1.org.

Team Two Project Assignments v1.0

Pr	#	Status	Due	Project	Leadership
A+	1	✓	2015	Industrial Robotics	Caterpillar, Siemens, ABB, Boston
A+	2	✓	2015	Automation Software	SAP, Microsoft
A	3	✓	2015	Strategic Digital Transformation Dashboard	DashFlows
A+	4	✓	2015	Treatment for Alzheimer's	Two teaspoons Coconut Oil daily*
A	5	✓	2015	Clean-Diesel and Diesel vehicles	Audi, VW, Mercedes-Benz
A+	6		2016	Tricorders	XPrize Foundation
A+	7		2016	Communicators	Samsung, Google, Apple
A	8		2016	Video watches	Samsung
A+	9	✓	2016	Voiced Universal Translators	Google
A+	10		2016	Rapid Charge Batteries	
A+	11	1 million miles	2016	Driverless Cars & Trucks	Google
A+	12		2017	3D Printers - Food, Clothing - Tools & Parts	Feetz, Disney, Mark, AutoDesk, StrataSys
A++	15		2018	Automate Production Economy AND Configurable Matter from Energy	See #WPProjects below AND Imperial College London
A+	16		2018	GDP Import & Export Automations	Per Country
A+	17		2022	Cold Fusion	Canada
A+	18		2023	Robot Maids	Honda, Moriyama
A++	19		2035	Self-Powered Replicators	Team 2

Many Hands make Light Work

Untold billions of years ago, an enormous explosion of energy that we call the "Big Bang" synthesized our material universe from energy, and with it, created all of the things that we see around us today.

Imperial College of London Physicists will begin recreating that process in 2016 and, when perfected, that technology will be able to replicate our houses, food, and other needs on request. To mitigate the risk that this science will take quite a long time to mature, we will automate our production economies manually as well.

Automating our current Manual Production Economies in Mining, Farming, Fishing, Construction, Manufacturing, Food Production, and Distribution may sound like a tremendous amount of work. Realize that, first; many of the projects needed are within the grasp of an advanced high-school robotics club, and second; we are seven billion people working from 207 countries that can support the work needed worldwide.

I easily identified 150+ automation projects and assigned each of them, one per country - ensuring that countries with the highest GDPs were challenged with more difficult assignments accordingly. I could have assigned projects to states, provinces, and major cities as well, John F. Kennedy gave the Space Program to Houston for example, but there just were not that many high-level projects.

Although it's true that a high-school robotics team could accomplish some of these tasks within a handful of months, other tasks are formidable. Automating the creation of a park, marina, road or building directly from a computer plan, are examples of sophisticated projects that will benefit from assigning sub-projects to many teams - within communities, perhaps even through regional robotics competitions, in other countries. And so on – all project teams then report status up to their World Peace Agendas via a coordinator.

Countries are very welcome to self-assign, reassign or request more projects. There are many blank spaces left free for volunteering and reassigning the leadership roles of suggested projects as well.

Countries, states, cities, companies, clubs, and even individuals are encouraged to sign up to help. Countries are asked to fund resources as requested - and as able.

And finally, if each higher-GDP country and large company were to sponsor resources needed by efforts led in developing countries that could get them contributing very actively as well. I have suggested sponsorships for each country with an imperfect understanding of each country's relationship – and although that may be a good thing for World Peace generally, some juggling may be required for public safety and practicality.

The intent of delegating tasks worldwide is to include as many of us in a working solution for World Peace as possible. The Transition is a plan supported by the world's greatest technology, companies, and countries for a global win-win – so commit to an assignment and count yourself among the team.

Assigning Projects

Some effort has gone into matching top exports from each country with the needs of automation projects.

Cold Fusion requires micro-particulate Palladium and Nickel with facilities similar to those in a fission nuclear energy classroom, lab, and plant. Rapid Charge Batteries – need Lithium and specific lab needs too; Construction projects include manufacturing, installation, and all solutions must work in sequence - wiring comes after roofing, which comes after framing, after foundation, and so on, and all must assemble per a computer plan's spec.

Assigned Project Leads are responsible for resource delegation, performance and they can draw resources as needed worldwide - which should increase exports for those contributing nations as well.

All major projects have a one-year timeline for a release 1.0, and all systems must be interoperable with one another by no later than release 2.0.

(Grossman, 2014)

Companies that will assist automation of ERP systems are SAP and Microsoft - with open plugged-in solutions from robotically controlled assemblers and other projects provided by third parties and World Peace Transition Teams. Most delivered solutions will need to be triggered and managed by ERP software, and so I suggest that this is a deliverable of every project team as well.

In v1.0 of the Automated Production Economy Project List below, Canada is assigned Cold Fusion despite the Soviet Union's 250,000-resource-strong nuclear industry resource base. Russia is preferred to lead robotic automation projects because of their expertise in drone technologies and engineering. Can Russia contribute to Cold Fusion as well? Absolutely and enthusiastically - yes; either in part or in full, and we ask only for a collaborative discussion with a Canadian counterpart who has been asked to maintain Status Reports before you proceed. All documentation should adhere to Google Translate and conventions laid out in Chapter 10, and then you are good to go.

Here is the initial per-country project assignments list. There are many missing projects, but this list covers a broad start. You could take an inventory of 100 top selling items at local grocery stores to get another list of projects, and popular staple products vary widely from country to country. A U.S. database of international trade information is available for query by the public at https://dataweb.usitc.gov/scripts/user_set.asp. Pharmaceuticals are not complete on this list so unassigned countries can sponsor the following list (Drugs.com, 2014) or other important regional medicines.

Aristotle said that a Good Life is full of the things you need, and not necessarily the things you want. For this reason, products like Tobacco (which kill 20% of smokers), Marijuana, opium and illegal drugs and substances are excused from the list. Alcoholism too is a tragedy for individuals, families, and even whole communities but these products made it onto the list with the caveat that a world with everything provided cannot be a death sentence for the young and addicts. Addictive provisions will be monitored and restricted where public health and productivity concerns warrant it.

This list will be revision controlled, and Change Logged on the CSQ website as more projects are added.

Several countries have provinces or states that are the population and GDP of most countries – examples include Texas and California. Countries are assigned project leadership and may request or volunteer resources to or from other project teams.

Team Two – Automating our Production Economy

Country	GDP	Project List v1.0
Afghanistan	21,618	Sponsor Uruguay Flower potting and shipping
Albania	12,904	Sponsor Lithuania Canning containers
Algeria	208 m	Automated Education Pre to K12
Andorra	3,249	Sponsor Pakistan LED Lights & Fixtures
Angola	121,692	Auto Harvest – Lettuce & Cabbage
Anguilla	284	Sponsor Brazil Coconut and Oils
Antigua and Barbuda	1,241	Sponsor Iran Medical Imaging – CAT Scan
Argentina	611,726	Auto Harvest – Wine Fresh Peppers Picking
Armenia	10,431	Sponsor Croatia Butter Production
Aruba	2,589	Clean Water Plants
Australia	1,531m	Rapid Charge Batteries Automotive
Austria	428,322	Ground Transportation Mgmt. Paint Systems Management
Azerbaijan	73,557	Hi-Rise Crane Operation Coffee Makers & Grills
Bahamas, The	8,420	Sponsor Sri Lanka Missing Supply Contingency Planning
Bahrain	32,898	Plastic Recycle
Bangladesh	153,505	Auto Harvest – Carrots & Potatoes
Barbados	4,228	Sponsor Iraq Garden Build, Plant, Sprinkler, Maintain
Belarus	71,710	Auto Dredger

Country	GDP	Project List v1.0
Belgium	524,806	Produce Transport Chocolate
Belize	1,624	Sponsor Denmark Washer & Dryers
Benin	8,307	Sponsor Belarus Ice and Crushed Ice
Bermuda	5,574	Auto Ground Maintenance
Bhutan	1,781	Sponsor Malaysia Dish Washers
Bolivia	30,601	Auto Harvest - Corn
Bosnia and Herzegovina	17,852	Traffic Control Systems Boots - Rubber
Botswana	14,778	Sponsor Morocco Carpets & flooring covers
Brazil	2,243m	Auto Peacekeeping – Air Confections Candies
British Virgin Islands	916	Sponsor UAE Automated School Build
Brunei	16,111	3D Boat Building
Bulgaria	54,481	Farm Tires
Burkina Faso	12,547	Sponsor Ghana Table Honey
Burundi	2,549	Sponsor Finland Locomotives
Cambodia	15,250	Auto Food Distribution 30% of all emissions is food transport (ask for help from SAP & MS)
Cameroon	29,568	Auto Harvest Baby Foods
Canada	1,838m	Satellite Water Divining, Collection & Distribution Cold Fusion Drone Jets & Air transport
Cape Verde	1,861	Sponsor Argentina Auto Defibrillator maker
Cayman Islands	3,474	Czech Republic Auto Fish Finders & Sonar
Central African Republic	1,585	Sponsor Uruguay Lamb Meat Processing
Chad	10,640	Sponsor South Africa Violins & Chelos
Chile	277,043	Beef Processing Leather - Beef

Country	GDP	Project List v1.0
China	9,181m	Robotic human hand & Trainer software Housewares, dishware, cooking utensils
Colombia	378,148	Auto Pharma - Pill/Package/Transport
Comoros	622	Sponsor Spain XG Guitar, hardware, and cables. Classical Guitars.
Congo	32,691	Satellite Resourcing & Mining exploration – Lithium, Gold, Diamond, etc.
Cook Islands	330	Sponsor Italy Cooking Tools - choice
Costa Rica	49,621	Fruit and Vegetable Distribution
Côte d'Ivoire	28,593	Assist France Dining Room Furniture
Croatia	57,869	Automated Dump trucks
Cuba	78,694	Surgery – Gall Bladder Removal
Curacao	3,148	Sponsor Portugal Automated Planting with Seed
Cyprus	24,057	Smelting Automation
Czech Republic	208,796	Robot-3D Printer handover
Denmark	336,701	3D Printed or Assembled Buildings
Djibouti	1,456	Sponsor Venezuela Shoes - Womens
Dominica	498	Sponsor Australia Motorcycle production
Dominican Republic	60,612	Auto Bathroom Tub and Shower Installation
Ecuador	94,473	Surgery – Eye Banana Picking and Processing
Egypt	255,199	Robotic Process - Poultry
El Salvador	24,259	Auto Harvest Onions & Mushrooms
Equatorial Guinea	18,532	Automated Forestry Harvesting & Milling
Eritrea	3,438	Sponsor Kazakhstan Aluminum Foil
Estonia	24,880	Stone Cutting & Transport
Ethiopia	46,017	Auto Harvest Carrots and Beets
Fiji	4,034	Sponsor Qatar Flex monitor screens & Active Glass Touch Monitors
Finland	267,329	Fishing fillet, package, freeze

Country	GDP	Project List v1.0
France	2,806m	Auto Road Build/Repair Cosmetics Wine
French Polynesia	641	Sponsor France Shellfish and Clam Production
Gabon	16,970	Sponsor Luxemburg Diet Foods
Gambia, The	902	Sponsor Argentina Automated Seed Harvesting
Georgia	16,127	Auto Diesel Engines
Germany	3,730m	Auto Vehicle Delivery Driverless Cars Bavaria – Beer
Ghana	47,830	Automated Marine Engine Mounts
Greece	242,230	Grocery Web Interface
Greenland	2,418	Sponsor Iran Organic House-cleaning products
Grenada	831	Switzerland Medical Imaging – MRI
Guatemala	53,797	Construction Siding Coffee
Guinea	7,219	Sponsor Slovakia Synthesizers and Keyboards
Guinea-Bissau	1,036	Sponsor Poland Distribution Warehouse Makers
Guyana	2,990	Sponsor Ireland Hifi Music Microphones & Stands
Haiti	7,691	Sponsor Libya Shoes - Men
Honduras	18,569	Auto Grocery Bag Manufacture
Hong Kong	274,027	GPS Tractors Holiday and Gift Wrapping
Hungary	133,424	Auto Fence building B eer
Iceland	15,330	Pipelines – Oil & Gas
India	1,937m	Automated Community Build
Indonesia	868,346	3D Food Printers
Iran	492,783	3D Clothing, Tricorder
Iraq	195,517	Water/Waste Pipe Tunnelling
Ireland	232,077	Auto Ground Keeping

Country	GDP	Project List v1.0
Israel	291,567	Auto Airplane design
Italy	2,149m	3D Cement Systems Wine
Jamaica	14,270	Travel Booking
Japan	4,898m	Household Robot Grand & Upright Pianos Pharma Diabetes - Lantus Solostar & related
Jordan	33,594	Blue Crude Plant Automation S/M/L
Kazakhstan	224,415	Automated Cement Mixers
Kenya	54,443	Construction Glass
Kiribati	175	Sponsor Germany Off road vehicles
Kosovo	6,837	Sponsor Morocco LED Stage Lighting, Scaffolds, control panels
Kuwait	175,831	Plastic Container - Water
Kyrgyzstan	7,226	Sponsor Ecuador Laundry – detergents & fabric softener
Laos	10,760	Sponsor Sudan Music Drum Sets
Latvia	30,886	Small Engine Marine
Lebanon	47,221	Construction Framing – Metal
Lesotho	2,230	Sponsor Hong Kong Zambonies & Ice Makers
Liberia	1,946	Sponsor Chile Ice Rinks
Libya	74,597	Auto Peacekeeping - Sea
Liechtenstein	5,647	Sponsor Angola Woodwind Instruments
Lithuania	46,403	Robotic Process – Rice
Luxembourg	60,131	Automated forklifts
Macau	51,753	Construction Brick
Macedonia	10,767	Sponsor Bulgaria Sport Canoes & Kayaks
Madagascar	10,612	Sponsor Uzbekistan X10 Home Controllers & Software
Malawi	5,146	Sponsor Kuwait Flower Seeds & Pots
Malaysia	312,434	Auto Kitchen Appliance Build Server Production
Maldives	2,836	Sponsor Greece Soccer Balls, Volleyballs, Baseballs, etc.

Country	GDP	Project List v1.0
Mali	10,943	Sponsor Kenya Hockey Sticks & Raquets Tennis, RB & Squash
Malta	9,971	GPS Fishing Routes
Marshall Islands	189	Sponsor France Medical Imaging X-ray
Mauritania	5,516	Sponsor Bangladesh
Mauritius	11,938	Food Distribution Software
Mexico	1,259m	Auto Steel Manufacture Avacados, Mangos Tequila
Micronesia	333	Sponsor Japan Auto Toilets and Installer
Moldova	7,970	Sponsor Azerbaijan Music Drum Sets & Symbols
Monaco	6,559	Performance Management
Mongolia	11,516	Sponsor Macau Woodwind Instruments
Montenegro	4,417	Sponsor Romania Personal Hygene Grooming Products
Montserrat	59	Sponsor China Automated Rec Center & Gym, with Coffee Shop and Dance Hall
Morocco	103,836	Cold Fusion assisted Solar
Mozambique	1,105	Sponsor Norway Feminine Hygiene Products
Myanmar	63,031	Safety - Human Detectors
Namibia	12,580	Sponsor Tunisia Music Sundry, Guitar Picks, Drum Sticks, Capos, Strings, stands
Nauru	153	Sponsor Japan Music Amps Solid State & Tubes
Nepal	18,179	Auto Wood Mouldings
Netherlands	853,539	Conveyers - mining & farming
New Caledonia	9,712	Sponsor Dominican Republic Bar Mix Bottles or Cans
New Zealand	189,025	Farming - herding
Nicaragua	11,256	Sponsor Guatemala Bass Violins
Niger	7,407	Sponsor Oman Violins, Chellos, etc.
Nigeria	514,965	Auto Med Supplies

Country	GDP	Project List v1.0
North Korea	15,454	Auto Med Supplies
Norway	522,349	Auto Oil Production Diapers – All Electronic Drum Sets
Oman	79,656	3D or Auto Blankets
Pakistan	225,419	Grocery Distribution & Delivery
Palau	240	Sponsor UK Molusks Production
Panama	40,467	Auto Pier-Builder - Wood
Papua New Guinea	15,420	Auto Harvest Baby Powders
Paraguay	29,208	Automated Small Engine Manufacture - Marine
Peru	200,269	Robotic Sheering
Philippines	272,067	Technician robots
Poland	525,863	Tech Architects & QC Approvers Interoperability Standards Transportation Software
Portugal	227,324	Marina Management System
Puerto Rico	105,149	Construction Drywall – Manufacture & Auto Installation
Qatar	202,450	Automated Fish Net Harvest
Romania	192,094	Robotic Process - Fruit
Russia	2,096m	Drone Peacekeeping – Ground Auto Snow Removal Oil Alternatives - Paraffin Diesel and Blue Diesel Plants (perhaps using nuclear ship power) Vodka
Rwanda	7,601	Sponsor Cuba Automated Pest control
Saint Kitts and Nevis	743	Sponsor Turkey Pharma Remicade
Saint Lucia	1,336	Marina Management
Saint Maarten	1,021	Sponsor Sweden Pharma Enbrel
Saint Vincent and the Grenadines	709	Sponsor Netherlands Pharma Cymbalta
Samoa	691	Farming – Tropical Fruit
San Marino	1,802	Sponsor Singapore Pharma Neulasta

Country	GDP	Project List v1.0
São Tomé and Príncipe	342	Sponsor India Cars for Wheelchair
Saudi Arabia	748,450	Rapid Charge Batteries - Robotics
Senegal	15,152	Sponsor Panama Pharma Humira
Serbia	45,520	Auto House Building
Seychelles	1,445	Prepared Meal delivery
Sierra Leone	4,929	Sponsor New Zealand Pharma Spiriva
Singapore	295,744	Robotic Mining System
Slovakia	97,713	Construction Framing Wood Bottling & Returns
Slovenia	47,990	Construction Plumbing
Solomon Islands	1,073	Sponsor Belgium Pharma Advair Diskus
Somalia	1,399	Sponsor Thailand Auto Fuel Station Clean Diesel, Gas
South Africa	366,060	New Mining by Satellite
South Korea	1,304 m	Automated TV, & Monitor production
South Sudan	11,804	Sponsor Lebanon Satellite phones
Spain	1,393m	Automatic Park Builder
Sri Lanka	67,203	Robotic Motor options Auto Tea
State of Palestine	12,579	Sponsor Costa Rica Pharma - Nexium
Sudan	54,595	Construction Tile
Suriname	5,299	Sponsor Vietnam Guitars
Swaziland	3,523	Sponsor Algeria Equestrian Equipment
Sweden	579,680	Marina Build System – up to 80 boats
Switzerland	685,434	Surgery – Knee, Hip, Elbow
Syria	35,164	Auto Harvest - Nuts
Tajikistan	8,506	Sponsor Myanmar Pharma Diabetes - Lantus Solostar
Tanzania	44,698	Auto Toilet Paper & Paper Towel
Thailand	420,167	Robotic & Drone Bulldozer

Country	GDP	Project List v1.0
Timor-Leste	4,941	Sponsor Ukraine Pharma Copaxone Hockey Equipment
Togo	4,158	Sponsor Peru Pharma Rituxan & related Fishing equipment, tackle & boxes
Tonga	440	Sponsor Canada Pharma Abilify & related Hockey Sticks, Pucks, Goal Masks
Trinidad and Tobago	24,463	Tropical Fruit Pickers
Tunisia	46,883	Auto cabinet pre-fab, transport
Turkey	822,149	Satellite Driven Well & Irrigation Builder
Turkmenistan	41,851	Automated Kitchen cabinet installation
Turks and Caicos Islands	706	Sponsor Indonesia Medicine - Lyrica
Tuvalu	38	Sponsor United States Chronic Care Robots
Uganda	26,444	Auto Harvest Apples & Pears
Ukraine	188,350	Tunnel builders – Passenger
United Arab Emirates	599,000	Fishing fillet, package, freeze
United Kingdom	2,678m	Energy to Matter Conversion Sanitation – Waste Treatment Scotch – Single Malt & Whiskey
United States	16,768 m	GPS Threshers Harvest Tricorder Med Console Hawaii - Auto Harvest Macadamia Nuts California – Blue Crude & e-Diesel Cheese & Cream Products
Uruguay	55,708	Construction Roofing
Uzbekistan	57,210	Construction Electrical
Vanuatu	800	Sponsor Saudi Arabia Power WheelChairs
Venezuela	371,339	Subway car builder
Vietnam	171,222	Rural Sanitation Black Peppercorns
Yemen	34,714	Auto Door Hardware Rock Cutting to Plan for Home and Harbour Foundation
Zambia	22,384	Auto Window Hardware

Country	GDP	Project List v1.0
Zanzibar	1,161	Sponsor Nigeria Pharma Atripla
Zimbabwe	13,490	Sponsor Ethiopia Pharma Januvia

Adding World Peace Agenda Projects

This chart is a very high-level, initial list that has missed dozens, and even hundreds of component projects. It might even be safe to count on at least one more project per country before the list completes.

These are complete automation projects, no human hands touch the products from raw materials to boxed product. Integrations should be well thought out so that housekeeping robots are able to spot and trash expired produce or meats in refrigerators. List dependent projects in Charter and communicate your needs to them in Requirements.

Do not hand projects to Project Leads who do not know how to begin. You have handed it to the wrong resource and failure is imminent. The right project lead is the engineer who has run these projects a hundred times - and is someone who knows when and how to ask questions that guarantee immediate progress past roadblocks to a final successful v1.0, v2.0 and beyond. You have two years to release v2.0 with support for three to nine variances – so there is a busy two years ahead.

Liaise and seek sponsorship with leading companies and engineers and seek outside engineering help as needed.

Review the list above and use the forums at CSQ1.org or email info@csq1.org to suggest and track additions.

All projects need: assignment to a sponsoring country or business, a project manager with contact information, Charter and other project docs – both in-process or approved, budget and resources, and then the work can begin.

I have not tried to guess at what additional project teams, companies, and individuals will volunteer to take on component automation projects. I hope that the number is considerable and that a small army of engineers and master builders feel they are enabled and can take enough queues from these very basic sketches. I can tell you that the really capable engineers will find 80% of the directions they need here – and divine the rest during the build process.

Find the most recent version-controlled Master World Peace Agenda Project List at CSQ1.org.

If you are not assigned a project to begin with, seek funding, start with Maslow's Hierarchy of Needs, and build up from there. Water – Creation, Delivery Coordination, and Transport; Food, Clothing, Shelter, Sanitation, Security Communications, Transportation, Education, HealthCare, tablet computers, and so on.

Should TVs and video games be on the list? All of these questions need to be vetted so put them on your requirements list and build all deliverables in order of priority – whatever that priority might be.

Break down problems and then build up solutions. Wouldn't it be terrific to reach every High School Robotics team and have them collaborate with the coordinator of the technology that they decide they want to tackle first? Science Fairs, Exhibitions, Trade Shows, the sky is the limit, and I hope to see you and your project team results there.

Project and Operational Performance

- Strategic Digital Transformation and Performance Management in Three Easy Steps:
    1. Shared, Transparent Project Management Approach – simply referred to as PMLC, and Revision/Release Control.
    2. Status Reporting – Weekly at Project and Program Team Level with Communications Monthly and Annually with Steering Committees.
    3. Operations KPI monitoring
- Project leads, lead engineers and local governments present Status, Exceptions, and Operations Reports. The United Nations

coordinate project deliverables at a global level with their Global Goals projects:

- Red Status – the team needs help or Operations are falling short of Target KPIs
- Yellow Status – the team is struggling and may need help.
- Green is fine – active teams have what they need and are operating within target measures

Building Great Technology

When I was in Baccalaureate Math classes in High School, we had two members of the Canadian Space organization visit our class to talk about the importance of math in our future. I thought I had forgotten that 1979 – thirty-four-year-old memory entirely, but as I think back on lessons learned, I can recall their binary math demonstration in that class vividly.

The first instructor wrote down a sixty or eighty character string of 1's and 0's on the white board. The second instructor turned away as the first changed a dozen or more of the seemingly pattern-free binary, and then the second returned to the board and proceeded to correct all of the changes made easily and quickly. The instructors explained that the ability to correct errors in binary sequences was a critical part of any communication in space because of the low reliability and high loss rate of the transmissions received. More times than not, communications arrived scrambled and had to be corrected before they were useful.

The lesson here was always to embed a layer of self-correction into your work. I heard it echoed again years later while working with programming teams: a great programmer was someone who could create code that subsequent programming teams could not easily break during revision updates.

That was a great lesson.

Example Project - Industrial Robotics

Leaders: Caterpillar, Siemens, ABB, Boston ...

These are the devices that have been around since 1961 when Unimate, the world's first industrial robot, was first used on automotive assembly lines to weld die cast parts onto auto bodies for General Motors in Detroit. Today these devices work in series, one after another, to complete complex assembly tasks and then hand the task to the next machine that continues the build.

When creating a Project Plan for robotics, it is important to structure programs that are revision controlled for regular updating and for the easy addition of new sub-projects.

Consider the creation of a robot arm as a simple example. The arm must meet an array of articulation, extension, and strength needs. Designers can equip robot arms with many more options than a human arm. As you enter the Charter Phase of this project, you will layout the different sub-projects as you can think are important initially. The Arm's base for one; will it be a fixed base or will it need to move? Arm articulations? How many elbow joints will there need to be? How much weight must each joint sustain? What are the ranges of motion needed? A human wrist turns just 120 degrees; a robot joint can turn 360 degrees and several times if needed. The Arm's Reach – how far must the arm extend and maintain lifting strength? Number of Arms; arms costs from $22,000 per unit and so you will want to use as few as needed but also keep up with manufacturing's volume needs. An example is "50" cars or units per hour.

The Project of building the robotic Arm is a Program in itself. The Project of the "Hand" is another Program. Some "hands" weld, some lift, some rivet, some place seats and other place parts. Fingers too, need direction, strength, reach, and grip.

If the Project Hand is that of a "picker" of grapes for example – maybe two fingers are needed with video controlled positioning that also senses how much pressure fruit can endure at a maximum. Humans learn when they apply too much pressure, will your robot learn too? Perhaps twenty or 100 hands work simultaneously to pick more rapidly than two human hands ever could.

When the robot breaks or the accuracy of its work needs to be improved, maintenance tools or robots need to be available, and when those maintenance tools break, they too need to be repaired. In this way, a system is made sustainable, and the availability of the picking solution meets Service Level Expectations consistently.

Robot car projects work in this same way. With Sub-Projects spawning from larger Programs with hundreds of deliverables, each requiring the running of pragmatic requirements, design, build, pilot, implementation - or Project Management Life Cycle (PMLC). Thousands of tasks and engineers running in hundreds of sub-projects, mitigating risks and meeting needs one at a time, status reported weekly, and all rolled up to compose a successful driverless car hi-tech deliverable.

Limitless Complexity - laid out simply, so as to be buildable reliably. The same way you might eat an elephant-sized cow if you had to.

And then, when the first version of the v1.0 car is built, work begins on the release of version 1.1, 2.0, on and on.

Peer Review and Coaching

For simplicity and scalability, initial documentation data elements are suggested to reside in a web-driven database, and document repository called the World Peace Agenda. A database all sounds very complex, but it's not really and its intent is to simplify access to project data, science, lessons learned, and other helpful knowledge and resources. It allows tags and other smart organizations.

We do not want to slow down design teams, and we do not want to interrupt experts during their family meal either, so the World Peace Agenda Data Repository is designed to make a lot of support information readily and easily available - when experts are busy elsewhere.

It is a foreseeable eventuality that patent, security and privacy concerns will come to the forefront that will require us to extend the Schema of the World Peace Agenda. Listing Security and Privacy concerns to your Requirements Lists will address these issues in

Design. We also do not want our engineers to design a solution for World Peace only to have some reprehensible special interest group patent the tech and forbid its use as well. Considering the direction that this work has us heading in, I rather think that this behavior will adversely impact the patent requestor – but we do want to plan to minimize litigation interference.

Individuals and even whole firms do look actively for opportunities to extort capitalist system loopholes. In one example, according to CNN, "The Daily Beast" bumped Turing Pharma's CEO Martin Shkreli to their "Most hated man in America" position in September. He bought licensing for an important sixty-year-old life-saving drug Daraprim - that cost $13.50 a pill initially, and then hiked it up to $750 a pill immediately. Individuals and even whole firms do look actively for opportunities to extort capitalist system loopholes. Hillary Clinton called him on that decision, and Transition Projects will work to protect against any similar reprehensible behavior.

Would it be a surprise to find Human Resource and Procurement Departments asking for disclosure or otherwise discouraging work on World Peace initiatives? We saw this during the Open Source work on UNIX in the 2000s when some companies even sought to restrict recreation-time contribution to that work by its engineers. Wide calls for boycotts to offending companies was communicated throughout the internet then, and the policy of CSQ forums to not maintain a list of organizations that restrict their engineers in this way at any time.

I would want to be able to encourage my children, social leaders, peers, and any engineer or Hi-tech lead to be enthusiastic about taking on World Peace Initiatives. There are employers who ask employees to adhere to voting preference – even as a condition of employment.

Like most good advice about discussions of politics or religion, if you feel like others might discriminate against you, don't ask; and don't tell. Talk about the weather, teachers, and sports.

Chapter 8 – The Transition

Our Jobs are Automating

Jobs have been automating for several years now – and that's a good thing, but are our Governments ready for it?

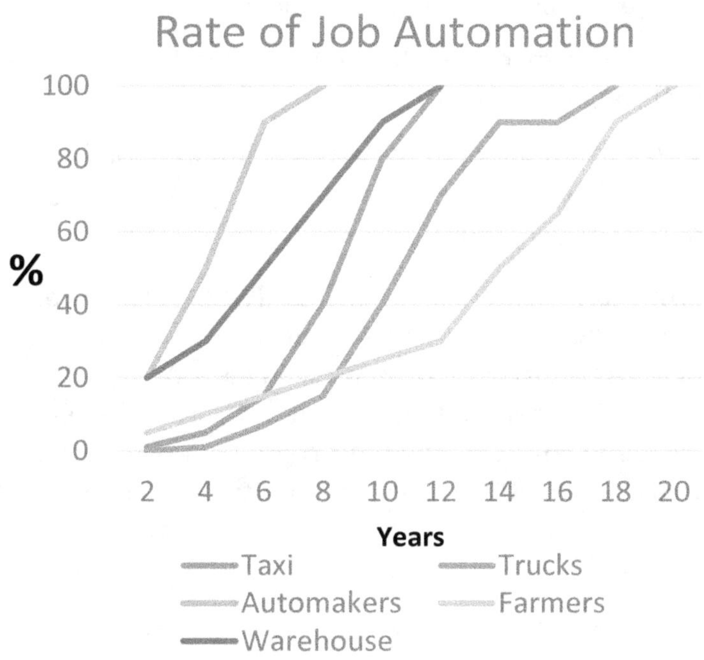

Most business and governments are scrambling to gain some understanding and insight on how to handle the problem that is beginning to touch every facet of society.

I used the example of Audi's 100% worker-free assembly lines above, and I encourage everyone to view the YouTube video of automated automobile production at Tesla as well. Automation will continue steadily on track to a countdown-to-zero. Zero workers – and, as I say, this is a very good thing for both productivity, and, for us and our countries as well.

The Rate of Automation chart above shows that most jobs are going to automate over the coming twenty years. The largest employer is the U.S. is the trucking industry and that group are expected to automate, completely - replacing all workers with computers, over the next three to twenty years.

If we did not have the plan to transition all of these displaced workers to a "Good Life", then we would have quite a lot of trepidation about this change. We could simply accept the automation as it comes to us with no planning, but there is too much to gain by staging the automation just a bit - as you might expect.

There are at least four important considerations to manage as we transition through the change

1. Rate of Automation targeting Wealth Creation first
2. New Safety Nets – for Engineers and Minimum Income
3. Wealth Distribution

In the end, technology solutions will exist to automate our production economies entirely. At that point, food and shelter will automatically come to us – but some of the technology solutions may take up to twenty years to fully implement across the planet. In the meantime, between now and the arrival of our technology safety net, perhaps 50% to 60% of society's jobs today will be automated.

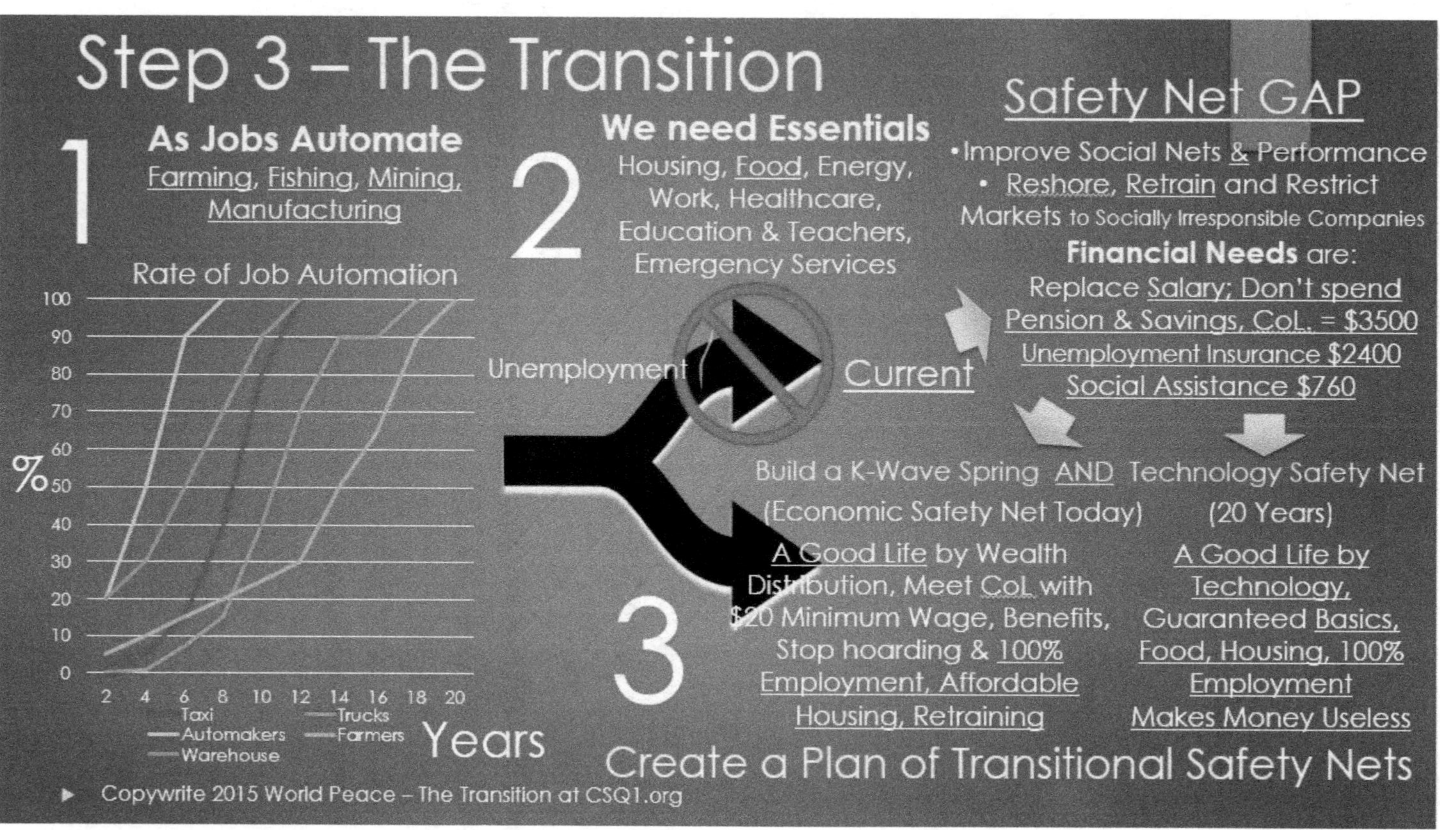
Step 3 – The Transition
1
As Jobs Automate
Farming, Fishing, Mining, Manufacturing
Rate of Job Automation
100
90
80
70
60
%
50
40
30
20
10
0
2 4 6 8 10 12 14 16 18 20
Taxi
Trucks
Automakers
Farmers
Warehouse
Years
2
We need Essentials
Housing, Food, Energy, Work, Healthcare, Education & Teachers, Emergency Services
Unemployment
Current
Safety Net GAP
•Improve Social Nets & Performance
• Reshore, Retrain and Restrict Markets to Socially Irresponsible Companies
Financial Needs are:
Replace Salary; Don't spend Pension & Savings, CoL. = $3500
Unemployment Insurance $2400
Social Assistance $760
3
Build a K-Wave Spring AND Technology Safety Net
(Economic Safety Net Today)
(20 Years)
A Good Life by Wealth Distribution, Meet CoL with $20 Minimum Wage, Benefits, Stop hoarding & 100% Employment, Affordable Housing, Retraining
A Good Life by Technology, Guaranteed Basics, Food, Housing, 100% Employment Makes Money Useless
Create a Plan of Transitional Safety Nets
Copywrite 2015 World Peace – The Transition at CSQ1.org

For the first time, many of our unemployed workers are going to be skilled, experienced engineers and workers displaced by profit-motivated offshore interests, procurement teams, and pension de-risking – as per the instructions we gave business grads at our top MBA programs.

If experienced engineers cannot re-enter the workforce because CFOs instruct Human Resource departments to protect the company from high salaries and pension risk, two problems arise. First, unemployed engineers will begin to spend their pensions and savings in efforts to make up for the shortfall between Unemployment Insurance, or Social Assistance, and their actual Cost of Living. Second, the members of our society best qualified to build an automated infrastructure have no ability to assist in the work.

Current Social Safety Nets

The current safety nets of most countries are not able to take on the volume of new loading created by jobs lost due to automation projects.

The Cost of Living (CoL) in the Toronto area is approximately $4,500 per month – counting mortgage costs. Statistics discussed in Chapter 13 show that only 30% own their home free-and-clear and that all other homes are 50% financed. Rents have soared with the housing bubble due to the new demand for rental accommodation during this past three years. All other spending has increased too - on energy, food, insurance, automobiles, on-and-on.

At the same time, Unemployment Insurance payout maximums are at around $2,400 per month, and once those protections expire, Social Assistance is available at $660 per month for a single individual. Personal and Retirement savings start to be used to shore up the gap.

If an individual's savings are $100,000, they have 50 to 70 months (2.5 Years) before their savings run out. If $1 million, then they have twenty years – which won't get them through their retirement years. The typical time required by an experienced worker to find gainful employment was nine months last year and that time window is much longer this year. The rate of those that give up looking for work is at an all-time high as are suicides, soup kitchens, social assistance recipients.

Dawn of the Engineers

I can assure you that 99% of Hi-tech businesses fail in Canada through lack of investment funding. That sort of revenue-based, non-asset backed lending is not on the table at even business centers dedicated to innovation and hi-tech startups. Any news reports to the contrary are repeating the stories of government spending program leads that don't also maintain KPIs of companies helped as well.

So - first things first, let's get our displaced engineers working. The safety nets are going to pay unemployment insurance premiums of $2,500 per month to these folks to sit at home and beat their heads against the wall looking for work while they diminish their savings for upwards of eighteen months on average. They might get lucky and find something in a corporate workplace, but in the meantime, let's keep them productive.

Instead of insufficient current safety nets, let's give qualified hi-tech leads and engineers - initially - Cost of Living equivalence and full pensions in return for chipping away at the technology projects needed to build our fourteen production automation projects. Just as valuable as their technical skills are their project leadership skills, so this resourcing plan should be able to maintain a substantial momentum for project work.

I spent my "downtime" writing a working solution to World Peace, I can't wait to see what my peers can do with a finished plan in their hands.

Transition engineers can begin to work on World Peace Agenda Projects through either project sponsor companies, or through enhanced Government Safety Nets with benefits, pensions, and all other resource needs met.

Wealth Distribution

Transitioning to a Good Life

An important task that will build a Good Life while our Technology Safety Net builds is to reduce the impact of jobs lost to globalization by re-shoring our production economies and engineering work.

Government and business investors and leads could assist wealth distribution, social accountability, and automation without burdening government budgets in one move, by calling on major businesses and asking that they sponsor World Peace projects. Businesses could do this through the rapid hiring of high-salary, pensioned, and benefit-enabled hi-tech and engineering staff immediately, and they could also provide resource equipment, offices, and lab needs as required too.

When we are choosing what to automate first, always select productions that represent the biggest export earning products for your country, and then expensive imports, before others. Do not slow automation work but do monitor the social project results to ensure that planned benefits and profits are balancing. Bottom-up plans do more heavy-lifting than any top-down solution ever will - given a little time. Setting smarter measures will usually correct failures in revenue recovery.

Remember from the discussion of Economic Controls above that we have a thirty-year window of K-Wave Spring to look forward to (a Capitalism-driven Good Life) once we distribute wealth and emerge from our current K-Wave Winter. Wealth Distribution, therefore, should give us more than the time we need to transition our capitalist economy over to a sustainable technology-driven Good Life fully.

Transitioning Countries One Part at a Time

Two countries stand out as having a more significant challenge in rolling out a Good Life than others. China and India, are home to almost 36% of the planet's population; with a staggering 2.5 billion people. That means that by the time that most nations are long switched over to a Good Life, these nations are about 1/10th the way there.

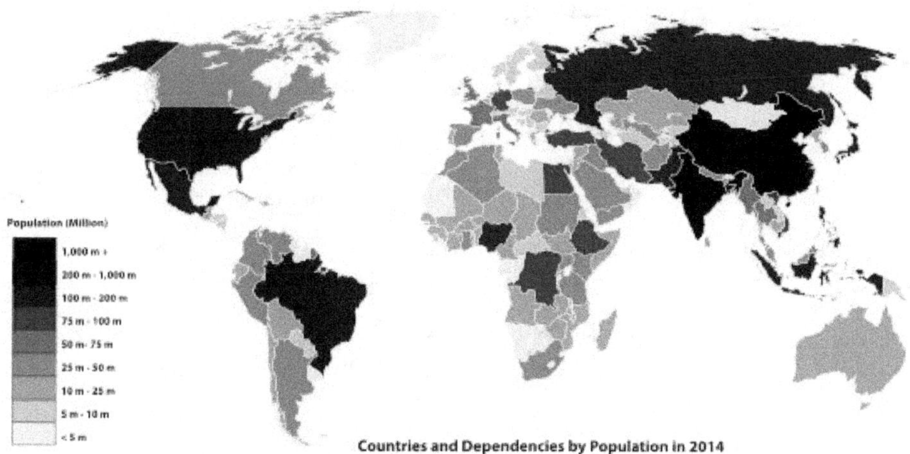

Countries and Dependencies by Population in 2014

World Peace is going to have to roll out in phases, and perhaps by region, in every country. For India and China, it will have to prioritize the neediest first – as they do today - and then advance quintile by quintile (in 20% lowest to highest income sections) and region by region; or perhaps octal by octal (10% lowest to highest incomes).

I was so surprised by the populations of many countries that I thought you might be too, and so I present a sampling of the top 29 here in the book. The global population map here is courtesy of Wikipedia.

Wikipedia and Google are doing a sensational job of presenting information about our planet this year. Research and supporting data for conclusions presented in this book were there almost whenever I needed them.

Rank	Country	Population	%
1	China	1,372,670,000	18.9%
2	India	1,280,256,500	17.6%
3	United States	322,042,000	4.43%
4	Indonesia	257,080,000	3.53%
5	Brazil	205,055,000	2.82%
6	Pakistan	191,165,000	2.63%
7	Nigeria	183,541,000	2.50%
8	Bangladesh	159,207,000	2.19%
9	Russia	146,406,280	2.01%
10	Japan	126,832,000	1.74%
11	Mexico	121,005,000	1.66%
12	Philippines	102,212,000	1.40%
13	Vietnam	90,730,000	1.25%
14	Ethiopia	90,076,012	1.24%
15	Egypt	89,684,300	1.23%
16	Germany	81,285,000	1.12%
17	Iran	78,717,200	1.08%
18	Congo	77,954,000	1.07%
19	Turkey	77,345,000	1.06%
20	France	67,087,000	0.91%
21	Thailand	65,104,000	0.89%
22	United Kingdom	64,800,000	0.89%
23	Italy	60,725,000	0.83%
24	South Africa	54,956,900	0.76%
25	Myanmar	53,897,000	0.74%
26	Tanzania	53,470,000	0.73%
27	South Korea	51,482,816	0.71%
28	Colombia	48,352,300	0.66%
29	Spain	47,705,000	0.66%

I could never have presented this plan to such a credible and validated level without the hard work of thousands of skilled engineers and experts that provided insight and context on every point.

Wealth Creation

Automate your production economy

Pharmaceuticals, Heavy Equipment, Medical Equipment, Hi-tech Software, Robotics, Automobiles, and other manufactured goods are examples of exports that generate a considerable profit for countries around the world. Find more examples by studying the GDP Exports of leading export countries around the globe. Countries like Hong Kong, Singapore, China, America, the Netherlands, Germany and others lead the world in the export of manufactured innovation.

By automating wealth-building industries first; high-value commodity exports in mining, fishing and farming at the same time; and imports that make your country more self-sufficient second, you begin to improve your financial position as you improve your safety nets at the same time.

Support and Sell

Needless to say, we need to establish markets with partner nations that need our automated products and consulting services. Governments should be assisting, and passively performance-managing sales relationships, until targets are missed.

Upon missing targets, assist sales actively until no longer needed. Monitoring is what a good government management team does.

All that is needed is for each country is to develop one piece of the puzzle and very soon 100 countries are contributing to a completely automated production solution for the planet. That might sound like a lot of coordination too, so for simplicity and practicality I will simply lay out the work ahead in the following few pages.

If one country is working on:

Household robots - we need to be supporting them with automation software, components and special attachments (human hands, etc.), component motors and controllers, and world leading manufacturing consulting expertise.

Technician robots – we can continue to manage our robots manually, or we can automate the maintenance of both our robots and technician robots.

Farming – Produce. If a client country is leading the automation of farm equipment, support them with controllers, robotic pickers and threshers, tractors, tires, oil products, robot arms, conveyors, and safety detectors. Automate transportation, warehousing, food preparation, packaging, freezing and transport systems.

Farming – Livestock. Improving conditions at chicken, beef, pork and other livestock farms is a constant concern for farmers and consumers alike.

Mining – pumps, drills, guidance systems visual, automated forklifts, conveyors, safety systems, infrared and other supporting automation systems are going to be required steadily now.

Fishing – GPS-driven drone ships that deploy and retract nets have been at work in our seas for years and will just get better. Next steps will automate the transport of catch to processing lines that filet and package automatically.

Manufacturing – completely automated everything from bottling to automobile manufacturing to packaging and shipping. Build the Model-T of our century – simple, cheap, reliable and automated.

Construction – Great buildings, houses, and communities are going to build themselves. We can automate handyman tasks and new construction as well. 3D Printer Construction, Cement Work, Sanitation, Crane Operations, Framers, Plumbers – and offer these capabilities at competitive pricing to construction projects.

Warehousing – fully automated conveyor and lift-driven warehouses are already at work delivering just-in-time inventory

Restaurants, Fast-Food and other – McDonald's is already rolling out fully automated kitchens in Pilot locations. These kitchens ensure optimal consistency and quality in an environment where only machine operators and supply loaders will have a role to play in the preparation of the food we consume.

Retail – is going to be interesting. What need is there of retail stores in the world where you can get the right fit, style, color, and price from a website application that assembles it or 3D prints it right in our homes - in just a few minutes? If our in-house equipment can't make it, a remote printer will create it and automatically transport it to our doorstep. If the vehicle that transports the product to us needs refueling on the way, it will either plug in and recharge or pump automatically at automatically refreshed clean-fuel stations along the way.

Tool Enhancements – 3D printing needs to be 300 times faster. Robots need to operate eighteen hours a day and recharge in just five minutes. Cars need to recharge in the time it takes to fill up gas at the pump; we need clean fuel (like blue crude), cold fusion, energy-to-matter-based food, and assembled composite, carbon fiber and metal-alloy replications as well.

Water – Automating the sourcing, desalination or treatment, transport, and distribution of Clean Water is a huge project. I live a stone's throw from a Great Lake in Canada, and it would still take a huge manual effort to get the job done. GPS Satellite Water Divining, Electrolysis from Sea Water, Water Pipelines, Water Container creation and storage, distribution, in-home storage and irrigation for home gardening too. All projects are fair game.

To name just a few project examples.

Running Projects that Solve Anything

Running thousands of projects is straightforward as described in Chapter 17 below and in the CSQ Common Sense 101 guide. Once reached for, they give you the examples and detail method needed to solve any problem no matter the scale nor complexity. You can also get online on our forums at CSQ1.org and ask a community member for help.

To keep it simple here, know that projects are run in the same way large or small; regardless of whether they are technical or social. A complex project is likely going to need several sub-projects and perhaps even sub-projects to those projects, so begin by creating - a Program.

A Program begins with a **Charter Document**. The Charter contains lists of all Stakeholders impacted by this project; other resources needed; dependencies on other project; working team members and steering committee members. Include rough timelines, scope, budget, risk plans, key measures of success and a description of managed releases planned (how often will this project will need updating); list all projects likely to be needed within the program and include a communications plan which links all status reports back to your World Peace Agenda.

Getting approvals for your Charter detail may take up to four weeks as needed to run workshops, speak to project owners and stakeholders, and then get the Charter document itself approved or signed off. With approved Charter In hand, you are ready to start.

Requirements Document - Projects and Sub-Projects within Programs, are run in the same way as well. Begin individual projects by assembling a list of needs, also called requirements or deliverables, and also document an inventory of existing items for your project at the start. This information constitutes the "Requirements Document". Your project is making a change, so identify all considerations and re-usable assets today such as people, technology, and processes. Ask stakeholders for important needs, KPI measures – for both the project team and final operational solution.

Engineering and operations teams are asked for this information by project teams routinely so they may already have a prepared list of requirements for your designers to consider. Teams that will offer requirements might include the database team, the server team, the network team, the security team, the enterprise applications team, business project management office and technology PMOs, the robotic team, and the 3D Printing Team - to name just a few.

Design Documents - include Detail Implementation Plans and instructions in the how-to **Build Book** for this version of the implemented solution. Upon completion of the detail design, and during the final implementation and systems testing, a **Run Book** can be developed as well.

Before the solution goes live, you will need **Training** on your Run Book for support staff and production teams within a **Pilot**; you will need an operational support team; technology administrators will need training and so will end-users of the new system. Security, technology, SLOs (Service Level Objectives) for Availability, Delay, Accuracy, Change – Moves/Adds/Change. Reporting stats, should all be spelled out in a list and reported on weekly, monthly, and annually in the **World Peace Agenda**.

Most technology and engineering groups are asked to participate in projects routinely and will have developed a list of their standards, naming conventions, and requirements of new project teams for **Change Management** controls. Change Management protects other production systems from your new additions.

These are good rules for any project, social or technical in nature. Example requirements for Wealth Distribution and include "no interruption of the basics of life" during the process. Food, shelter, energy, security, education and other needs of living must continue during any social project change.

Here we are, the Good Life.

Once we all have the basics of life and what we need, we now have the time and resources needed to roll out cookie-cutter solutions to other nations. Those members of our society that are not assisting with the work will have the time to take on the really interesting Science, Healthcare and Automation Technology Projects as described on the CSQ 100 Year Plan.

No longer will our best and brightest minds incented to reduce corporate risk. At this point, our best and brightest can explore space, science, improve the worldwide infrastructure. They can provide for the needs of others, explore the meaning of life, whatever they want really; which also better aligns with definitions of a Good Life and the Meaning of Life that we discussed at the beginning of the book and summarize in Chapter 26.

See Chapter 17, for example project documentation, and Chapter 25 - Transition Economics, for a detailed look at the balancing act needed to offset transition and safety net spending with these new revenues.

Chapter 9 – World Peace TEDTalk

In these first several chapters, I have explained many of the projects and reasons needed to roll out a plan to build a sustainable Good Life to seven billion people. The detail instructions and steps are in the chapters that follow, but the "TEDTalk" for World Peace can be presented now based on many of the important points already discussed.

A good summary explanation of the World Peace Transition Plan required basic instruction in economics, government systems, war, technology and other level-setting topics.

I didn't want to put off this summary too far into the book because the message here is that World Peace is within our reach today and is very straightforward as well.

A Good Life and World Peace is affordable, it is buildable and sustainable, and so without further ado...

TED
Ideas worth spreading

Our Generation's Moon Launch
- and Greatest Legacy

World Peace

THE TRANSITION – TO AN AUTOMATED SOCIETY

FOUR STEPS, THREE TEAMS, AND 19 PROJECTS TO A WORLD PEACE AGENDA

THE 5 YEAR PLAN & HOW-TO GUIDE TO MANAGING CAPITALIST ECONOMIES AND BUILDING A SUSTAINABLE GOOD LIFE AND FUTURE FOR OUR CHILDREN IN EVERY COUNTRY

Copywrite 2015 World Peace – The Transition at CSQ1.org

Cover - World Peace – The Transition

The Transition to an Automated Society and Economy is our Generation's Apollo 11 Super-Project. NASA spent eight years building their twelve projects and Worldville should take quite a bit less time given the tools at our disposal now.

Four Steps, Three Teams, and Nineteen Projects to a World Peace Agenda (WPA) Plan that we call the #WPProjects on Google+.

This book is a How-To Guide for managing Capitalist Economies and Building a Good Life and sustainable future for our children – in every country. In the same way that Alan Turing's computer cracked Enigma and put Man on the Moon, so too will it automate our society for all the right reasons.

In complex Technology Projects, a Plan is needed. Automation is happening already and it is going to continue in a very painful way if we do not plan safety nets for it as well. People will die; there may be wars; and so this Plan is an important one.

Automation provides benefit worldwide and therefore one or several projects are assigned to 180+ countries by this plan.

Side Notes: The word "Step" was selected as it signifies both action and a physical task as well. Step also infers that there is a natural sequencing too – so that Step One must complete before we can take on Step Two. In painful discussions, I have labored over suggestions for "4 Spokes", "4 Wheels", 4 Areas, 4 Activities and similar. Were there a 4 part lifter of heavy loads, I would have selected "4 Loaders". Know that I mean 4 Steps begin immediately, or have already started, and have little dependency on one another as they move forward simultaneously with gating and instruction only from Transition Economics in Chapter 8 and 25.

In this cover slide, the term "World Peace Agenda" is mentioned for the first time. A World Peace Agenda (WPA) is a Country Policy Governance Document used by every country to report the status of their individual efforts to build World Peace Transition projects listed here. See detailed explanations in Chapter 6 and 7.

A Knock at the Door in "Worldville"
pepper
WORKING
WORKING
WORKING
Copywrite 2015 World Peace – The Transition at CSQ1.org

A Knock at the Door in "Worldville"

The knock at your door is a delight as you have been waiting for this day for a lifetime it seems. Waiting on the doorstep is a package with a few-days-supply of canned goods, fresh vegetables, water, a cell phone, a tablet computer and a universal charger. A sheet of paper provides instructions to call with questions and it has directions on how to use your tablet or phone to order your groceries.

From this day forward, water, groceries, energy, and household supplies will continue without interruption for the rest of your life, for the lives of your children, and that is just the beginning.

The World Peace Transition Program, at the direction of your government, has installed your sustainable automated production economy and, this morning, it has brought the basic human right of a good life to your front door - and to your entire community.

A good life, in a family friendly community, without security concerns where life is respected; a life with equal opportunity, strong family and community values, universal healthcare, with advanced technology, 100-year longevity, and the assurance of liberty and the pursuit of happiness too.

Worldville's Pilots rollout beginning in 2018. No mere amusement park attractions, these are connected, automatically built and serviced sustainable villages capable of growth to many thousands of people.

Recognizing that the Problem Exists

- We are a world of 7 billion people battling :
  - Corruption, unequal access to opportunity, healthcare, education
  - 3.1 million children under 5 starve to death
  - 12.5% (842 million or 1-in-8) are undernourished
  - 23% of all deaths are caused by infectious and parasitic disease
  - Deaths from Obesity – 300,000 in the US per year
  - 750 million people lack access to clean water; 840,000 die from disease
  - War kills 60,000 people die directly; 200,000 indirectly
  - 800,000 per year die by suicide; many more attempt to end their lives unsuccessfully
  - 540,000 people die violent deaths in countries with no armed conflict
  - Misinformation on Energy, Warming, Population, Water Crisis, Real Estate Bubbles

0-4
5-6
7-8
9-10
11-15
16-20
21+

World Peace needs a Smart Plan

Copywrite 2015 World Peace – The Transition at CSQ1.org

Recognizing that the Problem Exists

... is the first step to solving it.

We are a world of 7 billion people battling:

- Corruption, unequal access to opportunity, healthcare, education
- 3.1 million Children under 5 starve to death
- 12.5% (842 million or 1-in-8) are undernourished
- 23% of all deaths are caused by infectious and parasitic disease
- Deaths from Obesity – 300,000 in the US per year
- 750 million people lack access to clean water; 840,000 die from disease
- War kills 60,000 people directly; 200,000 indirectly
- 800,000 per year die by suicide; many more attempt to end their lives unsuccessfully
- 540,000 people die violent deaths in countries with no armed conflict
- Misinformation on Energy, Warming, Population, Water Crisis, Real Estate Bubbles

Instead of getting easier, it's getting harder for more people and so ...

World Peace needs a Smart Plan

This summary of world problems is a quick introduction to a fact of life that many of we fortunate G8 citizens will seldom be victims of.

The ironies of 3.1 million children under five dying of starvation, 842 million are undernourished, while 300,000 die of obesity - showcase the differences between our lives here and there.

Step #1 - World Peace Starts at Home

Shared Values

- Society's needs:
  - A Good Life for All Individuals
  - Family Friendly Societies
  - Equality of Opportunities
  - Freedom from Fear
  - Security and Longevity
  - Recognition for Contribution
  - Gender Roles
  - Maslow's Needs Met

Step 1 + World Peace Starts at Home

Society's Needs and Shared Values:

- A Good Life for All Individuals
- Family Friendly Societies
- Equality of Opportunities
- Freedom from Fear
- Security and Longevity
- Recognition for Contribution
- Gender Roles
- Maslow's Needs Met

Abraham Maslow developed Maslow's Hierarchy of Needs in his paper "A Theory of Human Motivation". It parallels many theories of developmental psychology.

Each layer requires the lower building block for support - to exist. For example, one cannot reasonably be concerned with morality when unable to find food and water. Having our basic needs gives us the solid foundation needed to build a moral and creative society.

Basic Human Need was Aristotle's reasoning too when he stated:

> *"The Right Plan is the one whose ends, means, practical thinking and purposeful action resulted in a Good Life; a life full of things you need – and not necessarily a life full of everything you want. With a little luck, goods in body and soul, and making a habit of good choices which reflect moral virtues of temperance, courage, and justice, a Good Life should be sought and found. "*

Step #1 - World Peace Starts at Home

What is a Good Life?

Shared Values

- Society's needs:
  - A Good Life for All Individuals
  - Family Friendly Societies
  - Equality of Opportunities
  - Freedom from Fear
  - Security and Longevity
  - Recognition for Contribution
  - Gender Roles
  - Maslow's Needs Met

Age	Female	Male
0	A Parent is at Home	"
1-2	Childcare (or Mom)	"
5	Grade School	"
12	High School	"
17	University / College	"
20	Home & Marriage	Home & Marriage
21	First Child (optional) Family Cottage (Dacha)	"
22	Masters Studies (Optional) Or Full-time Mom Or Work	Masters Studies (Optional) Or Work/Take Income
23	Second Child	"
23		Work
24	Masters Studies (optional)	
26	Work (optional) Or Full-time Mom	
40	Work or Grandchild care	
41	Grand Children arrive	"
55	Mini-retirements	Mini-retirements
62	Great Grandchildren arrive	"
75+	Retirement	Retirement
85	Great Great Grandkids	"
100	Porch Swings & Community	"

Step One – Defining what is a Good Life?

The Life Goals Chart, taken from Chapter 1 above, explains that measures of a good life include the option to able to meet our great-great-grandchildren and have the option to be able to take advanced education. We should fulfill our gender roles by getting married, taking a home in our hometown – perhaps a cottage too if we choose, and then have the option to start a family as we carry on our Degree and Masters Level studies. This "crunch" of activity that signifies our starting a life, could begin at the age of twenty (shown here in blue).

When we get to different stages in life, appropriate housing will be there for us as needed, including nanny quarters should you wish to care for parents within our own home.

When we get to be fifty-five years old, we can begin mini-retirements until we retire at age seventy-five or sooner – or we can carry on a part-time basis if we wish.

This was the good life enjoyed by people living in many G8 countries in the 1950s and 1960s. Less so by the 1970s and finally the last of the G8 was Russia in 1986 when Perestroika dismantled this system. Countries that support this good life today include Norway, Saudi Arabia, Brunei and Cuba.

Today, affording a home in the town where you grew up, marriage, children, and degree studies needed by our employers, force young men and women to delay starting their lives for seven to fifteen years in many cases. Given the financial hardships of saving, maintaining work and home, these delays cause many young people to consider our system "broken".

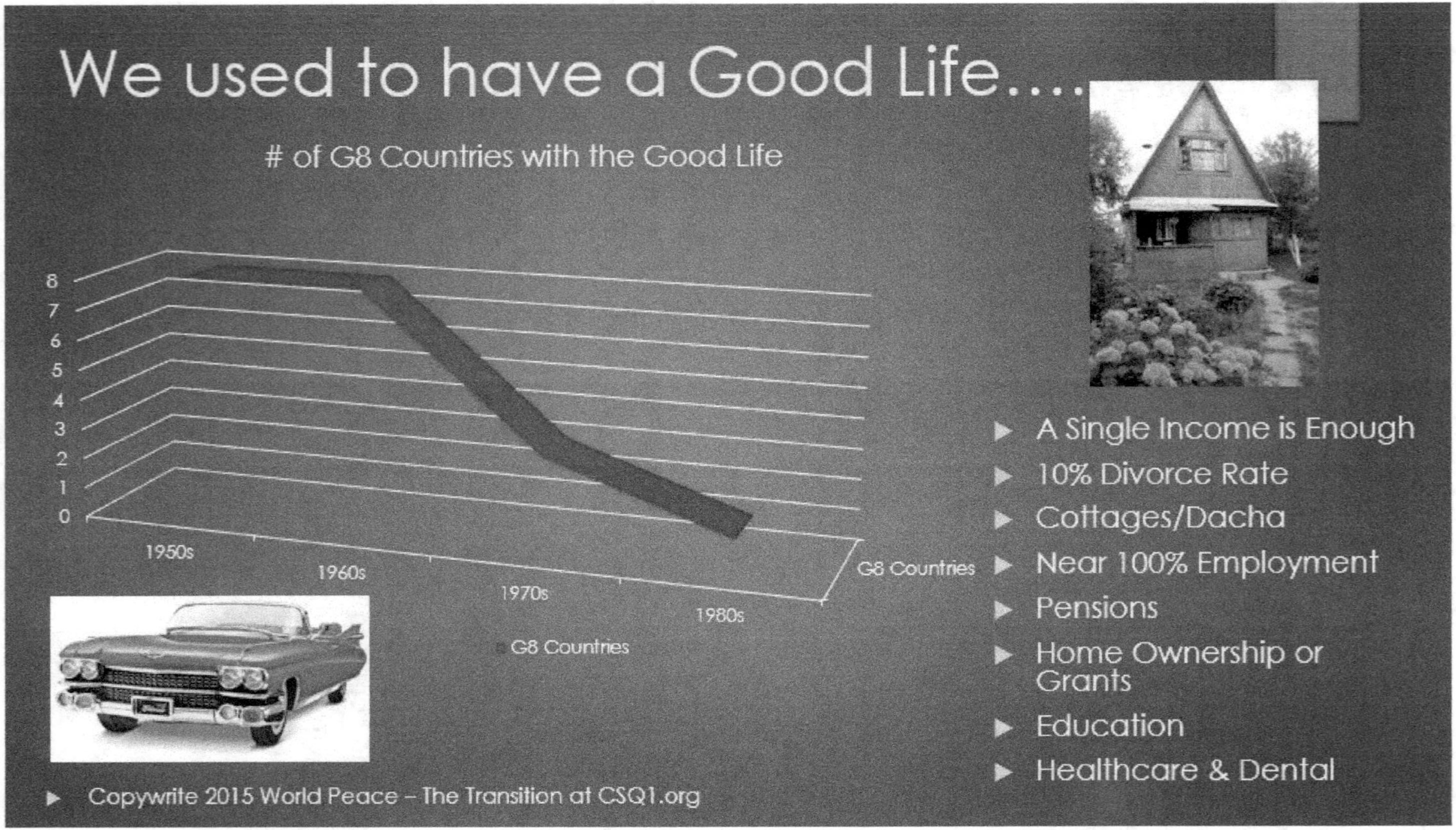

We used to have a Good Life....

of G8 Countries with the Good Life

8
7
6
5
4
3
2
1
0
1950s
1960s
1970s
1980s
G8 Countries
G8 Countries

- A Single Income is Enough
- 10% Divorce Rate
- Cottages/Dacha
- Near 100% Employment
- Pensions
- Home Ownership or Grants
- Education
- Healthcare & Dental

- Copywrite 2015 World Peace – The Transition at CSQ1.org

We Used to have this Good Life - in G8 Countries

In the 1950 and 1960s, the norm for young adults in North America was to get married and have a first child by the age of 20. My mother was living with my Dad at the age of nineteen and I came along when she was twenty, my brother was born eighteen months later. I had kids at thirty and my mom is seventy-one today so she is a big part of her grandkids life and she can't wait to get out and spend time with them next – which helps me quite a bit too. Especially as raising four girls through the teenage angst years got interesting.

My mom finished high-school and later went back for a then two-year registered nursing program which she worked at for twenty-five years before retiring last year. She would say that life was hard when she didn't live with family or within a family home, but that otherwise, her life was good.

When my friend was twenty in 1985, she was recently married and living in a state-assigned apartment in Moscow not far from her university where she and her husband attended Engineering and Masters Studies. She was the oldest out of all of her girlfriends when she had her first child at twenty-one and she had her second son at age twenty-three. As a married woman, she and her husband were assigned their own place with enough room for their two sons. Life was good – as a young adult living in Moscow.

As a young child she lived simply as the only-daughter of a divorced mom; sharing a large flat (apartment) with a second small family, and then she visited her family Dacha with her grandparents and great grandmother in the summer months. Things became more difficult only after travelling around the world and coming to live in Canada during recessionary times in 2008.

What happened to our Good Life?

1. K-Wave Economic Theory 101
   - In Capitalist Societies
   - Every 60 Years +-
   - Documented In:
     - Egypt
     - Mesopotamia 3500 years ago
     - In 18 cycles since 900AD
   - Every Capitalist Monetary Society "Troughs"
2. Our Leadership Changed

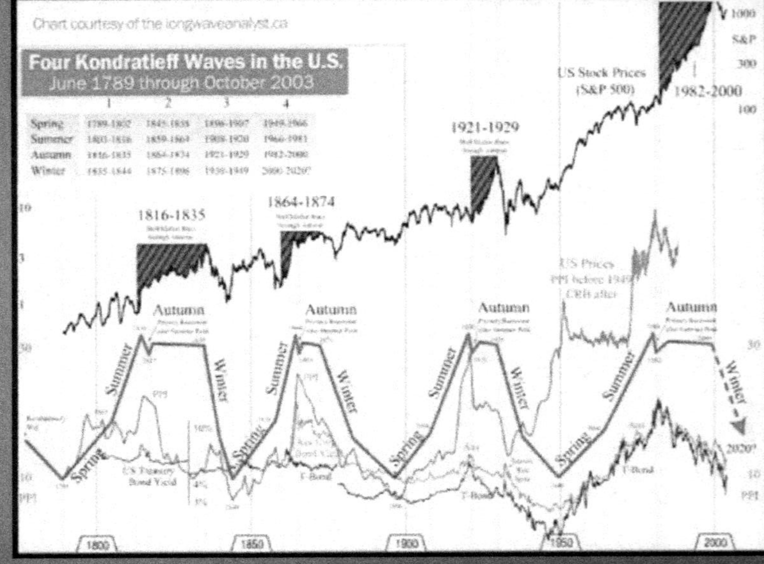

What Happened to our Good Life?

1. **K-Wave Economic Theory 101**
   In 1925, a Russian economist, Nicolai Kondratieff, published a book showing four longwave cycles of an economic phenomenon in Capitalist Societies that repeated on four occasions. Kondratieff Waves are also referred to as Longwave Economics. When Stalin asked him to prove that Capitalism was unsustainable after the crash of 1929, Nicholas continued his research until he concluded that Capitalism's peaks and valley were part of a cyclic system as follows:

In Capitalist Societies

Every 60 Years +-

Documented In:

a. 18 cycles since 900 AD were later confirmed in numerous thesis...
b. Egypt & Mesopotamia 3500 years ago by King Hammurabi of Mesopotamia. His economic controls for these waves are on display in the Louvre in Paris.

Every Capitalist Monetary Society "Troughs" – History calls these dips:

<u>Great Depressions and "Panics"</u>

2. **Our Leadership Changed**

What do Capitalist Cycles Look Like?

- **K-Wave Spring – Most recent 1950**
  - Wealth is distributed. Everyone has what they need; with their available income they can easily buy homes, property, and the things they need. Life is Good.
- **Summer - 1969 - 1982**
  - Wealth accumulates - in some hands more than others. Interest rates climb as the market for investment ramps up. Life is still good for 90%+

- Copywrite 2015 World Peace – The Transition at CSQ1.org

What do Capitalist Cycles Look Like?

Capitalist cycles resemble a Monopoly game. In the beginning, all players have equal money and opportunity to buy property and businesses. With time, some players get a little farther ahead; they start investing in houses hoping that their rental incomes will return well. Over time, with luck, houses turn into hotels and just one player wins all while all others have to mortgage and sell their properties and lose all money.

Kondratieff described the cycle in four parts that he called seasons:

- K-Wave Spring – Most Recent was 1950

Wealth is distributed. Everyone has what they need; with their available income they can easily buy homes, property, and the things they need. Life is Good.

- Summer – 1969 to 1982

Wealth accumulates - in some hands more than others. Interest rates climb as the market for investment ramps up. Life is still good for 90%+

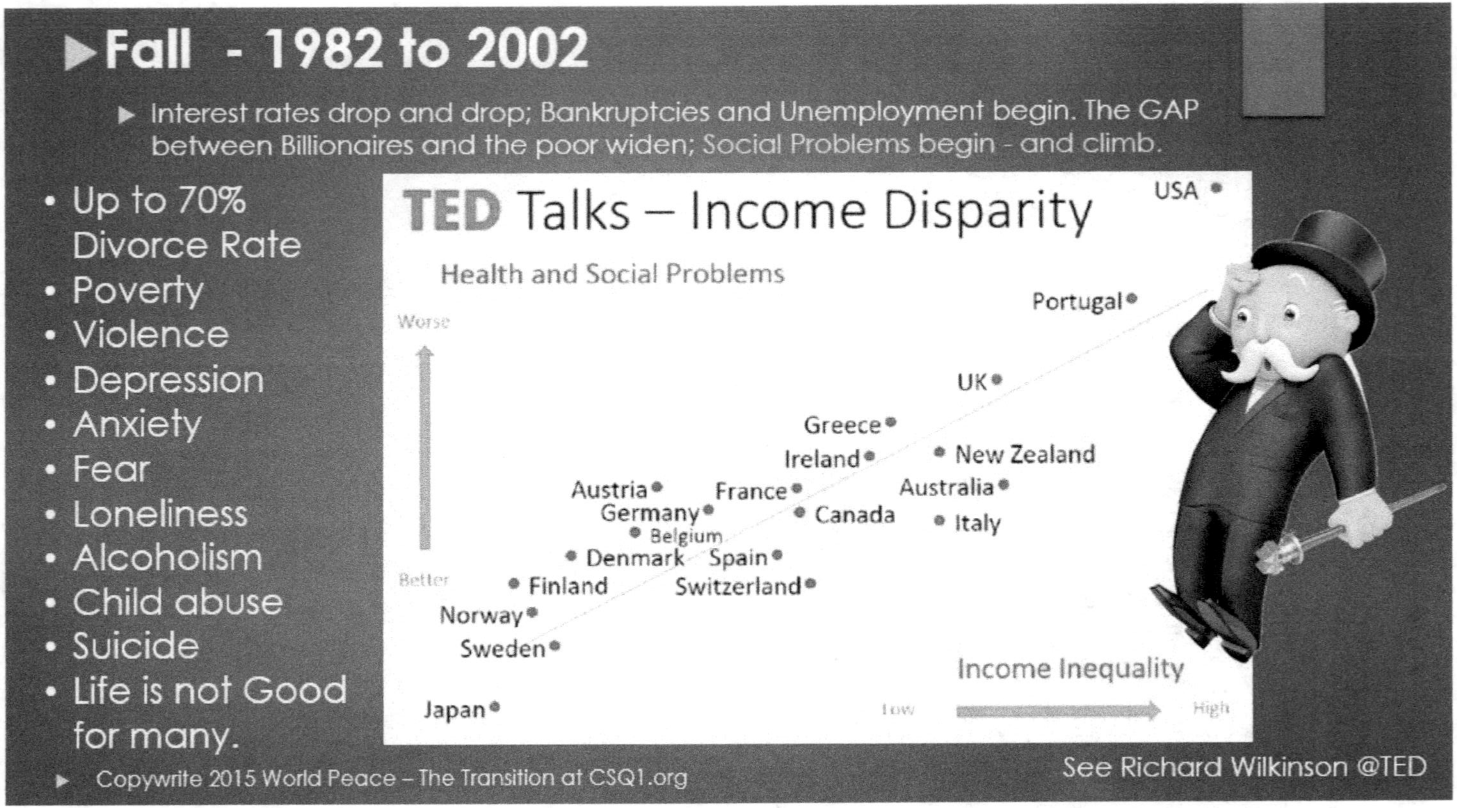
▶ Fall - 1982 to 2002
▶ Interest rates drop and drop; Bankruptcies and Unemployment begin. The GAP between Billionaires and the poor widen; Social Problems begin - and climb.
• Up to 70% Divorce Rate
• Poverty
• Violence
• Depression
• Anxiety
• Fear
• Loneliness
• Alcoholism
• Child abuse
• Suicide
• Life is not Good for many.
TED Talks – Income Disparity
Health and Social Problems
Worse
Better
USA
Portugal
UK
Greece
Ireland
New Zealand
Australia
Italy
Canada
France
Austria
Germany
Belgium
Denmark
Spain
Switzerland
Finland
Norway
Sweden
Japan
Income Inequality
Low
High
▶ Copywrite 2015 World Peace – The Transition at CSQ1.org
See Richard Wilkinson @TED

K-Wave Fall – 1982 to 2002

Interest rates drop and drop; Bankruptcies and Unemployment begin. The GAP between Billionaires and the poor widen; Social Problems begin - and climb.

- Up to a 70% Divorce Rate
- Poverty
- Violence
- Depression
- Anxiety
- Fear
- Loneliness
- Alcoholism
- Child abuse
- Suicide
- Divorce
- Life is not Good for many.

Richard Wilkinson's team from University of Nottingham collects social data on all of these statistics above for G20 countries and looks for significant correlations. This chart plots wealth inequity statistics against amalgamated stats on the Social Problems created by these inequities.

K-Wave Winters – The Great Depressions

- The cycle of doom concludes in public outcries for Wealth Redistribution – most often ending in a Trough **War and Revolution - if unmanaged**. Life is Hard for many.

Revolution

Or … New Wealth Arrives

IDEAL

Or … Wealth Is Distributed

And a new "Last" K-Wave Can Start Again

- **World War II**
- 60 Million Died
- 200,000 by 2 Nuclear Bombs
- **World War III**
- 4100 Nucs Actively Deployed today; 15,000 in Russia & US and 1,800 on high-alert
- **An extinction-level event**

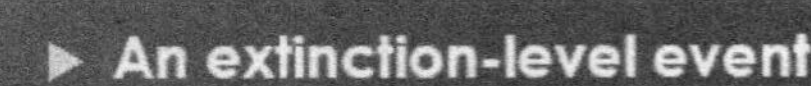

- Copywrite 2015 World Peace – The Transition at CSQ1.org

K-Wave Winters - the Great Depressions and Financial Panics

The cycle of doom concludes in public outcries for Wealth Redistribution – most often ending in a Trough **War and Revolution** - **if unmanaged**. Life is Hard for many.

During K-Wave Winters, income deprived masses – the 99% - as Occupy Wall Street protesters in New York City called themselves in 2008, have several options open to them to distribute wealth in societies.

1. Revolution - the French Revolution, the Russian Revolution, the Chinese Revolution, Iranian, Spanish Civil War, Taiping, Young Turk and many more tore down one system for another.
2. The Great Depression of 1835 was ended by the California Gold Rush, which multiplied America's gold reserves ten-fold. The 1786 and 1893 "Panics" were ended by new wealth from immigration – in North America.
3. War – World War II ended the Great Depression of 1929 by distributing wealth at the cost of 60 Million lives; 200,000 in just two nuclear bombings of Nagasaki and Hiroshima. World War III would be an extinction level event by nuclear winter, and any notion that first strike is possible is folly as well.
4. Wealth Distribution in the form of Universal Debt Forgiveness and a return to equality of opportunity was the economic control taken my Emperors in antiquity called "Jubilee".

And then a final K-Wave can begin again.

Controls for Sustainable Capitalism

What Happens without Financial Controls

- Interest compounds
- Year-over-year profit increases
- The Rich are earning 7/24, with enough time, the Rich will have everything; Poor have Nothing – just like Monopoly
- 0.3% of Wealth is controlled by 40% of population
- "Low Taxes" & Trickle Down Theory are failures
- No Distribution of Wealth can occur

Systems of Government

- Monarchs enacted Jubilee every 50 years – Fresh Start See "Code of Hammurabi – King of Babylon", Louvre
- Democratic Voters:
  - Voters were never taught K-Wave cause-and-effect economics management in school
  - We Voted for "Low Tax" political rhetoric
  - Competent financial leaders, Left and Right could not get elected.
- Copywrite 2015 World Peace – The Transition at CSQ1.org

Wealth Distribution Charts
US Federal Reserve Stats 2010

Low Taxes & Trickle Down =

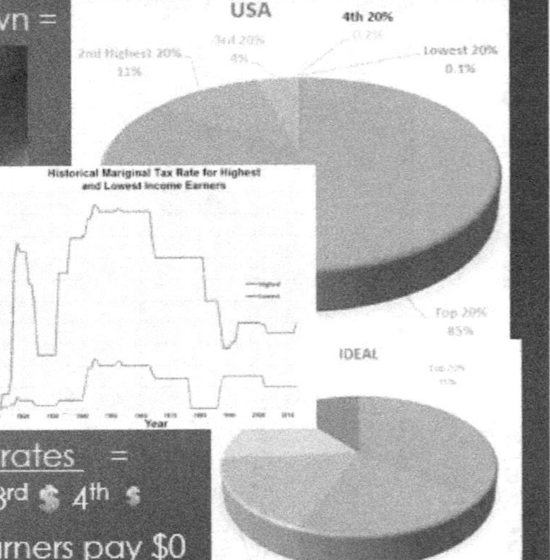

Graduated Tax rates =

Top 20% $ 2nd $ 3rd $ 4th $

Lowest Income Earners pay $0

With no schooling, We got what we voted for - Incompetent Financial Management

Controls for Sustainable Capitalism

Capitalism works but different economic controls are required at different phases in the cycle. Low Taxes and Trickle-Down Economics, enacted in a K-Wave Autumn, led to very poorly managed Wealth Distribution that created Social Problems.

Graduated Taxes for large business and wealthy individuals manage wealth distribution over time to avoid K-Wave Winters altogether.

What Happens without Financial Controls?

- ▶ Interest compounds
- ▶ Year-over-year profit increases
- ▶ The Rich are earning 7/24, with enough time, the Rich will have everything; Poor have Nothing – just like Monopoly
- ▶ 0.3% of Wealth is controlled by 40% of 360 million
- ▶ "Low Taxes" & Trickle Down Theory are failures
- ▶ No Distribution of Wealth can occur

Managing Capitalism by Systems of Government

- ▶ Monarchs enacted Jubilee every 50 years – Fresh Start. See "Code of Hammurabi – King of Babylon," Louvre, Paris
- ▶ Democratic Voters:
  - ▶ Voters were never taught about managing K-Wave cause-and-effect economics in school
  - ▶ We Voted for "Low Tax" political rhetoric
  - ▶ Competent financial leaders from Left and Right Parties, could not get elected.

With no high school training, voters got what we voted for - Incompetent Financial Management

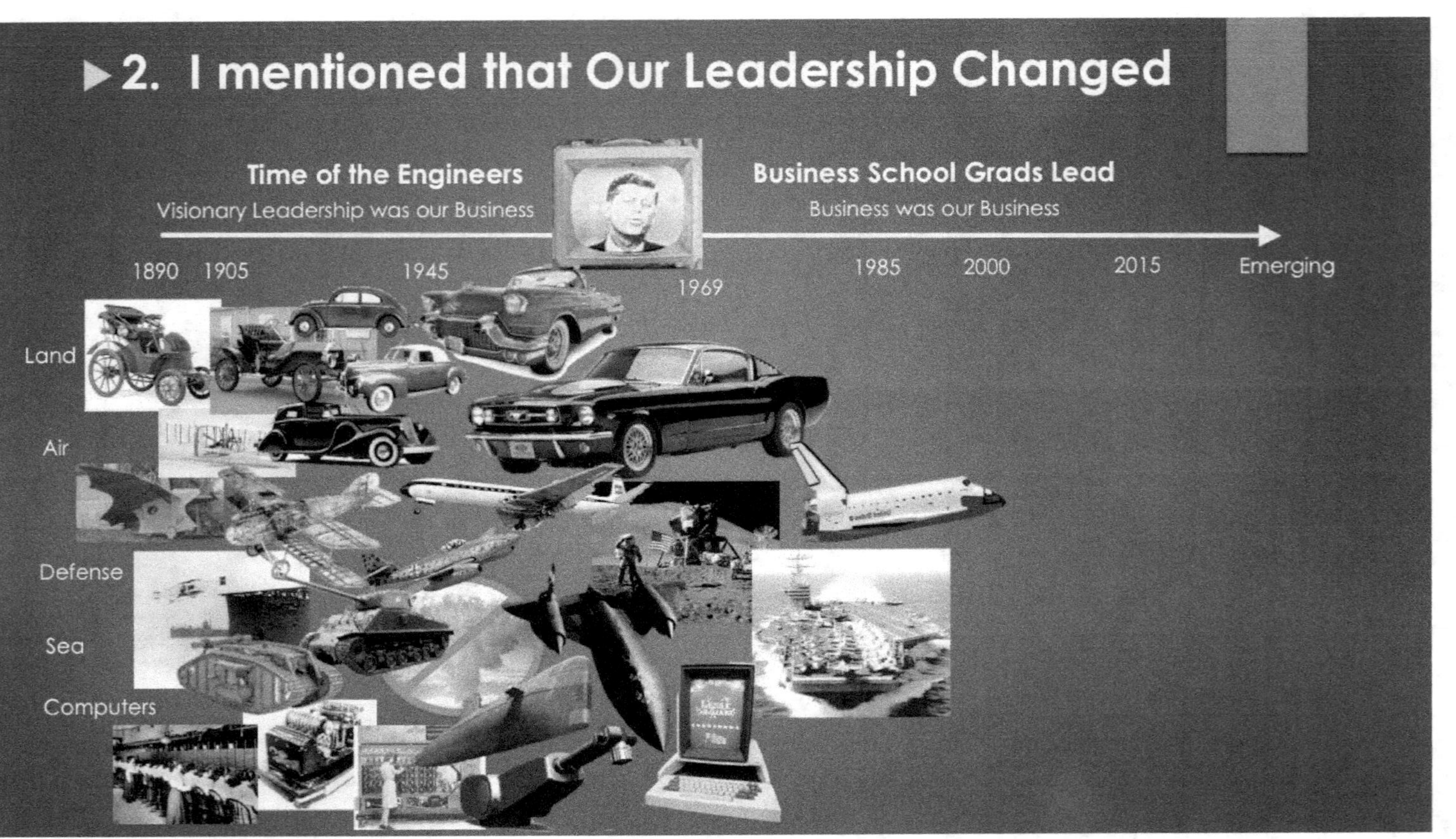

▶ 2. I mentioned that Our Leadership Changed
Time of the Engineers
Visionary Leadership was our Business
Business School Grads Lead
Business was our Business
1890
1905
1945
1969
1985
2000
2015
Emerging
Land
Air
Defense
Sea
Computers

Our Leadership Changed

The Time of the Engineers

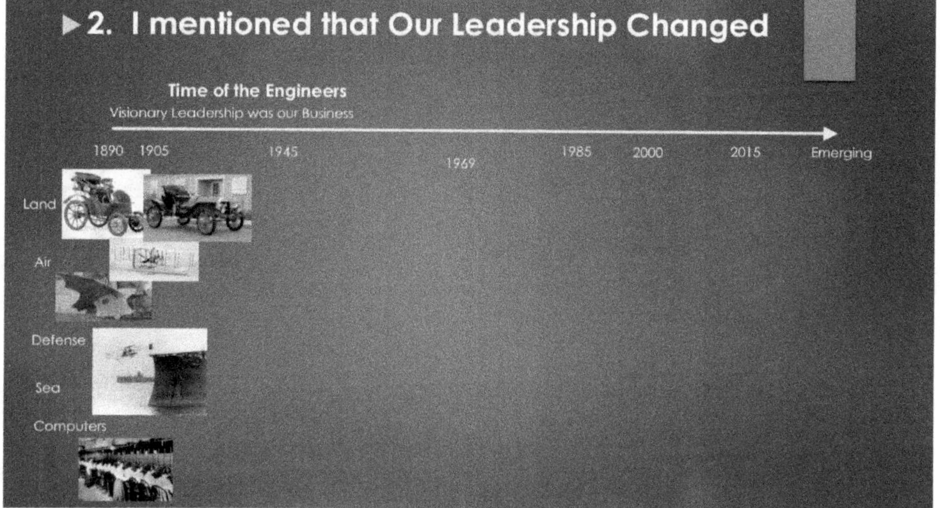

After sixty years of steam driven industrialization, North America and Europe stood at the dawn of the internal combustion engine era. Newly patented gasoline and diesel engines stood ready to issue in a new era of locomotion.

The first self-powered airplane was the steam-engine powered, bat-winged Éole. Clément Ader, a French inventor, flew twenty yards farther that the Wright Brothers and thirteen years earlier as well, on October 9, 1890, in Brie, France, on the lawn of the Château d'Armainvilliers. The Wright Brothers' Flyer flew first at Kitty Hawk, North Carolina and are notable in history in that they carried on their work.

From 1890 to 1900, hybrid gas-electric and gas cars were a cross between carriage and automobile. Ferdinand Porsche's 1900 hybrid gas and electric automobile resembled earlier carriage-based designs. This one caught my eye as it looks like an early Batmobile.

COURTESY: PORSCHE

By 1903, cars began to take on a more refined and modern look.

The **first electronic device** was the Relay back in 1835. By 1836, it drove the Telegraph and at the turn of the century, Telephone networks and switchboards grew quickly. The **first mobile call** was made in St- Louis in 1946 – that Bell cellular phone weighed eighty pounds.

The **first warships to carry aircraft** emerged shortly after the commercial production of the Wright Brother's airplanes began shipping to the army.

Technology in 1917

World War I between 1914 and 1917 brought an array of new weapons, aircraft, and artillery

technology advances including much faster and more reliable biplanes and the first tank.

By the end of World War II

Sophisticated cars, trucks, tanks and self-propelled artillery matured at incredible rates.

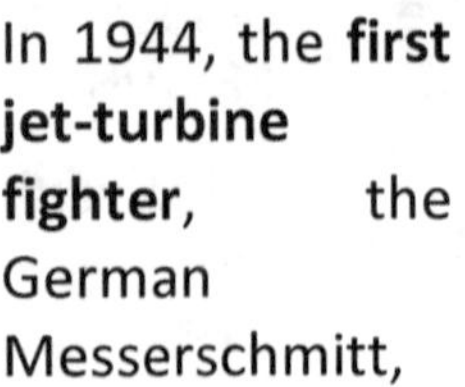

Germany's discovery of **Nuclear Fission** in 1939 was first weaponized by the United States in the Manhattan Project. Nuclear Deterrents have kept us safe since

In 1944, the **first jet-turbine fighter**, the German Messerschmitt,

was in mass production. This was the fastest fighter jet that fought in WWII.

In 1926, Enigma was the most sophisticated mechanical computer of its time. Animatronic inventions by noteworthy clockmakers and inventors including da Vinci were built for a century by this time, but none put to such wide production and use as Enigma.

Alan Turing's Turing Machine Computer deciphered Enigma messages for the Allies, reducing the war by two years and saving twelve million lives.

1945 to 1969 - the Cold War

De Havilland produced the world's first pressurized commercial passenger jet. The DH 106 Comet flew first on July 27, 1949, from its Hatfield Aerodrome in Hertfordshire, England.

General Motors, Ford, and Chrysler were making cars that defined luxury and reliability in Detroit, USA with cars like this 1953 Cadillac.

The USS Nautilus (SSN-571), the world's first nuclear submarine, sailed in 1954 after three years of construction. Nautilus broke many records due to her break from diesel engines. Modern nuclear submarines only have to unsubmerge when their food and supplies run out – usually every 90 days (three months).

Telstar 1 (America) and Telstar II (Europe) launched in 1962 and satellite broadcasts began on televisions throughout the U.S. and Europe. Telstar 1 & Telstar 2 continue to circle the earth today.

The fastest Jet ever built, the Lockheed SR-71 Blackbird, made its maiden flight in 1964. Its Jet engines could propel the craft to a Mach 3.2+ cruising speeds which would heat the aircraft's external surface well beyond 500 °F (260 °C). In flight, the internal temperature of the Quartz window shield was 250 °F (120 °C). An ejecting pilot could be subjected to 450 °F (230 °C) so a pressurized flight suit similar to an astronaut's was required.

Ford's 1965 Mustang Fastback was a very popular muscle-car bought new for around $2500.

President John F. Kennedy led the charge to put a man on the moon in 1962. Eight years later, Neil Armstrong set foot on the lunar surface for the first time in 1969.

The first Space Station, the Soviet Soyuz, launched in 1971, was followed by seven more by 1977. There have been sixteen space stations in all and only the International and Chinese Space Stations are operational at the time of this writing.

Timeline of Space Stations

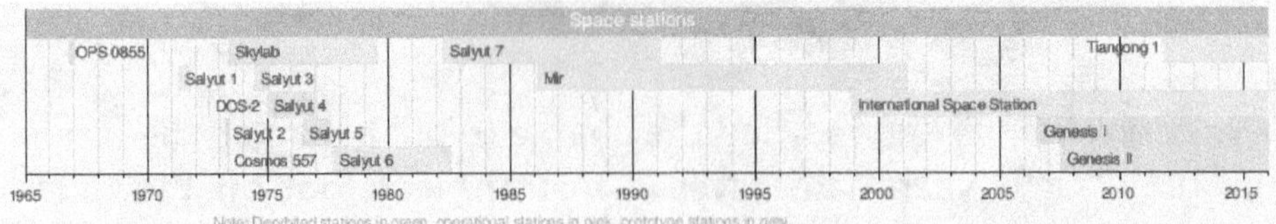

Designs for the largest nuclear warships in the world, were complete by 1972. The USS Nimitz Class Aircraft Carriers are small cities that desalinate water, cook, run heavy equipment and hydraulics, and propel this five-football-field-sized watercraft 35 knots (55 km/hr) in sustained operation with a crew of 6,000 via just two nuclear reactors. The reactors need core replacements only every 25 years.

Technology deemed worthy of being on one of these marvels of technology can be counted among the best in the world. Waste from these super-ships is converted to plasma and then to recyclable glass.

Plasma is the forth state of matter – from a solid, to liquid, to gas, and then finally to ionized gas or "Plasma" at three thousand to nine trillion Fahrenheit. Plasma temperatures can exceed the heat of even nuclear reactions. Even "cold plasma" creates a lava waste product that is explosive in the presence of water, so Nimitz converts its waste to recyclable glass. Measuring these temperature in Fahrenheit is like measuring the galaxy in feet.

The final incredible addition to the cold-war technology explosion was the realization of the 1969 Space Transportation System Plan which designed the first ever reusable Space Shuttle "Orbiter" which was launched in 1981.

In 1973, the Xerox Altos became the first mouse-operated GUI computer ever developed.

Business School Administrators Lead

The Age of Engineers came to an end in the 1970s. To gain an appreciation for the pace of change over the following 45 years up to today, we complete the slide as follows.

2. I mentioned that Our Leadership Changed
Time of the Engineers
Visionary Leadership was our Business
Business School Grads Lead
Business was our Business
1890
1905
1945
1969
1985
2000
2015
Emerging
Land
Air
Defense
Sea
Computers
Google

45 Years Later

The Mustang Fastback runs on gas and four wheels with minor changes that improve safety, reliability, computerized controls that add engineered obsolescence incenting owners to sell their used cars before a 100 working motors on the vehicle start to fail - but same gas engine, same basic fuel economy, the computer adds a few radio stations and this vehicle costs 20 to 30 times its 1965 counterpart. The Jetsons' clean, anti-gravity cars where slated for delivery in 1995 - back in 1960 – but they are no-where in sight today. In summary, some evolution, but no revolution.

Robotics and computers advanced in capability, but nothing revolutionary – again progress continued at an evolutionary pace; as profit permitted. Computers were more miniature and faster, mapping & GPS got better, but no groundbreaking advances really.

MISSILE
COMMAND

=

WORKING
WORKING
WORKING

Google

!=

Metallurgy improved, and industrial robots became more dexterous and reliable – but rarely to a level useful for robotic home maintenance. Robotic ineptitudes were made painfully obvious during the Japanese Earthquake of 2013 when technicians had to sacrifice their lives to shut down damaged nuclear reactors despite Japan being currently the most advanced of all nations studying robotics and automation.

Miniaturization, Improvement, Profit

During this Era of the Business Leads, the good life vanished from the G8 altogether with Russia being the last community committed to supporting the Good Life as described in slide three above. In 1986, Perestroika put an end to seventy years of communist land grant practices and a new era of capitalism and trade began.

Throughout the 1940s through 1970s, the sense that we all had to work together as a community or perish – was gone now. A growing sense of selfishness emerged in society during this time. A better-than Good Life could be obtained easily for a time during the 1980 to early 2000s, and the needs of our neighbors were set aside because we wanted ourselves and our children to be above.

No veteran ever risked their lives because they wanted to ensure that individuals could use their fought and won freedoms to prevent the freedoms of their children and millions of other countrymen. They risked everything to ensure that their children, and everyone else, had security, opportunity, liberty, and the pursuit of happiness – which is why many constitutions ensure this.

In the U.S. and other high-GINI (high wealth inequality) democracies, the growing ranks of the poor had no vote and no collective bargaining power, so their needs grew in keeping with inequity in society. Militias emerged, incarceration rates rose to record levels, on and on - just like in Monopoly – and just like in our analysis of Good Lives in G8 Countries above.

2. I mentioned that Our Leadership Changed
Engineers were Marginalized and Business Ethics Grads took to the Boardrooms
Time of the Engineers
Visionary Leadership was our Business
Business School Administrators Lead
Business was our Business
1890
1905
1918
1945
1969
1985
2000
2015
Emerging
Land
Air
Defense
Sea
Computers
Google
Copywrite 2015 World Peace – The Transition at CSQ1.org

Emerging Technology

A strengthening in technology stocks after 2010 gave rise to interesting technology sponsored by company leads at Google, Tesla, Audi and a handful of VC firms that also benefited, fueling a surge in development and $20 billion demand for 3D Printing, driverless cars, and others.

What Can Era Differences Teach Us

We discussed that war-driven or peaceful wealth distribution restarts capitalism and a good life every 60 years for the past 4000 years. This restart occurred most recently at the end of World War II. The Cold War, therefore, took place in a K-Wave Spring and Summer.

By the 1980s, in K-Wave Autumn and then Winter, the continued goal of capitalism and profit left the poor minority with no vote and little opportunity to curtail their growing ranks pre-emptively. In extreme capitalist democracies, social problems are left to build to crisis levels when basic human rights do not include a Good Life.

The difference in human development between these two eras is so great; so extreme, that we can draw three clear conclusions credibly.

First, the leadership of engineers during times of need, with their oath of service to the greater good, made the grandest advances in civilizations historically. Engineers led during times when money is meaningless, and profit was not a consideration, and it was also during these times that society and technology also advanced most rapidly.

Engineer-led advancement was true in steam eras, in combustion engine eras, and in computer eras. Google's driverless cars were born of 40-year-old hi-tech engineers with unlimited resources – for one example. Rudolf Diesel was an Engineer, as were Henry Ford, Howard Hughes, and others.

Second, Democratic voters do not understand and do need training in, Economic Controls, Business's Accountability, and the role of voters in supporting strong Societies.

Third, business and capitalism rarely create worthwhile projects - and then only at a very slow rates of speed. Non-SME financial and administrative overseers were important until the tail started to wag the dog as admins began gatekeeping access to funding and resources. Boards and Executive teams filled their ranks with non-SME votes, leaving the socially accountable voice of engineers and medical professionals diminished and even marginalized.

I attended a Dragon's Den styled panel review of 20 of Canada's Most Innovative companies yesterday in Toronto. Five of these companies were building worthwhile projects and ten companies did not deserve to be there. Not a single 3D Printing Company was among them. The judges were Merchant Bankers (admins and salespeople) and assorted non-SMEs with questions routed in financial gating aimed at prevented investment. Eight of the ten panel members composing and defending the selected companies, had no business being there.

The Cold War makes this case most strongly due to the absence of an actual day-to-day war. For 30 years, engineers' research and leadership were uninterrupted and the urgent need to create a nuclear and space-based deterrent kept them at the lead of society from 1945 to 1970. These factors resulted in the greatest 30 years of mankind's development in history.

And the opposite can be proven as well. During times of peace, when profit can take the greatest importance again, society benefits least - and regresses dramatically too.

Plato, Cicero, Socrates, and Aristotle said the same. Money was originally designed to facilitate our production economy, but as soon as it is used for profit - above and beyond borrowing needed to fund startups and manufacturing capabilities like real estate debt sales for example – society pays more than it benefits.

There is no man in all things prosperous,
There is no man among us all is free,
For all are slaves of money or of chance.

Aristotle, Politic 322 BCC

Shakespeare's The Merchant of Venice, is given to this theme almost entirely.

Neither a borrower nor a lender be;
For loan oft loses both itself and friend,
And borrowing dulls the edge of husbandry

Shakespeare, Hamlet, Act I, Scene III, 1600

There are many more quotes than I could hope to add in defense of Usury laws and prudence in borrowing and debt.

As a society, we all pay for the incompetent gating and encumbrance of worthwhile innovation companies and projects. Ensure that approvals are generous and that gating panels are engineers. The Return on Investment of that approach will greatly surpass our current methods.

Team 1 – Recessionary Policy

1. **World Peace Agenda** in Law and **Worldville** Project Approvals (Our Moon Launch)
2. Fiscally, this is not the Time to Zero national debt, **Governments and Business must finance**:
   - **Technology Projects** - per World Peace Agenda and CSQ 100 Year Plan
   - **Re-shore** automated Production and Engineering
   - **Sell** to other nations effectively high-quality GDP Exports
   - **Divorce & Housing** Projects
3. **Profitably**, these are not the years for record profits and bonuses. Just like in War-Time Economics, government leaders should not need to be on the phone to every business to assure rapid hiring and **100% employment** with benefits.
4. **Accountability** - Ethics at Business Schools must include CSR and balance a Good Life AND Profit. Investors, Boards, and Executive Teams must balance engineers with admins and prefer "Right Plan" Projects. A Licensing Model so Business Officers lose their license when society is harmed; today, socially irresponsible decisions are bonused.
5. Right and Left **Political Parties** must be excellent financial managers. Both must fast-track Wealth Distribution, Safety Nets, Human Rights, and Wealth Creation.
6. Increase the **Minimum Wage** in support a Wealth Distribution Plan for each country. Norway has 3% unemployment, #7 GINI, #1 HDI, with $20 per hour minimum wages. The USA calculates to $15 per hour and Canada $20.

Copywrite 2015 World Peace – The Transition at CSQ1.org

Team 1 – Recessionary Policy

1. World Peace Agenda in Law and Worldville Project Approvals (our Moon Launch)
2. Fiscally, this is not the Time to Zero national debt, and **Governments and Business must finance**:
   - **Technology Projects** - per World Peace Agenda and CSQ 100 Year Plan
   - **Infrastructure Automation** Projects
   - **Re-shore** automated Production and Engineering
   - **Sell** high-quality GDP Exports to other nations
   - **Divorce & Housing Projects –** see Chapters 12 & 13
3. **Profitably**, these are not the years for record profits and bonuses. Just like in War-Time Economics, government leaders should not need to be on the phone to every business to assure **rapid hiring** and **100% employment** with benefits.
4. **Accountability** - Ethics at Business Schools must include CSR (Corporate Social Responsibility) and balance a Good Life AND Profit. Investors, Boards, and Executive Teams must balance engineers with admins and prefer "Right Plan" Projects. A Licensing Model, so Business Officers lose their license when society is harmed; otherwise today, socially irresponsible decisions are bonused.
5. Right and Left **Political Parties** must both be excellent financial managers. Both must fast-track Wealth Distribution, Safety Nets, a Good Life, and Wealth Creation.
6. Increase the Minimum Wage in support of a **Wealth Distribution Plan** for each country. Norway has 3% unemployment, #7 GINI, #1 HDI, with $20 per hour minimum wages. The USA calculates to $18 per hour and Canada $20.

Team 2 builds Step #2 Technology Projects

- In 1962, JFK asked NASA to put a man on the moon. "Impossible" said his engineers.
  - JFK commissioned NASA to build 13 Projects with a budget of $5.7 billion; and a successful moon launch completed in 1969 (8 Years Later).
- The CSQ Plan asks our brightest companies and leaders to complete 14 Projects that Automate our Production Economies, Pilot in 2018, and sustain World Peace by 2035.
- **Worldville 1** Program Leads – Russia, China, and Germany
- **Worldville 2** Program Leads - United States and Japan

"If every instrument could accomplish its own work ...chief workmen would not want servants, nor masters slaves."

Aristotle, Politic 322 BC

- Copywrite 2015 World Peace – The Transition at CSQ1.org

CSQ 100 YEAR PLAN
PLANNING STRATEGIC LONG TERM GOALS
WHAT WILL BE THE LEGACY OF OUR GENERATION?
DISCUSS THIS PLAN AT CSQ1.ORG/MAG
50 years out
75 years
100 years
A+ Human Cloning
A+ Reverse Replicators - which convert matter to energy and back again.
A Consciousness Transfer Technology
A+ Anti-Gravity
A+ Warp Drive - Negative Energy Technology
A True Artificial Intelligence
A Transporters
B Impulse Engines
A 2055 The Singularity - Computers Programming
A+ 2055 End of the Monetary System - in G7 No taxes
Long-Term - 40 Years out
A+ 2035 Start 'The Transition' from Money to Social Utility Government
A+ 2035
In-home Self-Powered Replicators for Food, Gold, Clothing AutoDesk, StrataSys
Social Problems Decrease
Incomes Corrected Wealth Distribution Correcting
Mid-Term 10 to 20 Years
Mid-Term 8 to 10 Years
Short-Term 3 to 5 years
A 2020 Holodecks - w/o feedback Microsoft
A+ 2022 Clean Energy – Cold Fusion
A+ 2023 Robot Maids in every home Honda, Moriyama
A+ 2024 2 day, 4 hour Work-Weeks @ Full Benefits Government
A+ 2025 Fully Diagnostic Medical Imaging Fujifilm, GE Healthcare, Siemens, Philip
A 2025 Holodecks - with touch feedback Microsoft
A++ 2018 Worldville 1 & 2 #WPProjects
A++ Leadership
A+ 2018 Configurable Matter from Light Imperial College London
A+ 2017 3D Printers for Clothing, Food, Goods Featz, Disney
A+ 2016 Voice controlled Universal Translators Google
A 2016 Video watches Samsung
SAMSUNG
A+ 2016 Communicators Samsung, Google, Apple
A+ 2016 Tricorders XPrize Foundation
XPRIZE
A 2015 Clean-Diesel vehicles Audi, VW, Mercedes-Benz
Very Short-Term – 1 to 2 years
A+ 2015 Treatment for Alzheimer's
A 2015 Performance Management Software DashFlows
Microsoft
SAP
A+ 2015 Automation Software SAP, Microsoft
A+ 2015 Robotics Caterpillar, Siemens, ABB, Boston
A+ 2015 Close Markets to Socially Unproductive Companies Government
A+ 2015 Re-shore Engineering Government
A+ 2015 Re-shore to Create Wealth Government
A+ 2015 4, 3 and 2 Day work-weeks equal to automation profits Government
CSQ Compliance & Certification
Wealth Distribution 2015
Bottom 40%
Top 20%
CSQRESEARCH: WORLDPEACE - THE TRANSITION
COPYRIGHT 2015 EDWARD TILLEY ALL RIGHTS RESERVED

Team 2 builds Step 2 - Technology Projects

- In 1962, JFK asked NASA to put a man on the moon. "Impossible," said his engineers.

  President Kennedy commissioned NASA to build 13 Projects with a budget of $5.7 billion, and a successful moon launch completed in 1969.

- The CSQ Plan asks our brightest companies and leaders to complete 12 Projects that Automate our Production Economies and sustain World Peace by 2035.

Automate our Production Economies for Pilot in 2018 and rollout to a sustainable World Peace by 2035.

Our Generation's "Moon Launch" are Worldville's 1 & 2 – which need 14 Projects to complete.

- Worldville 1 Program Leads – Russia, China, and Germany
- Worldville 2 Program Leads – United States and Japan

"If every instrument could accomplish its own work ...chief workmen would not want servants, nor masters slaves."

Aristotle 322 BC

Team 2 – Automate Production ...

1. Each Country Builds a Part:

   A. 3D Printers and Robotic Assemblers of

   - Food – proteins, vegetable, vitamins, organic
   - Blankets, clothing, 300 times faster (set KPIs) ...
   - Robot Maids, Pickers, Miners - with ...
   - Rapid Charge Battery Systems
   - Cars, Homes

   - AND -

   B. Build a Replicator with Energy to Matter conversion – this is "The Easy Way"

2. Clean Power via a **Cold Fusion** Portable Energy Source

Copywrite 2015 World Peace – The Transition at CSQ1.org

Team 2 – One Project to Automate our Civilization

1. Each Country Builds a Part:

   A. **3D Printers and Robotic Assemblers** of:
   - Food – organic proteins, vegetables, pharma.
   - Blankets, clothing - 300 times faster
   - Robot Maids, Pickers, Miners – with Rapid Charge Battery Systems
   - Transport, Homes, Community Centers

   B. **Build a Replicator - Energy to Matter Conversion** and then Plasma "De-Replicator" see PyroGenesis

2. Clean Power – **Cold Fusion** (working in Germany in Dec 2015) **,**
   Fuel Cell and Blue Diesel Portable Energy

Reassembling products that we use every day atom-by-atom is not going to an easy technology project deliverable. Consider that all elements created from energy by the big-bang here on earth billions of years ago, were then reshaped through geologic forces, decay, pressure and heat. We took those raw materials and made alloys or plastics, and finally fabricated them into products.

Hard Projects - like this one and the Moon Launch - can take a decade or more. That work begins at Imperial College at the University of London in 2016.

Robotic Automation Projects Mitigate the Risk of Energy-to-Matter technology delays by assigning automation projects.

- 200 Countries are assigned hi-tech projects in Chapter 7
- Some projects are advanced high-school robotics or competition-level projects, and others are sophisticated robotic builds.
- Countries with higher GDP are assigned more difficult projects and are also asked to Sponsor smaller countries as well.

Automations Target Top Exports
North America
Motor Vehicles & Parts
Capital Goods
Clothing & Shoes
Europe
Machinery & Motor Vehicles
Motor Vehicles
Machinery & Transportation Equipment
Manufactured Goods
Asia
Copper
Electronic Equipment and Machinery
Precious Stones
Oil and Gas
Middle East & Central Asia
Petroleum & Petroleum Products
Oil
Petroleum
Crude Oil
Apparel & Foodstuffs
Africa
Gold
Crude Oil
Oil
Diamonds
Coffee
Cotton
Iron Ore
Petroleum & Natural Gas
South America
Transport Equipment
Natural Gas
Soybeans
Copper
Petroleum
Copywrite 2015 World Peace – The Transition at CSQ1.org

Each Country are Assigned #WPProjects

G+

Country	GDP	#WPProjects
Belgium	524,806	Produce Transport, Chocolate
Brazil	2,243m	Auto Peacekeeping – Air, Confections Candies
Canada	1,838m	Satellite Water Divining, Collection & Distribution, Cold Fusion, Drone Jets & Air transport
Denmark	336,701	3D Printed or Assembled Buildings
Finland	267,329	Fishing fillet, package, freeze
France	2,806m	Auto Road Build/Repair, Cosmetics, Wine
Germany	3,730m	Auto Vehicle Delivery, Driverless Cars, Bavaria – Beer
Hong Kong	274,027	GPS Tractors, Holiday and Gift Wrapping
Indonesia	868,346	3D Food Printers
Israel	291,567	Auto Airplane design
Italy	2,149m	3D Cement Systems, Wine
Japan	4,898m	Household Robot, Grand & Upright Pianos Pharma Diabetes - Lantus Solostar & related
Mexico	1,259m	Auto Steel Manufacture, Avacados, Mangos, Tequila
Micronesia	333	Sponsor Japan Auto Toilets and Installer
Monaco	6,559	Performance Management
Netherlands	853,539	Conveyers - mining & farming
Norway	522,349	Auto Oil Production, Diapers – All, Electronic Drum Sets
Pakistan	225,419	Grocery Distribution & Delivery
Philippines	272,067	Technician robots

Country	GDP	#WPProjects
Poland	525,863	Tech Architects & QC Approvers, Interoperability Standards, Transportation Software
Russia	2,096m	Drone Peacekeeping – Ground, Auto Snow Removal Oil Alternatives - Paraffin Diesel and Blue Diesel Plants (perhaps using nuclear ship power), Vodka
Saudi Arabia	748,450	Rapid Charge Batteries - Robotics
South Africa	366,060	New Mining by Satellite
South Korea	1,304 m	Automated TV, & Monitor production
Spain	1,393m	Automatic Park Builder
Sweden	579,680	Marina Build System – up to 80 boats
Switzerland	685,434	Surgery – Knee, Hip, Elbow
Turkey	822,149	Satellite Driven Well & Irrigation Builder
United Arab Emirates	599,000	Fishing fillet, package, freeze
United Kingdom	2,678m	Energy to Matter Conversion Sanitation – Waste Treatment Scotch – Single Malt & Whiskey
United States	16,768 m	GPS Threshers Harvest Tricorder Med Console Hawaii - Auto Harvest Macadamia Nuts California – Blue Crude & e-Diesel Cheese & Cream Products

- Copywrite 2015 World Peace – The Transition at CSQ1.org
- **200 Countries with at least One Project**

Each Country is Assigned a #WPProject

In sequence by GDP, each country in the world is assigned automation projects in all of our basic needs leveraging top exports from every country. If a country is a world leader in one of basic need for a Good Life, then as much as possible they were assigned this task. More complex tasks were given to countries with highest GDP and smallest GDP countries were paired with sponsor nations inversely so that smallest and largest GDP countries assist one another.

In this way, all countries can contribute to our generation's greatest accomplishment, an automated economy. An end of slavery, and the start of a universal human rights. High-schools, States, Companies too, engineer tinkerers, all are welcome to contribute.

The #WPProjects List is high-level in this v1.0 release. Dozens of projects are missed and it might even be safe to assume that many countries will be asked to add another project or two quickly. Discuss #wpprojects on Google Plus and Twitter.

How long to build one of these projects? It takes twelve months for a v1.0 working release of each technology. 24 months for a 2.0 release with interoperability with all other systems. By v3.0, there will be at least three to ten versions available, which means that an automated house builder will be able to build three types of home – perhaps townhouse, bungalow, two-story-home; and then three styles of townhome and others as well.

To build this list, I took the top 100 grocery items sold locally; then the top 100 Pharmaceuticals, then the top 50 household items – and so on.

Every region will have special requests, and we ask that you contact the project lead for a technology in the host country if your team wants to work on parallel or dependent projects. This is encouraged, as is swapping projects with other countries if this initial plan can be fine-tuned. Simply request an update be made to the master WPA #WPProjects Plan with consent by the host country attached in our forums.

Automate Primary, Secondary & Tertiary Production Economies

Primary – Farming, Fishing, Mining

1

- The farmer plants, grows and harvests grain

Secondary Manufacturing

2

- The Miller takes grain and makes flour.
- The Baker needs flour to make Bread

Tertiary – Transport, Distribution, Wholesale, Retail, Banking, Real Estate, Insurance

3

- The Wholesaler and/or Retailers pay for packaged bread
- The Consumer purchases the Baked Goods using a credit card or cash.

- Copywrite 2015 World Peace – The Transition at CSQ1.org

Automating Production Economies

Today we plant and pick our crops manually. Tomorrow we will automate Farming, Fishing, Mining, and other resource Commodities. Where do we begin? We look at our country's GDP Exports today and also look at where you want it to be in 5 years, and then we automate the most important exports first. In this way, we grow both wealth creation and wealth distribution, at the same time.

As we automate Primary Production, begin to automate Secondary Manufacturing as well - using the same prioritization technique above of GDP export planning. Target Imports and seek to minimize dependency where any hardships that arise are manageable by your suppliers and customers.

Automated Wholesale, Retail, and Distribution are planned to implement at a rate that generate profits and rely upon Social Safety Nets less.

If you are concerned that your priorities and tasks are not well understood nor well executed, look to CSQ Certifications and CSQ CSR Training Worldwide designed to assure due diligence was exercised in the automation choices made first, so that social benefits and the greater good are served as best can.

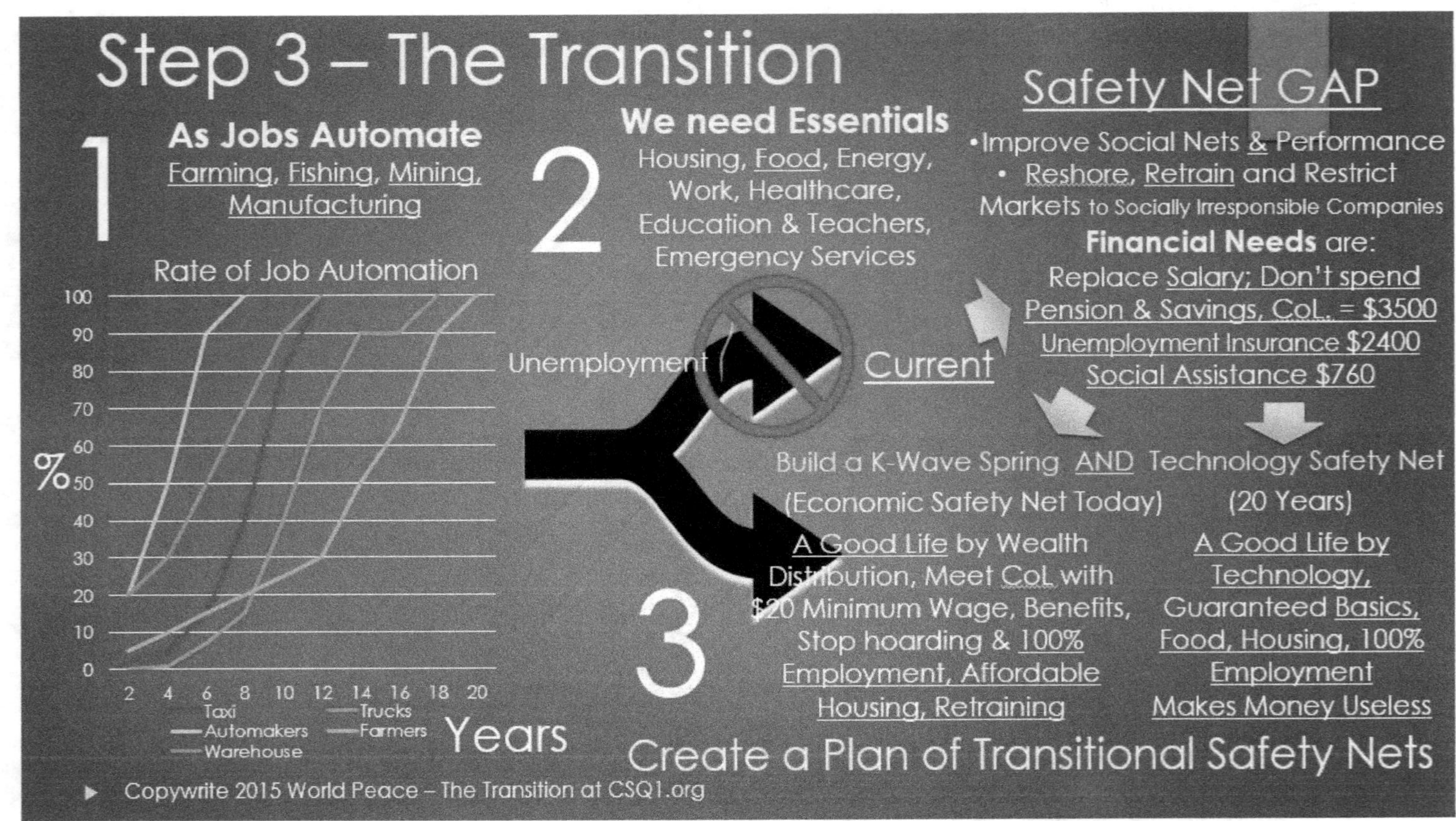
Step 3 – The Transition
1
As Jobs Automate
Farming, Fishing, Mining, Manufacturing
Rate of Job Automation
100
90
80
70
60
50
40
30
20
10
0
%
2 4 6 8 10 12 14 16 18 20
Taxi
Automakers
Warehouse
Trucks
Farmers
Years
2
We need Essentials
Housing, Food, Energy, Work, Healthcare, Education & Teachers, Emergency Services
Unemployment
Current
Safety Net GAP
•Improve Social Nets & Performance
• Reshore, Retrain and Restrict Markets to Socially Irresponsible Companies
Financial Needs are:
Replace Salary; Don't spend Pension & Savings, CoL. = $3500
Unemployment Insurance $2400
Social Assistance $760
Build a K-Wave Spring AND Technology Safety Net
(Economic Safety Net Today)
(20 Years)
3
A Good Life by Wealth Distribution, Meet CoL with $20 Minimum Wage, Benefits, Stop hoarding & 100% Employment, Affordable Housing, Retraining
A Good Life by Technology, Guaranteed Basics, Food, Housing, 100% Employment Makes Money Useless
Create a Plan of Transitional Safety Nets
Copywrite 2015 World Peace – The Transition at CSQ1.org

Step 3 – Manage the Transition

Our Jobs are Automating. This is already happening, and it is important that these automations must continue and even accelerate if we are to a reach a sustainable Good Life.

Automation has impacted Autoworkers, Warehouse workers, and many facets of manufacturing and shortly distribution too.

The largest employer in the U.S. is the trucking industry, and that industry will begin to lose jobs rapidly as automated driving cars and trucks begin to chip away at manual jobs and the inevitable countdown to 100% automation.

Second, our priority must always be to providing the essentials of a Good Life as these changes slowly begin to benefit us all. Engineers and Physical Scientists need to transition first, for example, some Medical transitions may be unnecessary or will happen toward the end of the transition, as tax collection and reporting become less important – those jobs transition next – and so on.

Third, our Safety Nets today are insufficient and must be transitioned to a Two-Tier Safety Net approach, where initial Transition Safety Nets fully cover cost-of-living (COL) until Technology Safety Nets are implemented over the next 20 years.

Create the field of Transition Economics, leveraging Transition Economy conventions and automation rate balancing.

In the discussion of Safety Nets – we cover engineers first and then empower them to build automation with priority.

Then we automate Industry by Industry - in priority of profitable exports first – and coordinate with other nations so that Canada automates Mining, as Mexico takes on robotic food production; then the UK takes on landscaping and ground maintenance; and Japan leads household robot-maid production – and so on.

Step #4 – WPAs
World Peace Agendas

#WPProjects *build* a Good Life for All

World Peace *is a Project*

World Peace *is a Human Right*

- Advanced Technology Anti-Gravity, Home Robots, Perfect Medical Imaging
- Social Safety Nets both Financial and Technical
- The greatest evolution of Mankind since Hunter-Gatherers turned to Civilizations 10,000 years ago. A High-Performance Meritocracy
- Our Best & Brightest explore Space, Science, Engineering, HealthCare and Contribution is Recognized
- Democracies do not war with Democracies

Team 3

The United Nations reports World Peace Agenda WPA Status for all Nations. For addition to their 2016 Agenda

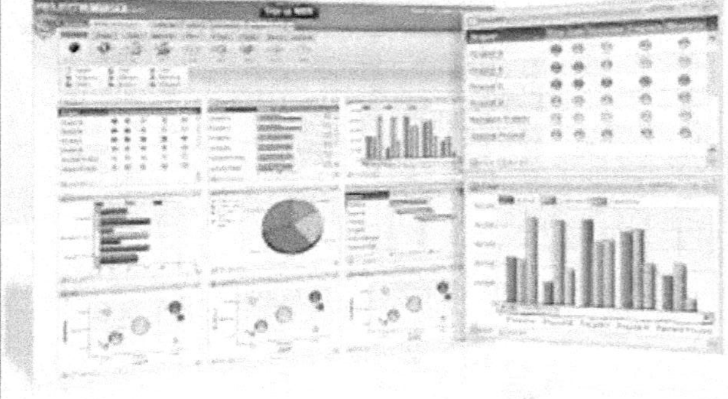

- Copywrite 2015 World Peace – The Transition at CSQ1.org

Step 4 – World Peace Agendas - Rollout Worldwide

Step 4 introduces Team 3 in the form of a central administrating body that consolidates and assists the World Peace Agenda of all countries worldwide. Initially, CSQ Research manages the list and then the United Nations may assume this role consistent with their international mandate today.

The World with Global World Peace Agenda (WPAs):

- Social Safety Nets both Financial and Technical are monitored
- The greatest evolution of Mankind since hunter-gatherers turned into civilizations 10,000 years ago
- A 7/24 Call Center is available and scales up with need
- World Peace Agenda Review Board
- The World Peace Agenda Repository
- Democracies do not war with Democracies
- WP-TV News and Documentaries
- Our best & brightest explore Space, Science, Engineering, HealthCare and the Contribution & Merit is Recognized
- A High-Performance Meritocracy evolves
- A Good Life is the minimum Human Right of everyone

Next Steps – Action Right Projects & Vote !!

Vote

Democratic voters and Monarchs elect Engineers & HiTech Builders first; then Medical; Legal; Sales; then Business

Wealth Distribution

- Graduated Tax on Wealth & Big Business
- 100% Employment & Education
- Affordable Housing
- Onshore Engineering

Wealth Creation

- Automate Exports Retraining

Investment

- Automation in 3 Tiers of Economy
- Clean Energy & Waste

#	A+ Priority Social Policy	Canada			
		Conservatives	Liberal	NDP	Grn
1	Long Term Planning	No	No	No	Yes
	Wealth Distribution				
2	Bench Strength of MPs for Project Execution	Maybe	No	No	No
3	Minimum Wage Targets	No	Maybe	Yes	Yes
4	K-Wave Economic Controls - Graduated Tax	No	No	No	Yes
5	3 Day Work Weeks or Income Safety Nets	No	No	Yes	Yes
6	100% Employment	No	Maybe	Yes	Yes
7	Housing & Anti-Usury Law	No	No	No	No
8	Onshore Engineering	No	No	Yes	No
	Wealth Creation				
9	Increase Quality Exports	No	No	No	No
10	Reduce Imports	No	Yes	Yes	Yes
11	Sellers Manufacture Locally	No	No	Yes	Maybe
12	Profit becomes Investment	No	No	Yes	Maybe
	Transition				
13	Tech Project Safety Nets	No	No	No	Yes
14	Automation of Import Export	No	No	No	No
15	World Peace Agenda & CSQ Plan Projects	No	No	No	No
	If Yes=2, Maybe=1, No=0	3%	13%	47%	53%
	Leaning	Right	Left	Left	Left

- Copywrite 2015 World Peace – The Transition at CSQ1.org

Next Steps – Vote and Choose Projects Well

In Democratic Nations, advise politicians that your vote is contingent on Social Programs that support your Country's World Peace Agenda.

- Vote for and become, Engineers & Hi-tech Builders first; then Medical; then Sales; say No to Business Leads, say no to small thinkers and short-sighted emotion.
- Vote for Policies in:
    - **Wealth Distribution - graduated Taxes on Wealth & Business** and financial controls "with teeth"
        - 100% Employment
        - Universal Healthcare including Dental
        - CSQ Compliance for your Projects and Invest in CSQ Tech
        - Affordable Housing
        - Families stick together and decrease Divorce
        - Onshore Engineering
    - **Wealth Creation**
        - GDP Exports improve like Norway & the Netherlands
        - Retraining occurs as needed to support the automation
        - Political Leads in North America are behind in their economic controls.
        - Status is maintained at CSQ1.org and reported at WPTV Magazine. Check this regularly and get your political party active in its support.
- Invest and volunteer time in projects that Automate Production, Manufacturing, ERP, automated transportation, 3D Printing Technology, Clean Waste (Plasma – see PyroGenesis), Clean Fuels (Ballard and Audi), Community Events, and Performance Management.

GAPs in the UN's Global Goals

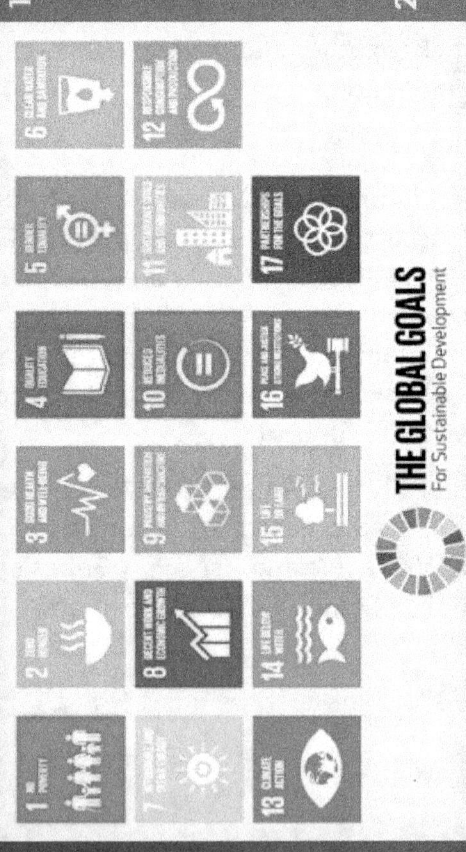

1. **Wealth Distribution** and a Good Life
   - Create Two Safety Nets
     - Engineers working automation projects
     - Minimum Incomes sufficient to have a basic Good Life (Human Rights)
   - Enforce CSR Business Accountability
   - Graduate Tax
   - Minimum Wages with targets
   - Affordable Housing, Debt w/o Usury
2. **For Wealth Creation**
   - Fund automation by Transition Plan
   - Manage WPAs with local ownership

Prefer Sustainable Bottom-Up Plans

Top-Down Plan - To Feed a Man a Fish, provides for what he needs today.

Bottom-Up Plan - Provides this man with the tools and teaching to fish for himself

Sustainable Bottom-Up Plan – automates to feed him and all of his neighbors too

Copywrite 2015 World Peace – The Transition at CSQ1.org

GAPs in the UN's Global Goals

The United Nations announced its first Plan since 2000 in 2015. Global Goals discusses many topics in the Transition Plan, but it misses important points I have termed "GAPs" here in the book.

- ▶ **Top-Down Plan** - To Feed a Man a Fish, provides for what he needs today.
- ▶ **Bottom-Up Plan** - Provides this man with the tools and teaching to fish for himself
- ▶ **Sustainable Bottom-Up Plan** – automates to feed him and all of his neighbors too. Always prefer a Sustainable Bottom-Up Plan

Projects not discussed in Global Goals include:

1. **Wealth Distribution** and a Good Life
    - Create Two Safety Nets
        - Engineers working automation projects
        - Minimum Incomes sufficient to have a basic Good Life (Human Rights)
    - Enforce CSR Corporate Social Accountability
    - Graduate Tax
    - Minimum Wages with targets
    - Affordable Housing, Debt w/o Usury
2. **For Wealth Creation**
    - Fund automation by Transition Plan
    - Manage WPAs with local ownership

CSQ CSR Compliance & Certification

CSR **Corporate Social Responsibility**

Benefits include…

- Peace of Mind for Voters & Policy Organizations
- Higher staff job satisfaction, productivity, and attendance
- Higher revenues and sales volumes
- Assurance of Compliance for Ethical Investors
- Oversight on Ethical Business Practices, Bonus, Compensation, High Performer Recognition
- Mature Status Reporting and Project or Process Template Libraries
- The CSQ Logo for your website and marketing
- Strategic Business Planning

- Copywrite 2015 World Peace – The Transition at CSQ1.org

CSQ CSR Compliance and Certification

We often hear politicians explain how wonderful are their governments and what a difference that their programs will make – and then nothing happens or negative results comes of those investments with time.

Businesses are before The Hague in 2015 for the first time, and CSR Leads are now scrambling to correct glaring gaps in social accountability. Changes to policies include due-diligence for lending, investment, and operations.

CSR – Corporate Social Responsibility

CSQ's CSR Compliance & Certification monitors CSR targets and reports progress. When CSQ Certified Programs are promised, they get delivered. Benefits include:

- CSR Goals are established clearly and met
- Peace of Mind for Voters & Policy Organizations
- Higher staff job satisfaction, productivity, and attendance.
- Higher revenues and sales volumes
- Assurance of Compliance for Ethical Investors
- Oversight on Ethical Business Practices, Bonus, Compensation, High Performer Recognition
- Mature Status Reporting and Project Process Libraries
- CSQ Ethical Business Practices Logo for your website and marketing
- Strategic Long Term Business Planning

WP-TV

TV Shows & News Programs - Media Stories & Status Updates

- World Peace Canada
- World Peace Italia
- World Peace America
- World Peace UK
- World Peace Nederland
- World Peace Australia
- World Peace Denmark
- World Peace Sverige
- World Peace Türkiye
- World Peace Brasil
- World Peace Norge
- World Peace Deutschland
- World Peace Russia
- World Peace India
- World Peace China
- World Peace Japan
- World Peace Suomi
- World Peace Schweiz
- World Peace Mexico
- World Peace Espana
- World Peace Latino America
- World Peace Portugal
- And more…

Copywrite 2015 World Peace – The Transition at CSQ1.org

Final TEDTalk Slides explain WPTV and WP Magazine at
http://csq1.org/mag

World Peace – The Transition

Graduate Today …

- ✓ Society, Business, Leadership & Global Economics
- ✓ Strong Families
- ✓ Individual Achievement and Happiness

Forums at CSQ1.org

All material are protected by US & Canadian Copyrite. Owner Edward Tilley

WORLD PEACE
THE TRANSITION
EDWARD TILLEY
CSQ
COMMON SENSE 101

Copywrite 2015 World Peace – The Transition

Chapter 10 – The World Peace Agenda

As countries begin to build CSQ World Peace Projects to a Transition plan that builds a sustainable Good Life, our present systems of government and capitalism continue. They even continue with a restored sense of optimism because we have a plan of safety nets that responsibly manage the needs of engineers and citizens as automation continues to widely change today's manual production economies.

As work progresses, each country maintains their own World Peace Agenda (WPA), and all of these WPAs are then centrally administered so that working teams, stakeholders, and engineers and voters can track the progress and leverage contributions and lessons learned by international efforts elsewhere.

Positive change is afoot - but none of it happens without proper planning and respect for the individual needs of participating countries, scientists, and program leaders.

An important construct of that sharing is adhering to the unique requirements introduced by each country and corporate participant. Needs include accommodation for cultural diversity, religions, language and many ages ranges, many educational experiences, security, personal privacy, and even different levels of progress are encouraged whenever possible.

The values that built each country are not going to change because a coordinated plan, or policy, for World Peace is going to be uniformly documented in a standardized Agenda.

Every government routinely focuses policy including Transportation, Infrastructure, Immigration, Economic Development, Foreign Policy, Economic Policy, participation in the United Nations, and so on – to mention just a few examples.

Simplicity and consistency are important considerations of the World Peace Agenda. CSQ Certification is also a central quality control consideration because the Plan has to work well in the face of hundreds of individual subprojects running simultaneously.

Communicating Tasks, Timing, Dependencies, Risk Plans, Communications Plans and Progress Is made easy so as to ease adoption and avoid confusion.

For this reason, the World Peace Agenda is suggested to become Government Policy - just like any other, with an elected lead minister assigned from parliamentary leads or other engineering leads within each government.

World Peace Agendas are represented at Central United Nations Status Meetings monthly but roll-up weekly in a single one-hour meeting attended by highest level stakeholders in every country. A UN Monthly Review meeting should be conducted by the Economics and Technology Review Committee as well.

Asking current Ambassadors to the United Nations to take on this role is not recommended unless the lead has the availability and also an engineering and science project leadership experience set.

Each Country assigns Designates to the UN World Peace Agenda Review meeting to drive progress. This is not simply a reporting role.

We already know that many of the core projects of World Peace's Transition are already ongoing in the world today.

The World Peace Agenda will ensure that technology and social projects not impacted by conflicts of interest or apathy; nor simply because no-one recognized the importance of what they had and then squandered that work by not supporting it.

I mentioned that the U.S. had World Peace for Individuals in the 1950s and 1960s – through a system of Wealth Distributed Capitalism after World War II. Russian citizens had this too, through a system of land grants that gave them cottages (Dacha) as well, right up through to the mid-1980s. Brunei and Saudi Arabia have this today through financial grants and national incomes given to each citizen.

We want to build on these structures ensure that a Good Life is building worldwide steadily.

Governance and assistance on the setup of a World Peace Agenda are available from CSQ1.org's Consulting Teams initially – and then United Nations admin teams will likely continue support as more and more countries come online.

World Peace Thermometer and Clock

The deployment of a Good Life worldwide is initially proposed to be tracked with a % Complete Thermometer.

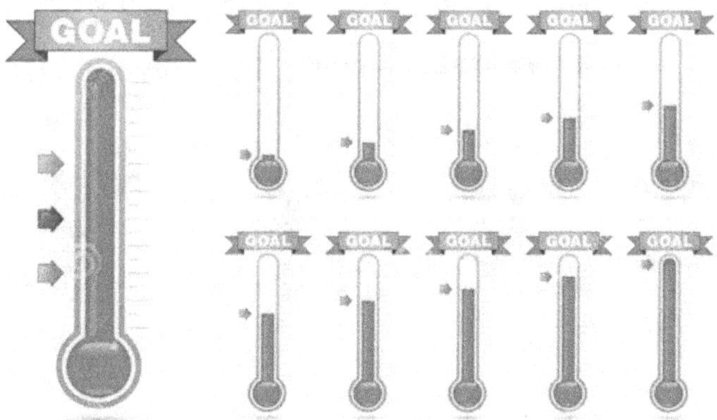

And a Countdown Clock indicates Target minus ...

It's almost **World Peace**

7022 days **22** hours **45** minutes **6** seconds 359

As World Peace Transition Projects in Technology begin to make progress, Social Projects begin to increase the number of Good Lives through Wealth Distribution initiatives. Technology makes a Good Life sustainable and even builds it eventually, but Wealth Distribution accelerates the process by up to twenty years and averts the risk of a nuclear WWIII too.

Communicating Status and Needs

Weekly, Monthly, Annually Status are important to the success of World Peace Agenda.

Monday	Tuesday	Wednesday	Thursday	Friday
Working Team Status Meeting	Confirm Red, Yellow, Green with clients	Program Report Rollups	One Hour Exec Review of RED Projects	Update and Forward Globally Once per Month
Annual Progress Summaries for all Projects & Teams				

Summary Status Reports by Country

Status Reports

Overall Status: Yellow

Social Status: Yellow

Technology Status: Green

Healthcare Status: Yellow

Click here for Detail Status Reports...

Geographic Status Reporting

Crowded Projects

Can you get too many teams working on a solution? Rarely. The physics teams working on Energy to Matter Projects utilize very specific and very expensive resources called supercolliders which are available in small windows of time in a small number of centers around the world. Resource constraint will restrict the number of teams working on this solution due to this constraint.

Most other science, technology, humanitarian and economic projects are less constrained and might only be gated by funds available for budget requests. In theory, many hands make light work, and in practice, administration and workgroup collaboration tools (Webex, email, wiki sites, document repositories, and similar) will be all that is needed to standardize and enable working teams to optimize their productive working time from anywhere around the globe. Teams that document their progress well within their World Peace Agenda will need to answer repetitive questions by other teams infrequently – so there is an incentive to document status well for this reason.

The cost of enabling engineers is often a fraction of tax programs offered to industry and is encouraged wherever possible.

There is always the chance that a group is going to sign-up to work on a program and then play video games all day every day. These cases will be few and short-lived and really, the real work is quite a bit more interesting than distractions as well.

Procedures to get Funding and Launch

The World Peace Agenda Repository is the place to find Application Procedures and templates to get your World Peace project funded and launched quickly.

Important KPIs like time-to-start, templates for Charter, Requirements & Inventory, Design, Detail Implementation, Status, Transition to Operations and Revision Management documents are all here to help you get started quickly and productively - with detailed examples and administrative help available to working teams.

Select a project, describe your team and approach, adhere to the project templates for build and status reporting - and get started.

Budget permitting, the **World Peace Agenda Review Board** is the administrative group that will get you up and running. Once running, they then monitor performance and ensure you get the tools you need, as well as the safety-net needed, to permit you to dedicate your time to the work.

Meet targets and this will permit you to continue working toward delivering a Good Life for everyone.

Acknowledgments and Recognition

Most important to the success of a sustainable automated production society are the acknowledgments and recognition lauded upon the engineers of seemingly endless hours of hard thinking, work, and trial and error.

A TV Awards Program is in the works to acknowledge the efforts and progress of the "rock stars" of World Peace. Programming to parallel and rival the Oscars and the SuperBowl Pre-Game are also up for consideration. Nothing is too good; nothing is over the top.

Operational Performance Measures

KPIs for operational performance follow a consistent look and feel whether systems are in technology or social in nature. The operational performance needs of each system are important to understand up-front so that these needs can be solutioned during design discussions and then met reliably in Operation.

Performance KPIs for Technologies include:

	SLAs	Level 1 Support	L2	L3	Vendor Support	$ & Resources
Availability	7/24 5 min down max.	Ticket Logs	7/24 Knowledge-base Analyst	Architect	5x12 5-hour MTR;	3 L1 Staff 3 L2s
Delay	.1 sec screen refresh	Ticket	7/24 Knowledge-base Analyst	Architect	9-hour MTS	2 L3
Moves/Adds /Changes (MACs)	5-day bug fixes; 5-week Feature enhancement – 6 months; Next Release - 6 months	Ticket				

These Application KPIs are then expressed for all technology infrastructure sub-platforms individually for Servers, Databases, Storage, Desktops, Remote and In-field Technical Support in the Design Documents of individual Projects and at the Call Center.

Similar KPIs for Safety Net, Wealth Creation, and Distribution Projects are similarly needed and documented.

WP Dashboard

Compatible with Wikipedia standards for political systems and demographics, the Dashboard gives everyone a drill-down glimpse into the progress of Transition Plan initiatives in every country – including the performance reports for individual country stakeholders and politicians as well.

The World Peace Agenda Call Center

Call Center staff can vet questions from the public and working teams via HotLines and the web-based Status Reporting Hub.

Project Metadata:

Look to the World Peace Agenda for worldwide and country status updates to contain important data on the projects detailed in Chapter 8 and to the specific Project and Program Processes and Status Reporting per Country in Chapter 17.

Metadata includes the following (and others will follow as required by analysts):

Form 1: Charter

Project Name & Number
Your Team Leader(s) – Primary and Secondary
Scope (which CSQ 100 projects is this team targeting)
Stakeholders
Steering Committee
Working Team
Risk Plan
Needs – working space, servers, salary needs, etc.

Form 2: KPIs

Form 3: Project Documentation

Form 4: Status

Form 5: Approvals

And that begins the World Peace Agenda process.

Chapter 11 – Technologia in Voting

Democracy is a system that permits the majority of registered voters to decide which political leaders will set the direction of the nation for the next four or five years. Like any system, elections can be swayed, influenced, or simply perfect and simple, depending upon their design.

Hearing generalities from candidates that capitalism, socialism, or communism are good or bad; or, Left or Right policies - are good or bad; "Big Government" is Good or Bad; these are not usually very helpful summaries. Sometimes socialistic policies are best – perhaps in a healthcare discussion. Sometimes Communistic policies are best – in a discussion of public housing. Sometimes capitalistic policies are most appropriate – when discussing open and free market economies.

Human Rights and the pursuit of liberty, however, are always good considerations as they speak directly to the underpinnings of a Good Life. Democratic Process makes wars unsustainable – so that is a very good thing to make use of and add to a requirements list as well.

For everything else, we voters are asked to try to boil down a hundred considerations, based on imperfect definitions from others, until finally we vote for one party over another. I can tell you for a fact that a lot of vote election workers at the polling booth who they should vote for – and this is a real shame. Many of us appear to need a set of criteria and checklist to help us vote well.

Our father voted for the right; my ancestors were Democrats; or some voters have other little systems, for example – "if three candidates discuss two policies consistently, I will know which is the best leader and cast my vote in that way." Some voters elect what their union or even employer tells them; so on and so forth – there are many influencing considerations.

Most Democratic systems of government leave losers of elections without income support, so for many candidates – everything, is riding on winning. There is not another job to go back to, so saying and doing whatever it takes to get into office and get elected, becomes a somewhat desperate goal.

Having sat on an election campaign team, I can tell you that campaigns can bog down with twisted discussions of media and message contortions. A politician gives a great percentage of their time and effort to re-election. Discussions of supporters are not often mature academic, point and counterpoint exchanges either. More often than not, untrained speakers use absolutes like never and only; they summarize better, best and worst imperfectly – and I am sure you can see yourself in this picture as can I see myself right beside you. People draw emotions into these discussions, and there can be flashpoints among contrasting viewpoints.

We can get so deeply wrapped up in these details; of the trees and blades of grass, that we can't easily step back to see the forest and big-picture anymore. Can you remember why our founding fathers enacted democratic elections in the first place? What did they run away from, and what did they know was important to protect against, and to work toward, when building a new country that could sustain their children's happiness, liberty, good lives and well-being?

Did America's Puritan forefathers envision that their country could become one of the richest nations in the world back in 1620? Yes, they did - by the way. Did they envision that their land of milk and honey would be without healthcare and minimum support for the poor once it did? Decide for yourself after reading a little bit about them.

Think like a Founding Father

I will direct a paragraph here to an excellent essay - The Puritans and Money by Leland Ryken. Puritans were middle-to-wealthy merchant middle class for the most part. They believed in taking benefit from god's blessings, and they also believed that poverty was no crime; that it had benefit as it "starved lust" too; but that the goal of poverty had no merit either. Puritans were Protestants, and they differentiated themselves from Catholics in this way. Puritans were scholars and knew well that the previous thousand years of social stalemate was directly attributable to poverty. Catholics believed poverty was Godly and should be strived for based on the teachings of the Bible of Emperor Constantine in the fourth century when the official religion of the Roman Empire became Christianity.

These founding father Puritans taught that poor people required part of the riches of others to help and comfort them (Latimer, 1858). They acknowledged that, although it did not have to be this way that it was rare to find extremely wealthy individuals who behaved in ways considered Godly. "The acquisition of wealth also has a way of absorbing so much of a person's time and energy that it draws him or her away from religion and moral concern for others." Richard Mather (Ryken, 2015). Puritans were students of Aristotle and these writings echoed ancient Greek thinking too.

There is no man in all things prosperous,
There is no man among us all is free,
For all are slaves of money or of chance.

Aristotle, Politic 322 BCC

The Puritans laid the groundwork for Harvard University within two years after landing in Massachusetts, and implemented universal education at home or in classrooms for every settlement of 50 or more, from grade school, high school, and right up to university - within just their first seventeen years. Latin was the spoken language at Harvard in those early days. High School students read Greek and Latin works including Aesop's Fables, Cicero, Ovid, Justin, Isocrates, Homer, Virgil, Horace, Perseus, and the Greek Testament.

Harvard Studies were modeled after the Universities of Cambridge and Paris. A University's technique to creating a system of thought, logic, and philosophy – referred to in Latin as Techno-logia – is its curriculum. Not to be confused with the modern word "Technology".

Harvard was unique in that its *Technologia* included both metaphysics' cosmology, ontology (conceptualization), epistemology (philosophy), and anthropology; and it also integrated the three "books of truth" – Nature (Science by Aristotle), Scripture (The Bible), and Logic (Philosophy by Plato). Harvard's Seal became Veritas, or "Truth", signifying the Puritan passion for Truth.

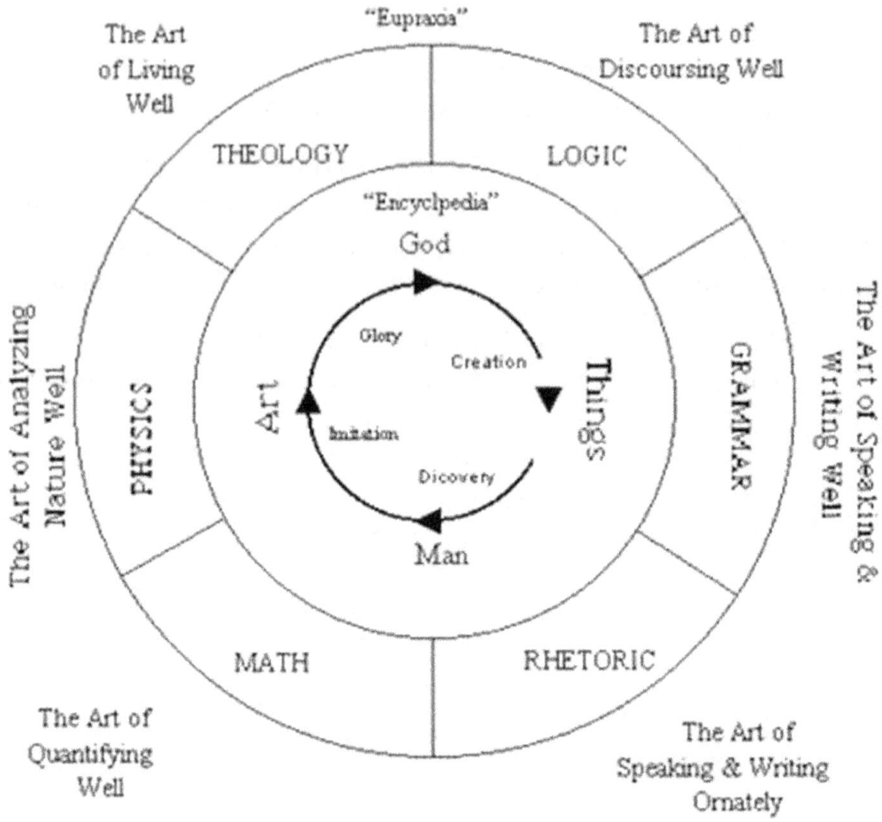

Needless to say, the Puritans were no simple farmers. Much more than colonial leads from France, Spain or Holland, the British Puritans were country builders and lawmakers of the highest standard. When they arrived at a vast New World, their first act before disembarking ship even, was to build what would later be called the world's first written constitution, the Mayflower Compact. A legal guarantee of democratic vote and cooperation among settlers for the good of the colony foremostly. My forefathers Edward Tilley, and his brother John were among the 41 Pilgrims who committed the Compact to law with their signatures in November of 1620.

The Mayflower Compact

IN THE NAME OF GOD, AMEN. We, whose names are underwritten, the Loyal Subjects of our dread Sovereign Lord King James, by the Grace of God, of Great Britain, France, and Ireland, King, Defender of the Faith, &c. Having undertaken for the Glory of God, and Advancement of the Christian Faith, and the Honour of our King and Country, a Voyage to plant the first Colony in the northern Parts of Virginia; Do by these Presents, solemnly and mutually, in the Presence of God and one another, covenant and combine ourselves together into a civil Body Politick, for our better Ordering and Preservation, and Furtherance of the Ends aforesaid: And by Virtue hereof do enact, constitute, and frame, such just and equal Laws, Ordinances, Acts, Constitutions, and Officers, from time to time, as shall be thought most meet and convenient for the general Good of the Colony; unto which we promise all due Submission and Obedience. IN WITNESS whereof we have hereunto subscribed our names at Cape-Cod the eleventh of November, in the Reign of our Sovereign Lord King James, of England, France, and Ireland, the eighteenth, and of Scotland the fifty-fourth, Anno Domini; 1620.

John Milton is considered one of the greatest English writers of all time (see Paradise Lost, Paradise Refound, and others). He was a Puritan in the Latin Office of the Cromwellian Protectorate in London, and he was just ten years old when William Shakespeare died, also in London, at age sixty-two in 1616. Shakespeare is widely accepted to be the greatest English playwright of all time.

Puritans took a dim view of show business's 3,000 person crowds in an era with some of the best new theatre of all time. They took this view, not because of the plays; rather they began closing play houses widely by 1642 in the complaint of bear-baiting, gambling, cock and dog fights, prostitution and other immoral purposes at the Globe, Rose, and other theatres. Read more info in "The Puritans and Education" (Herring, 2003).

The Importance of Education in Democracy

To read the Mayflower's ship logs and excursion transcripts is to learn that many of these were men of impeccable education and elocution. Their diction was superior to most Degree Graduates today – although I suspect that the needs of our immature computer interfaces have taken a toll on this generation that will be restored as our automation matures.

In 1620, under a monarchy system of government, King James I managed economic controls, security spending, and taxation based on a lifetime of training and lessons provided in the proper management of the economy, capitalism, military, law, and so on.

Today, our voters hold this same power and so they too need lessons in how to build and run a country. We leave ourselves open to corruptions when we cannot judge impartially for ourselves. We humans are well-intentioned and smart enough, but we will often make decisions based on how we feel about a subject when we don't have a process and experience to fall back on in logic; based on our experiences and our technologia – our personal techniques in logic.

Campaign marketing advertisements that promoted "Low Tax" and Trickle-Down Economics, are a good example for discussion. During a K-Wave Spring and start of a Summer, you can get away with these long-term irresponsible economic controls that will tend to benefit a few. Trickle-down in K-Wave Autumn and Winter, however, leave the greater good at risk of desperate acts and even wars triggered during these Great Depressions.

Trickle-down only ever works when there is proper monitoring in place to keep tabs on its impact on Wealth Distribution. Without these economic controls set in place, the poorest 20% not only slipped to below 11% of total wealth (or a similar planning target) – it was allowed to slip to 0.1% in 2010. In the US, that means 80 million people struggled to share 0.1% of the wealth; worse, the next 20% shared just 0.2% while 60% shared 99.7%.

How could this happen? In an extreme democracy, the top 51% are ok so, the vote and call for help from the minority poor is not heard.

It has taken 35 years from the start of this past K-Wave Autumn, for the poverty problem to grow to 49% and now their vote can finally be heard. As a voter, if you had been trained to manage wealth distribution back in high school, you would never have let it get this bad. Certainly a Monarch would have heard calls for revolution well before this level of income inequality.

King Hammurabi, for example, ruled a capitalist society by an economic control that forgave all debt every 50 years (Jubilee years) so that everyone was level-set before K-Wave Depressions could trigger rebellions and other social problems.

Reports by Berkeley economics professor Emmanuel Saezsay in his September 3, 2013, paper "Striking it Richer", documented that 95% of all new wealth from and after 2008 continued to go the top 1% directly without monitoring nor controls. So wealth distribution has not improved since 2010 reports at all.

To showcase the need for nationwide education; at the same time that this "Low Tax" election dogma was widely used in the 80s, it is very true to say that less than 1% of North American voters were aware that Capitalist Societies trough every sixty years. So really, how could voters ever be expected to vote for good management in that aspect of our capitalist economy? Education in K-Wave economics in high school, therefore, is crucial.

In school, you might have learned that although Wealth Creation and Wealth Distribution (the American Dream) are both important to a healthy society, was it also made clear that of the two, only Wealth Distribution has a proven correlation to decreasing social problems?

During history class, we learned that by permitting wealth inequity to become extreme, the French and Russian Monarchs exposed their non-combatant families to revolution and death. It never dawns on us that by permitting history to repeat itself today, we expose our families to suffering and death through a possible World War III. World War II killed 60 million people and surviving one of these wars was no picnic either.

A code book instructed Hammurabi and the other Emperors of Mesopotamia and Egypt that they followed as laid down by Emperors of the past thousand years. The code commanded countrywide debt-forgiveness every fifty years – just like our 2008 bailouts and much more.

Anecdotally, the code and other similar texts in the Torah, and in the Bible, also ensured freedom from indentured servitude (slavery), every seven years as well.

Were you ever taught that building anything large, complex and worthwhile, takes strategic planning and consistent forward project management? Of course, you were - Yes. In the TED Talks slideshow above, we see forty-five years of back-to-back four-year election cycles resulting in our heading away from a good life for our society with no long term plan of any kind.

Are you aware of a long-term plan of any sort for society? The Transition Plan is, in fact, a fundamentally important document for every government to maintain.

There will be another Spring Economy in three to six years, but it will only come at the hands of a hotly opposed wealth distribution project – or at risk of nuclear war and human extinction.

We do not have all of these rules laid out by our fathers any longer – and so this education shortfall risks democracy and even humanity itself.

In the absence of country leadership lessons in school, we voters relied upon the most leader-like candidates to manage this well on our behalf. If you are over 60 years of age, you can remember a time when everything that you heard in the media and on television news was considered truthful, so many people believed these low-tax instructions as they would have accepted a formal classroom training.

What is the old expression – fool me once, shame on you. Fool me twice, shame on me.

Democracy is an important construct of World Peace, and so we voters need to make intermittent adjustments that ensure it effectively manages our economy and detours us well clear of war and financial panics as well.

Voting Right or Left

Originally, when the concept of a right and a left political party came from the French Revolution, the right were the safer, traditionalist, status-quo, risk-averse group that tended to keep budgets well managed and the country strong in that way. These folks were pro-establishment and, therefore, sat on the right hand of the King.

Parties both left and right would be responsible for the good government of finance, infrastructure, and services for the country. There were, however, dozens and dozens of policy areas to take a position on beyond core policy areas; in immigration, foreign affairs, housing, finance policy, fisheries, mining and resources, environment, on and on.

So the purpose of the left party was to take on policies that had a more social and communal appeal. These policies would often have a higher cost or administration attached, and so the left party could be said to be taking on a somewhat higher risk for the country financially.

These social programs would be harder to afford if revenues from the country's exports were lower than anticipated. Early on there were no statistics to prove that social programs created wealth for a nation as we see today, so these left-leaning leaders were simply motivated to provide for the greater good.

In this way, voters had two options on Election Day – one party, which was safer and introduced little change and risk; and the other, which spent more on the greater good of the society. But again, no matter what, the core management of capitalism, budgets and finance were always to be managed well and responsibly by either left or right.

Financial measures included wealth distribution, wealth generation (GDP & Exports), diversity and quality in GDP are also very important, country and province/state budgets, debt, deficits and other important measures.

Elections tend to lean left in recession years. In extreme cases where the right has been seen to ignore its responsibility to wealth distribution, it is not unusual for voters to prefer socialist parties that recognize the importance of safety nets and making financial corrections rather than maintaining the status-quo.

Splitting the Left or Right Vote

In extreme democracies, poor voters only have a voice once their votes reach 51%. This change can take decades of slow, painful, and negative decline in society as Capitalism works its way through Autumn and Winter phases. Finally, in K-Wave Winter, usually enough voices from the poor can be heard to turn policies to favor wealth distribution.

But what happens when two parties split the Left or Right vote? Splitting the voice of the Left Party leaves the voices of the poor less heard, typically. If the poor fail to win a majority at enough ridings nationally, the chance of electing MPs and leaders, that have the interests of helping the poor reduces in a Split Left scenario.

A split left or right, is not an issue in a two Party System, but many countries have several Parties that leave the democratic system open to a wide array of abuses as showcased in the 2015 movie "Our Brand is Crisis".

Three solutions are available:

1) Influence your Right or Left MP to adopt World Peace Transition Agenda Policies (Chapter 6 & 7). As these policies are better for business and the economy both short and long-term, this is a good option. Germany, for example, has already adopted most of these policies.

2) Work toward a two party system where one puts forward World Peace Agenda Policies. Ideally there is oversight to ensure the two parties also fulfill their obligation to give voters a clear option.
3) With the caveat that I do not like this option due to the opportunity for abuse by poll takers: If a voter faces a split left or right scenario, use advanced polls to determine which of World Peace Agenda Parties are leading in your riding and nationally. Always align closely with the World Peace Agenda with CSQ Certification as well.

For Leadership, vote for Engineers and people like John F. Kennedy, that read books about World Peace routinely. Vote for successful big project track records in hi-tech, then engineering, then legal, then medical, then sales, and these people are the best choice to lead successful societies as well.

Which Policies to Prefer

Look for "CSQ Certification" on political party policy lists and their website as a symbol of alignment to World Peace Agenda Projects. We certify and then track these policies over time as well. Report abuses too at CSQ1.org. Until an expensive political marketing campaign successfully turns your public away from logic - and it would not be the first time, this should be a help.

In fairness, I will add that I developed these policies at a time when all available information said that these were the "Right Plan" next steps. The process that I followed was no different than steps I have followed hundreds of times before, so if something changes – these instructions will change through release management documented in change logs. In this way, World Peace thinking improves as it progresses.

What Policies do voters need to see from their Politicians?

i. Long-term Planning – what are your government's plans to address debt, housing, wealth distribution unemployment, safety nets, and wealth creation?
ii. Education – are kids learning technologia in voting, economic theory, trickle-down and low-tax liabilities, and what are the duties of good financial managers both the right and the left.
iii. Transition Economics in automation – target automations that grow GDP Exports while minimizing Imports.
iv. K-Wave Economic Controls – 4000 years of capitalist history shows that Wealth Creation is always important but that Wealth Distribution is crucial in Autumn and Winter.
v. Government Systems – examine right or left leanings for each policy individually. Some policies are social, some policies are communal, some policies are capitalist, some policies spend a little more than right-leaning policies but are also proven to generate more wealth that conservative policies.
vi. Affordable Housing – Usury is the immoral, and in many countries illegal, sale of debt to clients that will never be able to pay back the loan. Presently only 30% of homes in the USA are owned outright, and the rest of mortgage owners are victims of usury; it is rampant and on the rise with recent programs to encourage those that lost homes seven years ago back to the housing market with 3% down payments and very low-interest rates.
vii. Human Rights including the Right to Healthcare and a Good Life – or the American Dream.
viii. Business Ethics Licensing for Social Accountability.
ix. Technology – what is your government's plan to automate your production economy? Will your party commit to a World Peace Agenda?

Voter Comparison Charts

Use a simple chart like the following to work with your Political Parties in non-election years to shape policies that permit World

Peace Transition Projects. Once projects begin and thrive, so too should your economy.

#	A+ Priority Social Policy	Canada			
		Cons	Lib	NDP	Grn
1	Long Term Planning	No	No	No	Yes
Wealth Distribution					
2	Bench Strength of MPs for Project Execution	Maybe	No	No	No
3	Minimum Wage Targets	No	Maybe	Yes	Yes
4	K-Wave Economic Controls - Graduated Tax	No	No	No	Yes
5	3 Day Work Weeks or Income Safety Nets	No	No	Yes	Yes
6	100% Employment	No	Maybe	Yes	Yes
7	Housing & Anti-Usury Law	No	No	No	No
8	Onshore Engineering	No	No	Yes	No
Wealth Creation					
9	Increase Quality Exports	No	No	No	No
10	Reduce Imports	No	Yes	Yes	Yes
11	Sellers Manufacture Locally	No	No	Yes	Maybe
12	Profit becomes Investment	No	No	Yes	Maybe
Transition					
13	Tech Project Safety Nets	No	No	No	Yes
14	Automation of Import Export	No	No	No	No
15	World Peace Agenda & CSQ Plan Projects	No	No	No	No
	If Yes=2, Maybe=1, No=0	3%	13%	47%	53%
	Leaning	Right	Left	Left	Left

I completed this chart after a recent Federal Election. It is intended to be for example purposes only.

Educating Policy Makers

Educating democratic voters is important as discussed above. Training Policy Creation Committees, is another important discussion.

The constitution of many political parties document a process to forward policy by democratic election. Voting might take place without consideration for the credentials of the policy committee members in many cases. Policies, may discuss the economy, natural resources, manufacturing, etc., and each policy makes its way from riding offices throughout the country to regional review. When they survive a vote by a regional group, the policy next moves forward to a vote at a national convention.

Policy electors are not necessarily subject matter experts, and they are not screened for conflicts of interest in many cases. All policy reviewers should fully disclose conflicts that can include employment incomes from other members of committee. You should be elected a Subject Matter Expert before being invited to vote on policy as well.

A long time association and volunteering record with your party is terrific, but it does not however, justify your cancelling out the vote of an expert in a policy field. Invariably, there are far fewer experts in this world than not – so do not be offended if you are declined the vote in one subject area. If, however, you believe you are an expert and have been denied your right to participate as an SME, an appeal process should be available to you.

Chapter 12- Accountability

In the 2013 Oxford University Press publication "Firm Commitment", Colin Mayer explains in great detail why the corporation is failing us. Professor Mayor states that the corporation has created more prosperity and misery than ever imagined and that the balance is moving increasingly to the latter direction as it now threatens to consume us. These are failings that we now need to address as a matter of urgency.

Accountability in major corporations is one of the most pressing issues in our society today. CEOs, CFOs, Crown and State corporate procurement departments, Hedge Fund managers, and most every business school grad today - is taught, and rewarded based upon business ethics that externalize social costs.

The problem has been recognized internationally too in 2015 with Monsanto being brought up by the World Court in The Hague under charge of Ecocide and Crimes against Humanity in 2015. Major Lending, Investment, and Risk organizations are just beginning to implement CSR Leads and Corporate Social Responsibility Policy that are inserting new considerations into Due Diligence approval workflows for lending and support of socially irresponsible clients.

High Social Cost behaviors - such as eradicating pensions, offshoring (also called globalization and outsourcing), starvation-level minimum wages, tax and benefits dodges – have all become the norm. Any of these practices will also get you promoted and bonused at most any financially motivated company on the planet today as well.

Canada and the United Kingdom are in worse shape economically than others in recent years because they permit the wholesale offshoring of our engineering as well. This is a practice that the Netherlands outlawed twenty years ago and a process that America, Germany, and other nations, have started to correct already.

The Dutch legislated social responsibility in all aspects of their businesses and the reward was to give a country the size of Lake Ontario, an export GDP 20% larger than Canada and Russia.

When engineers and doctors harm society, they lose their license and can't practice: when business harms society, they get a bonus. When politicians permit these behaviors, we vote for them. We are accountable voters too of course; we need to be voting for engineers, but instead we elect business leads and career politicians that exacerbate these unproductive behaviors.

Good Process Never Takes Longer

Democracy is an underpinning to sustainable World Peace. A case can be made to say that G8 Government Legislatures wasted 45 years while eliminating a Good Life for a near-majority of their electorate. In those same years, in Denmark, Sweden, Netherlands, and Norway, 38 Million democratic electing voters preserved their Good Life. So, what was different between the G8 and these socialistic nations?

People of the rugged north, like people of the desert, realize that when they fail to feed a traveler or shelter him in a storm, that that person is being left to die as a certainty. Communities refused this outcome realizing that the next time it happens, it might be them or their children, who need help. These values were handed down from father to son, which embedded these thinking into their nation's basic values.

These countries defended a Good Life - and their socialistic policies ensured support of each citizen's pension, healthcare, education, and childcare – as a basic human right. This was hard work, with many detractors, costs, and attacks – but they defended their fundamental values, they profited from it as well, and they never lost the American Dream because of it.

As a member of executive teams, I have seen good leadership and process in the boardroom. I have also seen meetings with the same people descend into chaos in the absence of a good process lead as well.

Anyone who has sat through a government or committee meeting has seen this too. When filibusters, emotional outbursts, over speaking, and other tools of delay and persuasion are thought needed and important, you are simply experiencing a situation in which good problem-solving process is absent.

As elected officials and assigned leaders, your duty is, therefore, to adjourn emotional meeting discussions and work to regain a professional and productive momentum again. Seek the advice of strong engineering leads when efforts to gain a productive momentum fail.

Politicians who fail to do so, abuse the public trust and the business mandate that grants them permission to lead.

Designing Accountability

"Blame" is a very important legal term used when assessing financial and criminal accountabilities in a court of law. Blame looks to the past and assesses whether financial and criminal penalties should be levied or awarded to persons involved in unfortunate events that unfolded because of accident or intent.

For the most part, discussion of "blame" is best left in the courtroom as little more than lessons-learned can be gleaned by visiting events of the past in project work. Progress for the World Peace Agenda is expressed in status meetings and providing status reports on KPI targets as these are the best measure of performance.

If pension obligations appear to be shrinking year after year within our major businesses, where do those social obligations fall next? They shift to Pension Plans of that citizen's home country – and yet we hear nothing of these problems in the news. Present Conservatives and Republicans seem adamant to support business with no regard for social accountability whatsoever.

Once businesses, and business leaders, meet KPIs that include social accountability in quarterly and annual reporting, we are probably on the right track toward making businesses socially accountable.

Imagine what would be the impact if G8 countries agreed together to ban companies from their markets who did not meet minimum obligations for socially responsible business practices. I imagine that it would take just the smallest level of enforcement to turn the corporation back into the society-building asset that it once was. If the biggest markets on the planet all pulled products from the shelves of, no CEO could hope to keep his position.

The requirement of CSR should be a cornerstone of Business Ethics and KPI monitoring ensure that both profitability and social accountability are delivered reliably.

Insist on Forward Momentum

My career in transformation leadership in large organizations for twenty-five years provides unique insight into how things **can** change – but more importantly, it teaches you that everything **has** to change in a very specific way – for it to change well.

A single change process is a fact as sure as you are reading this sentence; I have watched this process govern hundreds and thousands of complex change teams simultaneously in mature change organizations - and I have watched immature groups falter without this process under just a fraction of a full workload.

It is not conformity to accept a fact nor to follow a proven process; you can think outside the box and still conform to a process that builds good solutions every time – and if that process is what gets your needs built quickly and measurably consistently, then you probably want to defend its use as well.

In career, I have run into individuals and organizations who were very straight-forward and said that they did not want a change to happen. Others made efforts to circumvent a good process as they rolled out or opposed a change or group decision in an indirect or even passive-aggressive way by agreeing and then doing something different. In parliament, and in a project, you are there at the request of an electorate who have good lives to get on with, and there can be little time for unproductive behavior.

In a busy change organization, most projects report weekly with Red status projects communicating their issues to senior project owners for a quick resolution in a one-hour weekly meeting. The process is a transparent, low stress, productivity engine that permits new initiatives easily and forwards lessons learned – so you will want to emulate that process in Parliament wherever possible as well.

Forward Momentum for Society begins with a long-term plan of smart social goals and objectives summarized as projects:

- Vote and Invest in Right Plan projects that build a Good Life – "the American Dream" - for everyone both rich and poor.
- Wealth Distribution – through Safety Nets and Graduated Tax
- Wealth Creation – through good GDP Export development, technology investment and the encouraging of self-sufficiency.
- Automate our production economy
- Ensure CSR Due-Diligence best practice to sustain business accountability
- Manage Transitions Well – builders vote; administrators assist
- Rollout the Transition Plan World Wide

See Chapter 6 Social Projects...

Simply add this list of projects to your agenda and also add this small list of bullet points to the requirements/needs list of every project within your charge. Although it is always best to coordinate efforts centrally, the power to gain momentum is greatly improved once all teams are thinking of ways to accomplish these targets most effectively.

See a case study example for Transition projects in this book below in Chapter 17. That process will always give you "Next Steps".

Business Accountability

Recalling the work of Dr. Mayer of Oxford University in the first paragraphs of this chapter.

Failure rates at major engineering universities are 20% to 30% despite the highest screening of any programs; I have never heard of an eMBA that did not graduate and yet MBAs are often thrust into roles that manage Technology or Engineering even within technology and engineering organizations, major airports, and similar.

An inverse correlation exists between society's forward progress and crowded executive administrator pools until finally, we begin to see major corporations pulled into The Hague under charge of Ecocide and Crimes against Humanity in 2015. Major Lending, Investment, and Risk organizations are just now beginning to implement CSR Leads – heads of Corporate Social Responsibility Policy - that embed new Accountability Policies into Due Diligence approval workflows.

As long as the trend of CSR accountability can persist, there is every chance of a turn-around in investor-driven change for the better.

Consider also a typical corporation operating executive team – say a pension fund, or major airport, which might include a CEO, CFO, CAO/HR, COO, CIO, SVP Sales, and SVP Marketing. There are one or two engineers to as many as five or more admins and yet all have an equal vote at the boardroom table. Human resources, finance, and C-level administration stewards have accelerated hiring and bonuses to their ranks based on their boardroom voting numbers as well.

If engineering, or social-leaning voices are strong, they risk being marginalized or not permitted at the table; to be replaced by Non-

SME generalists like MBAs, by the majority votes of other executive administrators.

From our discussion of Accountability and social responsibility above, this organizational design gives the strongest voice to those who are trained in Business Ethics classes - and not prevented by licensing - to externalize social costs with indifference to society's needs. Boards of Directors too, are often structured with far heavier administrative and financial votes.

Solutions to this accountability problem are implemented in Germany already. Here, workers vote for their executive team members. This means that Executives who have poor track records for socially irresponsible behavior, cannot be elected into the boardroom. Other controls might include the licensing of business leads in the same way that we license engineers and doctors today.

Whichever solution you choose, a choice must be made now. Business Accountability begins by revamping the Business Ethics programs within our biggest and richest business schools and in boardrooms everywhere.

Raising Successful Girls; Women in the Workplace

I am taking one for the team here, so please bear with my seeming very lofty for a few minutes as I try to leverage Sharon's advice for this section as follows.

To give advice, it should not be about one's way being the only right way. It should be about recognizing the human faults and flaws in others, and more importantly ourselves, and encouraging others to work on it by giving examples of mistakes or errors that we have made and how we corrected them. Humans, by design, are imperfect and have tendencies to be egotistical, however, upon saying that we can learn to look outside of ourselves, be empathetic, compassionate, altruistic, by being cognizant of others decisions, choices, goals, mistakes, tragedies, and even their dreams. We can only effectively guide or advise when we actively listen to what is being said and show a genuine interest in the person seeking our advice. I know I don't always follow my advice, but I try. I know my children will tell me what mistakes I have made with them, and that's ok. I need to listen to them and show them that even though I am their parent that I too make mistakes, but will take responsibility and ownership of those mistakes and work towards correcting them.

Sharon Gibney, 2015

I raised four girls who required, among other duties, that I attend twenty years of consecutive annual Ballet Recitals. There were the birthday parties, and the Barney dolls, and the Spice Girl Power videos, and-and-and. And I loved that phase of my life and will always be glad that I made it a priority to travel lightly and work very few weekends so that I could watch them grow up.

After twenty-five years working with and for women at all levels of corporate workplaces, and after witnessing a 70% divorce rate come to volition within just my lifetime, this is your father talking now, we need to introduce a few accountability lessons into the upbringing of our girls. As I am not your father, I will be brief.

Genders are equal, but they are not the same - and they are not raised the same either.

With my son, I talked to him about the lessons that my father talked to me about: When a neighbor needs clothes, you give him the shirt off your back; if he needs food and shelter, feed and harbor him – a man could starve or freeze to death without these accommodations. We are a Canadian family from Newfoundland, and this code was handed down through generations because to deny these things was to one day risk that you or one of your children is also turned away. It makes perfect sense when you think about it. My father told me just a small number of times, and the lesson stuck and became a part of my values as well.

Lessons included the commitment to family and to upholding a good family name. Honor your mother and father, and provide for a wife and children. You were the man, the protector, the provider. Honor, integrity, respect – you get the idea.

Girls did not get this basic training really. In my particular family, a girl was going to be educated, get a job and start a life. Her family name didn't matter too much as she would lose that to her husband's last name eventually. She was terrific, and computer literate, and empowered, equal and better to any man – but not responsible for upholding and providing for the family really. That was the man's gender role. More than great women and mothers, Matriarchs kept the generations of family together. My girls were encouraged to look up to women like their Aunt Madeleine and Nan – their grandmother.

I do not believe that this role of "Matriarch" was a lesson taught to girls in our society and I can recall few other woman among all of our friends that fulfilled this role in her family. My Aunt Madeleine and my daughter's Godmother, Sharon Gibney, were the strongest examples of Matriarchs that I knew. Girls have it in them to be much nicer people than men, but without this lesson, girls can become women who act on their wants with impunity from financial cost, social cost, nor the human suffering or immediate impact on their own family members. Whatever the root cause, "I don't care, I'm just going to do it" was the basic programming of every PowerPuff Girls episode that I heard my girls watch.

My girls had little interest in engineering, and none of the kids shared my likes in math. One took a degree in Biology, the other Languages and History - and she is now a TV producer, another took Marketing and Teaching, and finally the youngest started Computer Software Engineering. These were great choices and I couldn't be prouder of my kids.

Now, on to the business end of our discussion. A 70% divorce rate is the creation of my generation. I was born just after the baby boomers in 1963. A 70% divorce rate requires 50% to 70% more households and 50% to 70% more household incomes. That equates to a conservative 150% demand for jobs in a Winter K-Wave economy, in a country with globalization (offshoring) running unchecked.

The reason that divorce was not socially acceptable in the 1950s to 1980s in the U.S. and the G8, was to ensure sufficient household incomes through jobs. There was also the matter of family stability to consider. Divorced women could not afford to run a household financially and divorced men could not hold office nor senior positions in business nor government.

A 70% divorce rate means that 75% of society are unwilling victims of family instability as a child or adult. Divorce costs upwards of $12.5 trillion annually and 23% of divorced people will die younger in the U.S. (Dec 3, 2011, HuffPost). Divorce employs lawyers, real estate agents, and it fuels household goods spending, real estate bubbles, poverty, and unemployment and safety net spending as well. In comparison, smoking kills 20% and costs $289 billion annually. So Divorce has a much more harmful impact on a society than smoking.

When men ran things, it was 10%; when women called the shots, 70%. What is to blame? Infidelity? Statistics say 15% of men and 15% of women file this as the reason in their divorce papers – about equal. Spousal abuse? In 1960, homes with reports of spousal abuse were approximately 20% and divorce rates were 10%; today spousal abuse rates are 6% who report an occurrence over the past five years. Abuse is a social issue that has cleaned up quite a bit during this period. What accounts for the higher divorce rate?

According to the Center for National Health in the U.S., 80% of divorces are filed by women. The 20% filed by men aligns with the infidelity and abuse ratios. Women file 70% of divorces in the United Kingdom as well. Remember, women are equal - but not the same; there are important differences that have made our survival as a species possible.

Women prefer careers that are different from men. Women in Engineering total 12% and they overwhelmingly prefer roles in admin, sales, teaching, and healthcare. Women are caregivers in much higher percentages – we know this both intuitively and statistically.

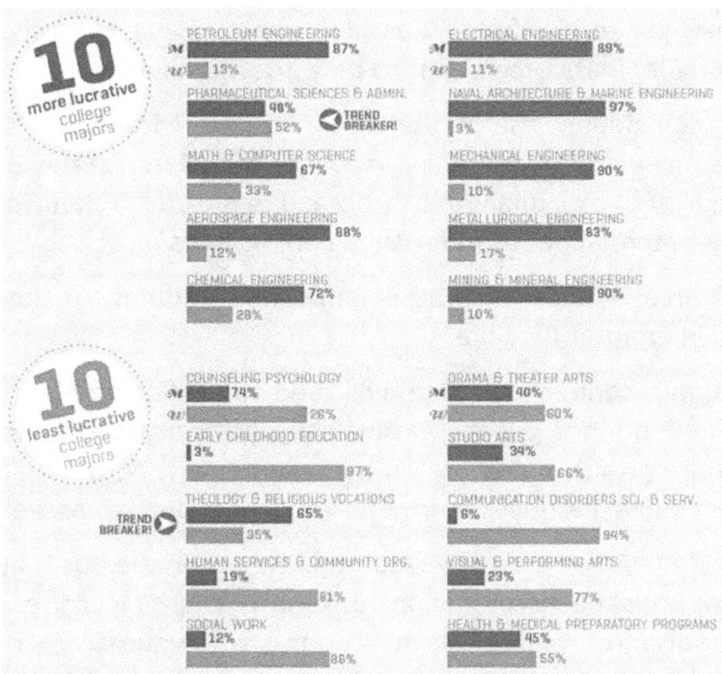

Anthropologically speaking, there is nothing more important, nor valuable to a species, than the individuals that give birth to its babies. In this regard, women have nothing to apologize for - ever. I lived in a house with three to seven females for twenty years, and I can count on one hand the number of times that I was acknowledged to be right, or that the girls admitted accountability for a problem. Men are providers of seed and security, so when one man dies, another can replace his important role. She bears immortal life and he keeps immortal life safe.

A 70% divorce rate is the creation of my generation, but not the creation of my gender. If female equality - like democracy, is to be a better system, mothers must be matriarchs who instill in both daughters and sons, the values to keep family first, and that foundations of respect and equality given toward one's mate must be recognized and returned.

Admitting that a Problem Exists - is the first step to solving it, and then we rely upon our process in Chapter 17 to roll out a system to repair it. Until we are ready to agree that families are to be a thing of the past, I respectfully, humbly and stalwartly insist that accountability in Divorce needs to be a priority now.

Was it a coincidence that in the First Council of Nicaea of 325 AD, a hundred scholars from many lands and beliefs, drawing upon thousands of books, unanimously agreed to include "What Good has joined, let no man put asunder" in wedding vows?

The "Divorce Project" is another important addition to the Social Projects in Chapter 6 above.

Women, in a connected group, will tend to prefer the company of equals and do not accept easily high-performers nor standouts within their group. If you are an average performer, a female boss will tend to give feedback more often than a male boss as well.

This concludes examples of Accountability here in the book, but it by no means ends the discussion on our forums. On the website you can agree, or disagree, and others may identify with your stories of other accountability concerns too.

Chapter 13 - Land Ownership

Russian Dachas are cottages assigned to families living in the major cities. Even the poorest Russian-owned one so that children, grandparents, and families could get away and enjoy the countryside, plant gardens, and change pace from a busy urban life during the summer months.

Trains connected the major urban centers to cottage country where dachas could be reached easily, often on foot by walking for just a mile or two.

Most North Americans aspire from a young age to own their homes, and then we work toward paying off a mortgage for most of our adult lives too.

Many people who were able to purchase their homes before the housing bubbles of the 1980s were able to pay off their home and cottages easily given the availability and relatively high wages of full-time salaries. In 1963, you could purchase three acres of Muskoka waterfront property, one of the most beautiful cottage areas in the world, for about $250. At the same time, salaries for steel workers was $300 to $450 per month, and this meant that most Canadians could easily afford a cottage.

When my 71-year-old best-friend Tom's dad was 17, back in 1965, he was a teenager working a student summer job and with his savings he had the choice of buying either a used car or two acres of waterfront property in Burlington, Ontario. Today that property would sell for millions of dollars and his summer income probably would be just about double what it was back then.

In 1982, incomes were up to $5,000 per month, and the average home price for a modest detached dwelling was approximately $35,000. That house would rocket up in price by approximately three times over the next three years and would never come down.

My parents told me from a young age to buy a home and pay it off as quickly as possible – and that made a lot of sense because mortgages could be paid off completely with five or ten years of concentrated effort. Today, only 29% of homes in the U.S. are owned free-and-clear, and the rest are paid off only 50% on average. Falling interest rates had to go so low to stave off depression in 2008 that raising them even a few percentage points would force the foreclosure of tens of millions of homes in North America.

Understanding lending costs

Usury is the practice of making unethical or immoral monetary loans intended to enrich the lender unfairly. Many countries enact Usury

Laws that govern maximum interest rates, maximum monthly payments, and maximum debt load as a ratio to income.

Banking laws in Canada presently permit Canadian Lenders to extend $26 for every $1 on deposit. This means that for the average detached home in Toronto which costs $1 million dollars financed at 5%, the bank will ask you to pay $5,816 per month; $1,613 will be the principle and $4,203 will be the interest payments. Borrowers will pay $50,436 in payments annually for the next 25 years, to return to the lender the $38,461 that they originally had to have on account. That is a working lifetime of steep payments in exchange for $38,461.

If at any time during that 25 year period, the homeowner is unable to continue payments, the bank can foreclose and sell the property to repay the full value of the mortgage commitment. Your forces your family to find shelter in a marketplace where rental rates increased 50% to 75% in response to the rising demand. There will be many additional monies wasted on moving and storage costs, and you might spend many years worrying about financial pressures as well.

No-one imagines that they will not be able to meet payments when they first apply for a mortgage, but the statistics show that most mortgage takers will have to refinance, downsize, or suffer foreclosure. 70% of all mortgages in the U.S. are never paid off by the original borrower.

My point in mentioning this example is that owning property is great, but it is terrific only when your system of government and economy supports it. The G8 economies supported home ownership in K-Wave Spring, Summer, and early Autumn, until around 1995, but that was 20 years ago. Today in K-Wave Winter, the statistics say that our economies and housing markets do not support home ownership.

At this time in a capitalist cycle, soliciting low-interest, low-down-payment mortgages, ceases to be beneficial to society and becomes Usury. The availability of easy financing creates bidding wars and pricing bubbles, and mortgages that statistics say will never be paid off. Most countries enact laws to protect its citizens against this form of Usury as well.

China and other countries that support land ownership are in the same situation where land ownership forces a very hard life for the great majority of its citizens.

Low-interest rates with just 3% down-payments are being offered in the United States again this year as people who lost their homes in 2008 become eligible to buy again. Does this sound like the beginning of just another bubble?

Economic Controls for Housing Bubbles

Economic Controls protect housing costs from soaring so that young people can purchase a home with a high probability of paying off their mortgage. In 1982, houses cost $32,000 with interest rates of 22% at a maximum. As soon as interest rates dropped to 12%, houses skyrocketed to $90,000. The same house climbed to $500,000 as rates dropped to 2%, so clearly falling interest rates are a bad thing. We probably hit the point when mortgages became usury at around 15%.

Housing prices become unsustainable starting with bidding wars between investors, immigrants, and new and existing homeowners. Today, highest-bidder-wins irrespective of all other factors. We also need to provide for those who prefer to rent short-term as well. Rent controls in New York City and "En Viager" contracts in Paris' (American Press, 1995), protect seniors from these market bubbles and Cost of Living up-swings.

Real Estate in some city centers and premium locations (beside water, on ravines, lots with a view, nicer neighborhoods, etc.) are more desirable than others too. Creating region-by-region price controls, or a cap on annual price increases, are necessary if children are going to be able to move close to their parents when they begin their families in time.

Another solution could include restrictions on who can offer to buy property so that 1) locals, 2) new home buyers, 3) previous homebuyers, 4) local investors – can bid only based on a ceiling price. If bidders make no offer, then bids are opened to refugees, immigrants, foreign investors, and so on. In this way, the cost of houses are protected from bubbles, and young people have a chance to begin productive lives in the neighborhoods in which they grew up.

A discussion of Housing Controls, therefore, must mitigate several risks:

- Beginning in late K-Wave Summer, **Interest Rates** can be permitted to run up, but if they do, they cannot then drop below 15%. Perhaps this is a good reason to cap interest rates as you consider another option - **a per-square-foot Price Increase Cap** that would prevent the rapid increase of house prices. I have never seen a cap like either, which probably explains current worldwide trends and a resulting phenomenon called Housing **Bubblenomics** (Aldrick, 2015).

  In his Telegraph article from November 2015, Philip Aldrick notes, "What's remarkable about this bubble is it has inflated without a recovery. *Markets have become completely detached from economic reality.*"

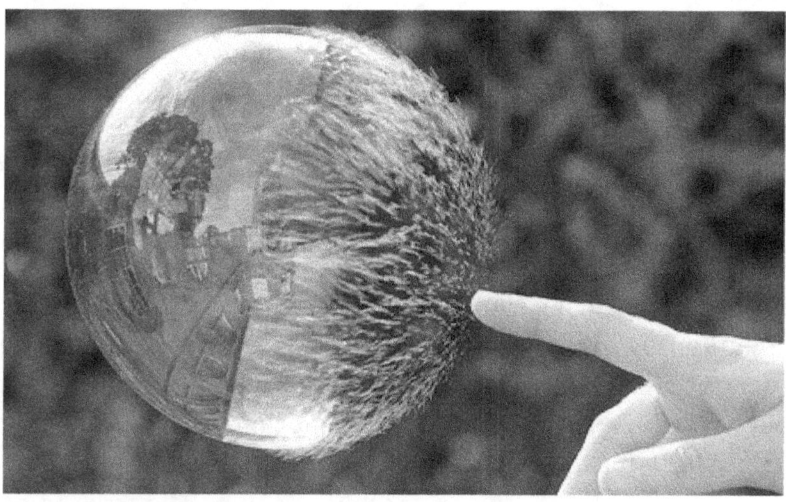

- For **First-time Homeowners** - lower-down payments, low debt-to-income ratios, low unemployment, and high minimum wage are required to service a mortgage. Perhaps lower interest rates are offset by higher investor rates.

- **Existing Owners** - Higher down-payments for existing owners with preference to owners that grew up in or live in the area presently.
- **Merit** - From our discussion in Chapter 19, perhaps Performance & Merit should be factored into which bids can be accepted and which not as well.
- **Investment Properties**, higher-down-payment percentages than owners.
- Finally, **Offshore Investment** – Should housing be considered an investment before all have shelter? If you see from my observations above that this question should be answered "No" as I did, then you might also agree that offshore investment in private homes should be discouraged. Foreign investment brings the Real Estate Bubbles and problems of other communities to your country. Perhaps these bidders could be permitted to compete only if local bidders are not available at mandated pricing. In this way housing costs stay low and new home construction is encouraged.
- **Switch** between Land Grant and Ownership models, as explained the next few sub-topics, to protect society from Usury.
- **Monitoring** is required to ensure rental properties are not soaring in monthly costs.

Clearly, our present systems are not brilliant. They encourage Bubblenomics and prevent children from raising families of their own in their own hometown. Current rules and systems are flawed, so do not blindly protect them but rather enact corrections as needed. Nothing is too big to fail and nothing is too important that it cannot be adjusted in pursuit of Human Rights and a Good Life.

Mortgage borrowing is stressful, it reduces our quality of life and longevity, and as with most non-business generating financial services, it advances society relatively little. It makes properties exponentially more expensive, our children cannot begin lives in their hometown, and it encourages purchasers to live beyond their means and reasonable debt loads.

In Mesopotamia, borrowers had to become the indentured slaves of moneylenders when they could not repay their obligations; and then, every seven years, they were released from their debt entirely. In many ways, the majority of today's mortgage borrowers are committed to a similar form of indentured servitude – but for a lifetime in a usury scenario.

Government Land Ownership

So – what was wrong with a government system that handed out land as we needed it in the first place? Opponents might like to argue or say that this is a communistic policy - in the same way that universal healthcare is a socialistic policy. The point remains that this is another system of land ownership which is widely used throughout G8 and G20 countries today.

Ownership of land is the norm in most G8 countries but, in its national parks, 100-year leases are granted only.

If we had an unlimited number of homes to give to citizens, land grants seem a natural option. Where we start seeing inequities, is when land grants award estates, or prime waterfront locations, which gives preference to some individuals and not others. This happens all the time today of course because families have been protected by class laws within a capitalist system just as were Royals or Races in other class-based societies.

Can we Switch - back and forth?

I mentioned that in K-Wave Spring, Property Ownership works well enough, and then I mentioned that in K-Wave Winter, Government Ownership works better. Remember that the important objective here is ensure that everyone has what they need and to ensure a Good Life for everyone. Can we switch from "Government Owned" to "Self-Owned Housing"?

Yes, absolutely; when the Soviet Union converted all of its states into Republics in 1986, all land grants were simply switched to an ownership model. Apartments in central parts of Moscow jumped over the next 30 years to cost millions of dollars. Nothing changed in the apartments; residents simply became victims of their own housing bubble. This example, and China's ghost cities also, showcase that Land Grants are the more sustainable system of housing.

Will landowners want to give their property to the state?

Yes and no. I suspect there are three groups as follows:

1) Free and Clear Home Owners - 30% currently

Those who own their homes free and clear might prefer to retain ownership of their home. A relatively low percentage of their incomes would go toward tax. These owners would maintain their properties themselves until robotic maintenance arrives, so that taxes could be lower.

2) Mortgage Owners – 70% currently

Owners with a short time left on their mortgage, and no concerns for employment, may elect to continue paying for and buying their home. Lower taxes would be levied in this scenario because homeowners are assuming costs and upkeep.

Others will decide to sign ownership of their properties over to government in the same way that the Russian Government did. Higher Tax would be needed to offset government admin costs - perhaps. Residents would be responsible for upkeep – until robotic maintenance projects complete.

3) Renters and New Home Owners

It seems reasonable to expect that as long as quality homes are offered close-by parents and grand-parents, most individuals will prefer to be given a home at a rate in keeping with the financial capabilities of society. A higher tax rate and resident upkeep would be required.

Transition Economics would have to dictate the rate of change that could be accepted in a housing system switchover. In democratic countries, 51% might like have to vote for these changes as well. These options would not be accepted easily until a high percentage of the population are disadvantaged or stranded outside the housing bubble – as in a K-Wave winter.

Recall also that our technology will shortly permit us to live anywhere we like, to build anywhere we like, without reliance on a power grid, or a daily commute into an office, nor a drive to a nearby mall to buy groceries or clothing. What difference will it make if the land is handed to us from our community's pool of available land – or if we go out and purchase it ourselves? Is our right to grind our way through a mortgage really so important to us?

Like any privilege, failure to maintain or upkeep assigned property before maintenance bots come onto the scene, could result in penalties financial or other.

As our automation projects make money less important, the residual value of an estate handed down to our children, also becomes less important.

What sort of home would I get?

This is a discussion of options and not recommendations. Discussion of a change in land ownership is going to shake any community to its core. I know myself that the notion of a government administrator deciding my lot in life – would worry me deeply.

Our robotic or manual home-building capabilities determine the quality and size of granted homes, as does current government incomes, available consigned properties, economy wealth, wealth distribution, and individual incomes.

If you are a brilliant engineer who builds hi-tech businesses, technology, and writes books that improve mankind's well-being, would, or should, your contributions be recognized with a larger and better parcel of land or home?

Our society rewards these considerations almost not at all today.

Would a Trust Fund Manager who earns $50 million per year by squandering 100,000 jobs to other countries - be rewarded with a larger parcel of land?

Our present capitalist system *does* reward this behavior today – with bigger houses, nicer cars, private schools, on and on. If you buy finished sneakers for a dollar and sell them for $100, we throw in a very handsome fleet of yachts as well. The net benefit to the planet is just about zero in this last example, as the value is delivered by the shoemaker and transportation can be automated right to the wearer's doorway. The same situation holds true for coffee growers who barely recover costs for their work and equipment while sales organizations reap enormous profits.

Consider that as money becomes irrelevant with new technology, are skills like "driving financial efficiencies" – especially the ones without regard for social well-being, important, desirable, or necessary? The answer is probably - less-and-less.

Would a cancer-care nurse who dedicated his or her life to attending to the needs of the suffering and dying be granted a more comfortable lake view and a nice home to come back to?

Today, our system rewards in-hospital nurses modestly and nursing home attendants are almost paid as minimum wage workers.

Through these examples, I am introducing a merit measuring discussion that focuses on using different measures to trigger social rewards. What contribution does this person bring to society? What is their performance? In the case of a Mother Theresa – this person is a leader as well? Does he or she inspire others by example to do amazing work for society with little regard for personal recognition? If yes, then a very nice lot would make a lot of sense. Just don't get me started about pro-sports salaries.

Whether you like the idea or hate it, this is just another very implementable change that makes a lot of sense and only takes a well-managed project and operation to achieve.

What is the right place for you to live?

Should people live alone? Should they live in Dorms? Do they live in apartments? Do they live beside lakes? Do they live on boats? Should special requests be considered?

I love these questions because there are many correct answers. When I was very young, the answer to this question was that I needed to live in a home with my family – ideally in my own room, with a yard for my dog, close to my friends and school. That was House One.

When I headed off to degree studies at seventeen, some of the best times of my life were lived in a dorm room with other students my own age; that was House Two.

When I had almost finished degree studies, I took a two bedroom bachelor flat with a buddy; House Three, and when I started working and wanted to start a family at twenty-five, I took a small three bedroom house with a big mature yard, on a quiet cul-de-sac, that needed a lot of weekend projects. House Four.

The woman of my dreams came along and our family grew to need a four-bedroom home with a pool within two years. A cottage would have been too much work so we owned a motorhome for a year or two until the kids convinced us that it was all about a pool for them.

Half of our neighborhood's kids agreed our pool was the best on the block and so my wife and I had to camp poolside at House Five to watch bathers for eight hours a day - some days. That went on for ten years until kids were eleven or twelve – and then the pool was very rarely used and we could have done without the extra maintenance work and energy cost as well.

Divorce came when the kids were thirteen and seventeen and we needed two homes temporarily. My kids headed off to University and a friend and me, a woman who was in a similar situation to my own, shared a three bedroom condo. Home Six. Our parents were getting older and we wanted to bring them closer to live with us, so a three bedroom bungalow with a nanny apartment or two, would have been ideal. House Seven.

As we get older, our needs will change too - until our kids either take us in, or we decide to live with other seniors within supervised apartments basically; House Eight. Like everyone, we will need 90% of our lifetime's healthcare within our last 5% of days most likely, and we hope to look back on a full life with grandchildren and maybe even a great-grandchild or two if we are lucky.

Everyone is important; some individuals represent a greater utility to society than others - and this must be recognized and rewarded. If we all keep two or three children families, it is entirely likely that a sustainable flow from home to home might actually work very well throughout our lives.

Value & Performance

The surest way to disincentive someone who does ground-breaking, terrific work with fantastic performance is to not acknowledge, recognize their value, nor permit their contribution.

Alan Turing, founder of the Computer Science and one day perhaps even World Peace itself, was ostracized and chemically castrated with the result that he committed suicide at the age of forty-one. Over four years, from 1940 to 1943, he built and made work from concept, an electro-mechanical computer capable of deciphering Enigma - the most sophisticated mechanical encrypting computer ever built – used to protect messages to German assault forces throughout World War II. His Turing Machine was an initial trial and pilot of the binary computer that all computers are based upon today.

Mr. Turing's genius and contribution are estimated to have saved the lives of twelve million people and reduced World War II by two years – as documented in the 2014 movie "The Imitation Game"

Albert Einstein (founder of the Theory of Relativity and Atomic Energy) was relegated to a patent clerk and ignored by academia for two decades; in 1989, Stanley Pons and Martin Fleischmann discovered Cold Fusion but academic peers condemned them to disgrace and took twenty-five years to rediscover formulations of nickel and palladium nano-powders that created cold fusion reactions.

Years lost; promise squandered; contributions ignored, and a future of clean energy delayed.

Everyday examples include resume pre-qualifiers that modern Human Resource interviewers and heuristics software use today. These tools and individuals are not SMEs (Subject Matter Experts) - and neither are MBA leads either, and this means that twenty plus year hi-tech leads with many hands-on roles and major project successes are considered poor candidates and, therefore, unhirable in a competitive job market.

Stereotypes of inflexibility, high cost, pension risk, non-relevant skills, distance of past training, ageism, racism, sexism, academic and other biases, and the dismissal of volunteerism and other life experience and community leadership measures - all work to increase a growing sense of arrogance in a greatly dumbed-down workplace.

The reality that measures of value are not always well considered is terrifying as a society because many of our best and most experienced, sit at home without productive output – but that is a solvable problem of course.

Housing KPIs (Performance Measures)

Key Performance Indicators – KPIs, are mentioned throughout this book again and again. It's important that we have shelter for children and parents; it is important to accommodate other needs as well; and just as important are needs of setting expectation well and meeting promises consistently.

If we need another home; it must be made available within "x" months but "y" months is too long; we must have a chance to preview a few ("x") options and we need not less than three months' notice of confirmation of a change. When automated software systems break down, we need a manual system to work in its place very temporarily. This takes the stress out of life – as with most good planning.

Good Planning with solid KPI exception reporting and release managed process improvement are the ingredients of the best self-correcting processes that you are ever going to work with.

Good Execution – I often say that the best programmers are those whose code cannot be broken by the next change that comes along. Performance can be imperfect either through training, overload, or changes from time to time too, so good testing rigor and contingency planning go a long way to maintaining a stress-free work life and productivity that meets everyone's performance needs - for Housing.

Corruption

Corruption happens in every system, the dishonest or fraudulent conduct or bribery of those in power. Monarchy was thought to be a protection from corruption, as these families were exempt from financial burden in perpetuity and history judged harsh abuses of their vows of service to the realm's greater good. Democracies, in contrast, could be persuaded relatively easily in comparison, as elected officials had to rely upon their incomes to provide for their family's needs.

How can this housing system keep corruption to a minimum and to the exception? There should be a reporting process. There should be a timely appeals process. Can complainants request changes to assigned committee leads? Ask an answer all of these questions in the Requirements and Design phases on your **Housing Project**.

Dante's Description of Hell

Those familiar with Dante's epic poem the Divine Comedy, are likely also familiar with his famously graphic description of hell. Dante's vivid descriptions inspired dozens of masters to render portrayals of suffering in an endless layer upon layer torture chamber. These exacting depictions described nine circles stacked in a cone, each catering to one of seven deadly sins: SALIGIA from the Latin, and in English; greed, lust, gluttony, pride, envy, wrath and sloth; plus two more levels for heresy, and "limbo" - a level where Aristotle and other great pagans live out their eternity as non-Christians. The decision to exclude great people that predated Christianity at the Nicaean Council seemed a little harsh really.

Vivid depictions of hell instilled terror in the superstitious, uneducated masses of Europe's dark ages and inspired absolute adherence to Catholic Church doctrines for a thousand years. Numerous abusers took advantage of that fear - in Inquisition and other witch hunts, and in the sale of Indulgences.

Indulgences were a sort of "Get out of Hell Free" card, but, of course, these cards were not free, they were for purchase or given for other services to an intermittently corrupt clergy. Dante paused to note many clergy members among the sufferers in his nine circles of Inferno as well.

The King James Bible of 1603 continued the fiction of an unclean limbo state from which dead loved ones could be "cleansed by the purchase of indulgences" into Heaven. The Puritans were very unimpressed by this practice and labored for 50 years to remove a profitable "Purgatory" from King Henry XIII's first endorsed English translation of the Latin and Greek Bible – "The Great Bible" of 1536.

The History of Bibles, Coran, Rigveda, Buddhist and other sacred texts is interesting learning and I encourage readers to look for more detail by searching for "Sacred texts of various religions" online - for a list of every text of every religion on the planet.

Here I simply want to note that the "Old Testament" were books written in Hebrew around 400 BCE (Genesis to Deuteronomy in "The

Law" (Torah), "The Prophets" (Nebiim), and "The Writings" (Ketubim). The Holy Roman Emperor Constantine assembled a New Testament from Palestine-area writings in Aramaic, and other languages, and then Hebrew scholars translated Old Testament texts into Greek between 250 and 325 AD (CE) in Egypt. The Rigveda of Hinduism, the oldest modern religion, was written between 1700 and 1100 BCE; the Coran was written in the seventh century; the earliest Buddhist texts are recorded in 858 AD.

The Gutenberg Printing Press was invented in 1450 and by 2007, all of these texts were freely available on the internet; and then fully translated into every language by 2012.

The "Vulgate" was Constantine's Greek Bible next translated into Latin late in the fifth century. The Catholic Church used the Greek Bible until accepting this Latin bible as their standard in 1545. The Church of England's Great Bible was a translation (Latin to English) of a translation (Hebrew to Greek to Latin), and modern English bibles are 90%+ based on this translated translation as well.

Why did Constantine make it his life's work to unite all of the various beliefs and religions of the Roman and Byzantine Empire sects of Judaism, Paganism, and many rival and openly disrespectful and even hateful factions of early Christianity? Peace. It is widely agreed that throughout the efforts of scholars to combine all of the regions' known sacred texts at the Nicaean Council, Constantine did not take a position on one belief or another as long as their negotiated outcomes rolled up to something they could all support in Peace, for the good of the Empire. His insistence was that all factions should believe in one thing – whatever that was – and stop finding unimportant reasons to create conflicts between one another.

During these years, clerics and university students of this time had to have a strong command of both Latin and Greek, as well as a level of education well above the average student's normal academic training.

Why ask you to read all of this background? Two reasons: First, because Dante's depictions of Hell give everyone just a hint of the pain that a mortgage owner endures for a significant percentage of their adult lives so, keep the minor achievement of paying off a mortgage minor - as a society. Mortgages should be a help that makes life better and easier, and not a thing that reduces our quality of life and foregoes our happiness and gives unfair advantage to Usurers.

Second: This very jam-packed history lesson, also tells us to respect and hold your own actions socially accountable, as you maintain the positive reasons that systems are created to make our lives better – be they religious texts, systems of government - or mortgages.

Great Emperors, like Constantine, cannot do this for us any longer, we must now teach ourselves and our children how to turn democratic crowds into an educated leadership.

Chapter 14 – The United Nations

Managing the humanitarian efforts of a very large planet of seven billion inhabitants often falls to the lot of the United Nations. Today's United Nations stands on the brink of being considered a world government. See UN.org.

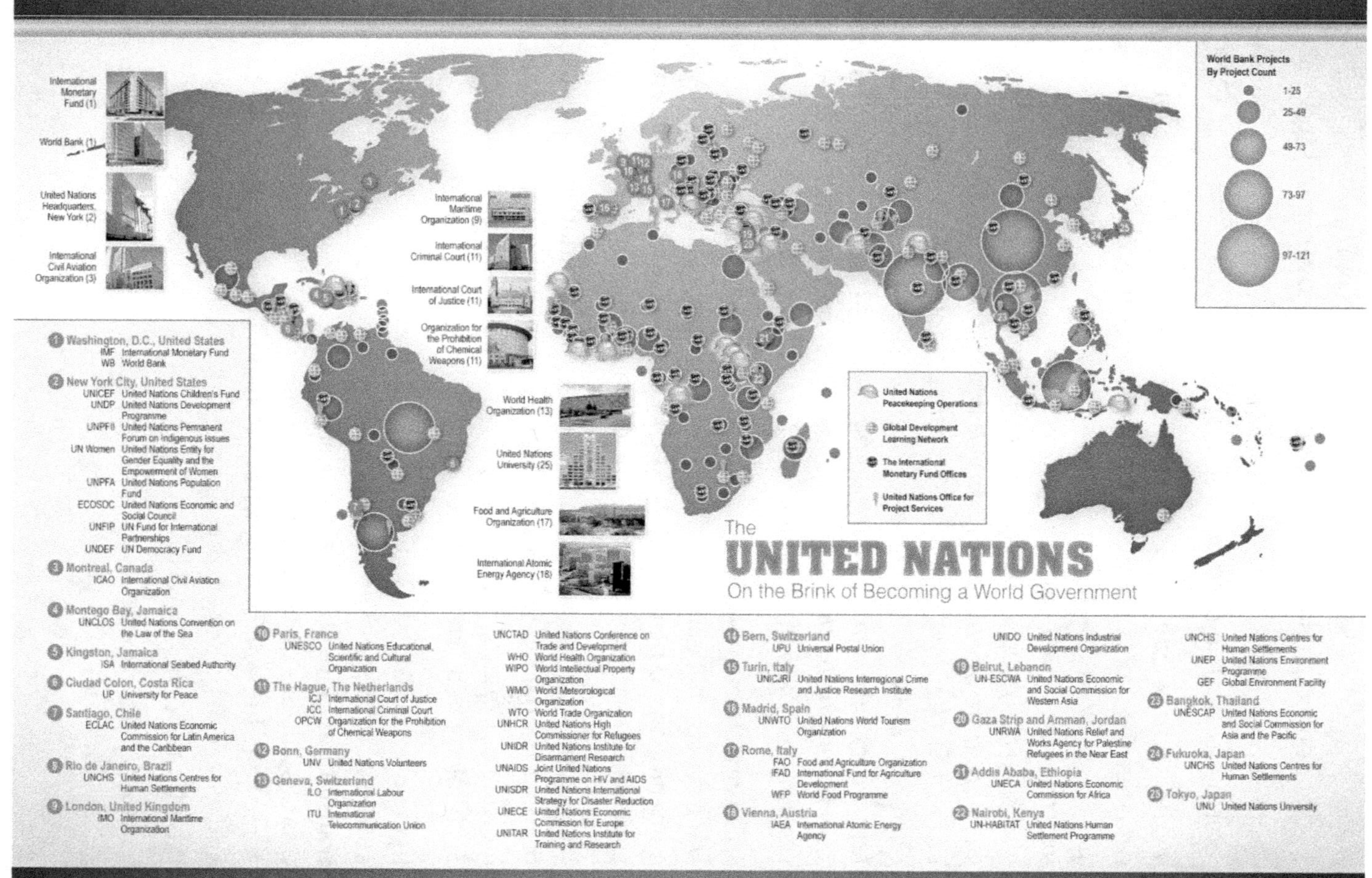
International Monetary Fund (1)
World Bank (1)
United Nations Headquarters, New York (2)
International Civil Aviation Organization (3)
International Maritime Organization (9)
International Criminal Court (11)
International Court of Justice (11)
Organization for the Prohibition of Chemical Weapons (11)
World Health Organization (13)
United Nations University (25)
Food and Agriculture Organization (17)
International Atomic Energy Agency (18)
World Bank Projects By Project Count
1-25
25-49
49-73
73-97
97-121
United Nations Peacekeeping Operations
Global Development Learning Network
The International Monetary Fund Offices
United Nations Office for Project Services
The
UNITED NATIONS
On the Brink of Becoming a World Government
1 Washington, D.C., United States
IMF International Monetary Fund
WB World Bank
2 New York City, United States
UNICEF United Nations Children's Fund
UNDP United Nations Development Programme
UNPFII United Nations Permanent Forum on Indigenous Issues
UN Women United Nations Entity for Gender Equality and the Empowerment of Women
UNPFA United Nations Population Fund
ECOSOC United Nations Economic and Social Council
UNFIP UN Fund for International Partnerships
UNDEF UN Democracy Fund
3 Montreal, Canada
ICAO International Civil Aviation Organization
4 Montego Bay, Jamaica
UNCLOS United Nations Convention on the Law of the Sea
5 Kingston, Jamaica
ISA International Seabed Authority
6 Ciudad Colon, Costa Rica
UP University for Peace
7 Santiago, Chile
ECLAC United Nations Economic Commission for Latin America and the Caribbean
8 Rio de Janeiro, Brazil
UNCHS United Nations Centres for Human Settlements
9 London, United Kingdom
IMO International Maritime Organization
10 Paris, France
UNESCO United Nations Educational, Scientific and Cultural Organization
11 The Hague, The Netherlands
ICJ International Court of Justice
ICC International Criminal Court
OPCW Organization for the Prohibition of Chemical Weapons
12 Bonn, Germany
UNV United Nations Volunteers
13 Geneva, Switzerland
ILO International Labour Organization
ITU International Telecommunication Union
UNCTAD United Nations Conference on Trade and Development
WHO World Health Organization
WIPO World Intellectual Property Organization
WMO World Meteorological Organization
WTO World Trade Organization
UNHCR United Nations High Commissioner for Refugees
UNIDR United Nations Institute for Disarmament Research
UNAIDS Joint United Nations Programme on HIV and AIDS
UNISDR United Nations International Strategy for Disaster Reduction
UNECE United Nations Economic Commission for Europe
UNITAR United Nations Institute for Training and Research
14 Bern, Switzerland
UPU Universal Postal Union
15 Turin, Italy
UNICJRI United Nations Interregional Crime and Justice Research Institute
16 Madrid, Spain
UNWTO United Nations World Tourism Organization
17 Rome, Italy
FAO Food and Agriculture Organization
IFAD International Fund for Agriculture Development
WFP World Food Programme
18 Vienna, Austria
IAEA International Atomic Energy Agency
UNIDO United Nations Industrial Development Organization
19 Beirut, Lebanon
UN-ESCWA United Nations Economic and Social Commission for Western Asia
20 Gaza Strip and Amman, Jordan
UNRWA United Nations Relief and Works Agency for Palestine Refugees in the Near East
21 Addis Ababa, Ethiopia
UNECA United Nations Economic Commission for Africa
22 Nairobi, Kenya
UN-HABITAT United Nations Human Settlement Programme
UNCHS United Nations Centres for Human Settlements
UNEP United Nations Environment Programme
GEF Global Environment Facility
23 Bangkok, Thailand
UNESCAP United Nations Economic and Social Commission for Asia and the Pacific
24 Fukuoka, Japan
UNCHS United Nations Centres for Human Settlements
25 Tokyo, Japan
UNU United Nations University

See http://www.fortruthssake.com/wp-content/uploads/2014/08/UNworld-government.jpg

Walter Cronkite, the late CBS anchorman and broadcast icon often referred to as "The most trusted man in America," stated in 1999:

It seems to many of us that if we are to avoid the eventual catastrophic world conflict we must strengthen the United Nations as a first step toward a world government with a legislature, executive and judiciary, and police to enforce its international laws and keep the peace. To do that, of course, we Americans will have to yield up some of our sovereignty. It would take a lot of courage, a lot of faith in the new order.... We cannot defer this responsibility to posterity. Democracy, civilization itself, is at stake. Within the next few years, we must change the basic structure of our global community from the present anarchic system of war ... to a new system governed by a democratic U.N. federation.

This portrait of a world government seems a natural eventuality if wars are to be eliminated and the non-combatant majority of humanity are to be protected – however, the right plan that ensures a good life has not as yet been discussed to my knowledge.

The recent announcement of a Global Goals initiative (UN, 2015b), the first Global UN plan in 15 years, has no provision for managing individual Country by Country World Peace Agenda progress in Technology and Transition Economics.

In the present Global Goals Planning, there is no consideration for an automated Good Life in each of the regions sponsoring and transitioning into the Goals either, and this would make the Global Goals' solutions unsustainable over time. For his reason, I will spend some time here discussing what the GAPs in Global Goals are today, and what can be done to ensure the long-term success for these well-intentioned plans as well.

At present, the goals include:

Wealth Distribution and a Good Life

1. No Poverty
2. Zero Hunger
3. Universal Healthcare
4. Education
5. Gender Equality
6. Clean Water and Sanitation
7. Affordable Energy
8. 100% employment
10. Reduce Inequality
11. Sustainable Cities and Communities
12. Responsible Consumption and Production
16. Peace and Justice – avoid World War III
17. Partnerships and Goals

Wealth Creation

8. Decent Work and Economic Growth
9. Industry, Innovation, and Infrastructure

Environment

13. Climate Action

14. Life Below Water
15. Life on Land

GAPs in the Plan

Top-down and Bottom-Up are both important approaches to making big social changes. Top-down plans to bring food and water to where they are most desperately needed is necessary Tactical Planning that the UN have worked at since their inception. The missing goal that has prevented World Peace by U.N. planners has been sustainability in fact – a failing of Strategic Planning.

Recall from Chapter 1, that To Feed a Man a Fish, is a Top-Down Plan that provides for what he needs today. Providing this man with the tools and teaching to fish for himself, and that is a Bottom-up plan because it can sustain him with food for a lifetime. If you next automate or give tools to all of his neighbors - to feed them too, this is a Sustainable Bottom-up Plan - because this man's starving neighbors will never need to take what you have given to just one man.

United Nations Teams have moved desperately needed supplies; at great personal danger, sacrifice, and cost, by manual solutions; for drinking water, wells, farming methods, medical supplies, peacekeeping and more. Wars, attrition, and more demand than capacity could rarely sustain these isolated manual solutions.
With a constant need and insufficient funding, there was only so much work that could be accomplished in the face of a much larger need worldwide.

Further along in this book, in our discussions of Alan Turing in Chapter 17 - Everything is Solvable, we describe how he came to realize that an automated machine was needed to solve very complex problems.

The World Peace Transition Plan is a concurrent Bottom-Up plan that builds a sustainable automated production economy. Why is that important? Because, once this plan is implemented, a Good Life can, and should, be considered a basic Human Right that can be brought to wherever needed. These results exceed the targets of the UN Global Goals teams today and the World Peace Plan is achievable within the next three to twenty years in a phased rollout.

To help the success of Global Goals, here are GAPs of Strategy:

1. Under Wealth Distribution and a Good Life –
   - Create Two Safety Nets – for Engineers and for others - as needed to support automation of profitable exports first – see The Transition in Chapter 8.
   - Graduate Tax
   - Fix CSR Business Accountability and get business leads contributing (see Chapter 12)
   - Implement Minimum Wage, Minimum Incomes and manage wealth distribution actively with 11% wealth given to lowest 20% percentile – see Social Projects Chapter 6.
   - Affordable Housing
   - Enable Debt without Usury
   - Accelerate Billionaire Bequeathals – Chapter 6.
2. For Wealth Creation
   - Fund Technology Project Teams that Automate our Civilization and Good Life (Human Rights). Projects
   - Manage Reporting of World Peace Agendas (WPAs) for Social and Technology Projects that align with the World Peace Transition plan.
3. Global Goals assumes a Top-down UN-Driven Plan - that might also deter adoption. The World Peace Agenda is a Country-driven, U.N.-monitored document that reports progress along to administrators in the UN for sharing lessons learned in other national teams.

This mix of ownership and autonomy is designed to encourage performance, and a scenario where the UN has to take over the driving of projects, might be a scenario that countries would work harder to avoid. Better every country should feel a sense of national pride in their contribution. The Canada Arm that ran on NASA's Space Shuttle, was of tremendous interest to young engineers here in Canada – for one example.

Mitigations for UN Policy concerns are constantly evolving including:

• Respect of country values

• The UN commands a Global Army of 110,000 serving on twenty peace operations led by the UN Departments of Peacekeeping Operations (DPKO). Only the US Armed Forces is larger.

• The Global Prosecutor, Judge, and Jury: The UN's International Criminal Court (ICC) officially opened its doors at The Hague ten years ago, in July 2002. The UN boasts that "the ICC has become a fully functional institution, with sixteen cases having been brought before the Court, six of which are at the trial stage. ICC judges have issued twenty-two arrest warrants and six arrests have been made."

• The Global Fed: The International Monetary Fund (IMF) and World Bank (WB) have wrought economic havoc worldwide for decades, burdening nations (especially the less-developed countries) with impossible debt and onerous economic policies. Over the last several years, a growing chorus of globalists has called for transforming and "supersizing" the IMF into the equivalent of a global Federal Reserve, with a global currency — SDRs, Special Drawing Rights — to displace the dollar.

• The Global Trade Cop: The World Trade Organization (WTO), which entered into force in 1995, has joined NAFTA (the North American Free Trade Agreement) in judging and overturning Usury laws and court decisions.

• The Global Environment: Through many environmental agreements, programs, and agencies — Agenda 21, the Biodiversity Treaty, UN Convention on Climate Change, the United Nations Environment Program, the Global Environment Facility, and others.

• The Global Firearm Cleanup: Through its Arms Trade Treaty (ATT) and its Program of Action (PoA) on Small Arms, the United Nations has been pushing feverishly for over a decade and a half to reverse the right of individuals to possess firearms throughout the world.

• Global Internet Control: the UN has been leading an effort to take over the Internet. Dictatorship-led and other countries in the forefront of this effort include China, Iran, North Korea, Cuba, Sudan, Tajikistan, and Uzbekistan, where Internet censorship and cyber spying on citizens are standard operating procedure.

Every problem is solvable, simply list the issues and work through each.

With 7/24 operations in fifty countries at any one time, the U.N.'s various committees manage Security and Human Rights for the unprotected Worldwide.

A World Government would likely play an important role in protecting the needs and safety of non-combatants in any conflict.

Functions of the United Nations

- International Monetary Fund (IMF)
- World Bank
- International Civil Aviation Organization
- International Maritime Organization
- International Criminal Court
- International Court of Justice
- Organization for the Prohibition of Chemical Weapons
- WHO – World Health Organization
- United Nations University
- Food and Agriculture Organization
- International Atomic Energy Agency
- United Nations Peacekeeping Operations
- Global Development Learning Network
- The International Monetary Fund Offices
- UN Office for Project Services
- UNICEF – UN Children's Fund

UN Departments

At the United Nations, day-to-day administration of issues that affect the whole planet is discussed, planned and coordinated in five Councils and Assemblies.

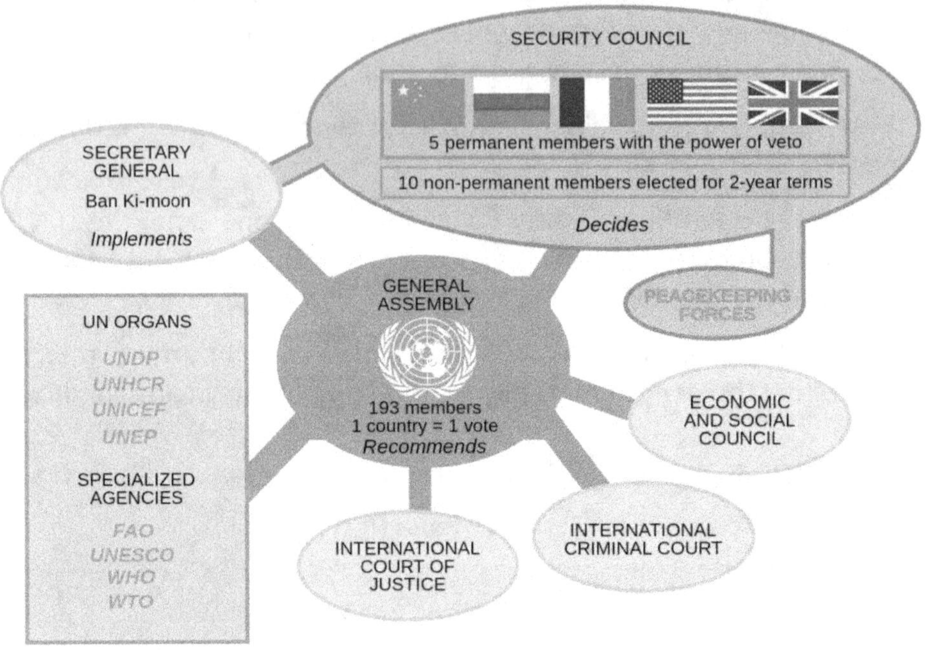

Statistics at the United Nations

The United Nations collects statistics for every country on the planet and then summarizes the data. A composite statistic of life expectancy, education, and per capita income indicators are presented in a Human Development Index Plan consistently so that countries in most desperate need of help can be most easily identified – here below by color. (UN, 2015a)

HDI Rankings for G8 Countries:

Rank	Rating	Country
1	.931	Russia, Moscow
5	.914	United States
6	.911	Germany
8	.902	Canada
14	.892	United Kingdom
17	.890	Japan
20	.884	France
26	.872	Italy

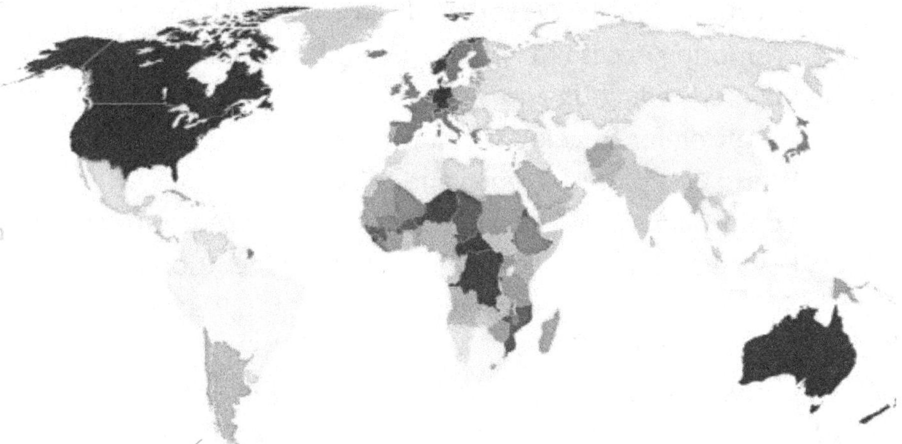

World map indicating the Human Development Index (based on 2013 data, published on July 24, 2014).[1]

0.900 and over	0.650–0.699	0.400–0.449
0.850–0.899	0.600–0.649	0.350–0.399
0.800–0.849	0.550–0.599	0.349 and under
0.750–0.799	0.500–0.549	Data unavailable
0.700–0.749	0.450–0.499	

The full-color Human Development Index is available on the web and at CSQ1.org

United Nations Annual Agendas

Planning for high-level discussions and direction by steering committees happens a year in advance due to the logistics of assembling so many committee members from all over the world. I had not seen U.N. Annual Agendas previously. Assuming that GAPs mentioned above do not add to next year's agenda, here are 2016's discussion topics as at the time of this writing - from the United Nations Website ...

The General Assembly discuss:

1. Cyber Security and Protecting against Cyber Warfare
2. The Threat of Transnational Organized Crime to International Security
3. Efforts to Control Weapons of Mass Destruction
4. Promoting Access to Renewable and Sustainable Energy for Poverty Reduction and Sustainable Development
5. Financing for Development
6. World Commodity Trends and Prospects
7. Comprehensive Review of Special Political Missions and the Future of UN Peacekeeping and Peace Operations
8. Intensifying Cooperation in Outer Space to Preserve Peace and Security
9. Improving the Situation of Non-Self-Governing Territories
10. Building Resilient Cities to Promote Climate Change and Disaster Risk Reduction
11. Realizing the Right to Adequate Shelter through the New Urban Agenda
12. Inclusive Urbanization for the Promotion of Equality and Social Cohesion

Economic and Social Council

1. Ensuring Universal Access to Water
2. Utilizing Youth Employment for Sustainable Development
3. Education in Post-Conflict Situations
4. The Impact of Sexual and Gender-Based Violence on Reproductive Health

5. Furthering Women's Participation in and Access to Information and Communication Technologies
6. Women's Empowerment and the Link to Sustainable Development
7. Social and Economic Development in Cities
8. Guaranteeing Indigenous People's Rights in Latin America and the Caribbean
9. Promoting the Sustainable Use of Natural Resources
10. Addressing Drug Trafficking and the Financing of Terrorism
11. The Role of Civil Society in Addressing the World Drug Problem
12. Evaluating the Impact of Global Narcotics Drug Control
13. Addressing Workers' Rights for Sustainable Economic Growth
14. Preserving Cultural Rights of Ethnic Minorities
15. Protecting Economic, Social and Cultural Rights for Refugees

Department of Development

1. Corruption and Environmental Governance
2. Improving Sustainable Forest Management Practices
3. Plastic Debris in the World's Oceans
4. Education for All: Strengthening Rural Education
5. Promoting Women in Science
6. Protecting World Heritage Sites against New and Emerging Threats
7. Agribusiness and Entrepreneurship Development for Poverty Reduction
8. Promoting Resource-Efficient and Low-Carbon Industrial Production
9. Mainstreaming Gender in Trade Capacity-Building Projects
10. Advancing Children's Rights in the Digital Age
11. Realizing the Rights of Indigenous Children
12. Ending Child Marriage
13. Enhancing South-South Cooperation

14. Empowering Youth for Development
15. Ensuring Women's and Men's Equal Participation in Democratic Governance and Peacebuilding

Human Rights and Humanitarian Affairs

1. Effects of Terrorism on the Enjoyment of Human Rights
2. Human Rights and Climate Change
3. The Realization of Rights of Persons with Disabilities
4. Transforming Refugee Camps into Sustainable Settlements in the Case of Protracted Displacement
5. Strengthening the Capacity of Refugee Host Countries
6. Addressing Temporary Displacements Due to Outbreaks and Epidemics
7. Encouraging the Eradication of Hunger through Cooperation with the Farming Industry
8. Improving Frameworks for the Supply of Food Aid
9. Responding to Food Insecurity in Yemen
10. Ensuring Universal Health Coverage for All
11. Combating Non-Communicable Diseases
12. Improving Health Care Services for Ageing Populations
13. Addressing the Needs of Palestinian Women and Girls in Gaza
14. Improving Coordination of Humanitarian Assistance and Relief for Palestinian Refugees in Syria
15. Strengthening Access to Education through the Human Rights, Conflict Resolution, and Tolerance (HRCRT) Policy

Peace and Security

1. Women, Peace, and Security: Women as Active Agents in Peace and Security
2. The Situation in the Central African Republic
3. Threats to International Peace and Security Caused by Terrorist Acts

4. The Security Council may adopt agenda items outside the topics listed here. Assigned delegates should stay abreast of current issues.
5. Measures to Increase National Reporting of Member States
6. Incorporating Gender-Sensitive Approaches in the Implementation of the Program of Action
7. Adapting to Recent Developments in Small Arms and Light Weapons Technology

Chapter 15 – Disarmament

To destroy all major cities of the leading nuclear power countries would require approximately eighty nuclear missiles with yields, or energy discharges, much less than the largest nuclear bomb ever tested in 1961; the 50 Megaton Russian "Tsar Bomba". Nuclear bombs that devastated Hiroshima and Nagasaki packed just one-twentieth that yield, between fifteen and twenty-five kilotons, by comparison.

Two decades after the end of the Cold War, approximately 15,700 warheads exist in nine countries around the world.

Country	Warheads
Russian Federation	7500
United States	7200
France	300
China	250
United Kingdom	215
Pakistan	110
India	100
Israel	90
North Korea	<10

Approximately 4,100 warheads are considered active, and 1800 are on high alert 24-hours-a-day. Statistics are presented here by the Nuclear Disarmament Resource Collection at NTI.org (FAS.com, 2015).

Nuclear Arsenals

Like guard dogs on sophisticated tethers, nuclear arsenals remain on standby indefinitely as a deterrent that ensures that no rogue leader nor country will ever dare risk the complete annihilation of our world through a nuclear attack. Nuclear war is a terrifying, potentially humanity-ending extinction-level event where those that survive the initial blasts succumb to radioactive fallout and nuclear winter in the years that follow.

A nuclear winter occurs when sunlight is blotted out by atmospheric dust similar to the meteor strike that ended the 300 million year reign of dinosaurs on earth 65 million years ago. That asteroid impact is estimated to have exploded at yields equal to approximately 100 nuclear bombs.

The firestorms that follow within hours of these blasts release 1000-times the energy of the original detonation and fill the skies with smoke and particulate. Years of darkness ends all growth, oxygen diminishes, and temperatures drop twenty to thirty-five degrees centigrade. The choking air will clear after two months, but the survivors will die, in all probability, of hunger, cold, and radiation poisoning.

Nuclear fallout is the radioactively charged debris and dust kicked into the sky by firestorms and then circulated by global trade winds. 20% of those who breathe in this particulate will die of cancer-related illness.

Today, eight democratically elected world leaders administer “buttons” that could initiate this war. Needless to say, highest levels of government place the highest level priority upon managing their nuclear capabilities.

Weapons, Militia, and Conventional Arsenals

Nuclear devices are not the only obstacles to peace. Unfortunately, handguns, rifles, armored personnel carriers and other military equipment remain in private households and community arsenals throughout the world. Whenever there has been a security concern, there is militia, and the perceived or real need for citizens to arm themselves.

In Iraq, the USA, and other war-torn nations, there are as many weapons as there are households.

Armies

One of the most dangerous times for any government administration is immediately after a war when celebrated Generals return from battle with trained and loyal armies. These strong leaders do not easily hand over the keys of power to their much weaker political leaders.

This was one of the leading reason for Monarchs to become God's appointed representative here on earth. Churches legitimized what was a fiction first designed to refer to the "greater good" of the kingdom or society, and over time, when religious fervor became very literal and entrenched within soldiers, major church groups used this as a control for the Monarchs of Europe as well. Churches demanded a tax, food, armies for holy wars, and then military support of their financial interests, and then fealty too.

Through land allowances and salary, militaries have ever been among the most reliable employers of men for thousands of years as well.

Religion in the Army

This interference by religious leaders was one of the reasons that King Henry XIII cast out the Catholic Church from England, and replaced it with another Church under his own control. Disconnecting from the Catholic Church was no simple task, and it took almost 100 years to establish the new religious order at the cost of many lives lost, even within the Royal family. Britain struck new military alliances with neighboring countries like the Netherlands, who also saw the value of separating and then uniting against tyrannical or corrupt religious emissaries from Rome.

No uneducated, superstitious, “god-fearing” peasant soldier would risk purgatory and hell for themselves and other family members in the next life, by revolting against God’s right hand on earth, the King. A God-Fearing Man was a term given to honorable, respectable, smart, family and community-valued individuals who pledged “fealty” to their King. Fealty was an oath of fidelity, service, and their lives if needed, in the service of their country. This oath protected Monarchs and even contemporary leaders from returning armies for thousands of years.

Chapter 16 – Rating your Government

Modern politics requires candidates and leaders to face the media and answer both fair and unfair questions about everything from policy to professional viewpoints, to private lives and even family connections as well. The education levels of many nations vary quite a bit from country to country and this will influence how the masses respond to tactics that can range from emotion-intense mud-slinging to credible, professional transparency. Transparencies here, are purely factual accounts of the KPIs met and missed by a government.

I refer to the high or low transparency requirements as pendulums that swing too far both ways from time to time. When an economy is good, cries for transparency tend to be a bit more relaxed generally and election years tend to see newspaper reports showcasing a potential police investigation into government. Attempts to corrupt the democratic system by taking advantage of rulemakings such as ex-pat votes, or students voting on campus, happen frequently and can be both very subtle and far too bold and brazen as well.

All Democratic voters are inundated with messages from re-election incumbents who explain what a wonderful job they have done for the country; what a terrible thing it would be if the other party was elected; and how and when you should cast your vote for them – and tell your friends and family to do so also.

Voting happens every four or five years in most democratic systems including Russia, China, the U.S., and so on. It's an expensive undertaking and a distraction from the business of running the country as well. This is the reason that a 90-day election window is typical in many nations. Canada's federal elections ran this year in October, and Provincial Elections ran two years ago.

Each candidate will be working very hard to ensure that they score 10-out-of-10 in their ability to manage the best run country in the world – your country.

There are approximately 207 sovereign nations, and half of the G8 countries are not in the Top 10 on the UN's list of the best places in the world to live (see Chapter 14's UN Statistics). This chapter sets out a method for measuring how good is your nation's Management Team and track record?

Rating the Management Team of your Country

In Chapter 6 – Socialistic Policies Mean Business, the United Nations administers a list of the best countries in the World to live. The number one HDI country out of all countries for the last several years is Norway, if we overlook Moscow, Russia's higher regional score. All other countries are scored based on consistently measured, weighted criteria in all countries of the world. This HDI list is intended to determine first, which countries are the best to live in, and second, which are in the greatest need of help. The list is not intended to tell which governments are good or bad necessarily – and so we are going to make a few adjustments so that we can leverage this resource.

Human Development Index, GINI, and GDP

The Human Development Index (HDI) Value is derived from life expectancy at birth, mean years of schooling, expected years of schooling, and gross national income (GNI) per capita. GNI summarizes gross domestic product (GDP) plus net receipts of employee compensation and investment income from abroad.

The HDI is not a perfect reference. Chapter 6's Wealth Creation sub-section showed Hong Kong's 15th highest HDI ranking in the world appeared to be completely unconcerned by 45% of citizens living in windowless shoeboxes essentially. Despite shortcomings in the measures of Wealth Distribution, we will leverage it alongside other measures like the GINI Index.

Let's compare the Netherlands and Canada for one example - the polite Irish custom is to pick on yourself first, so here we go.

The Netherlands is the "Rocky Balboa" of GDPs, so this comparison has wake-up call written all over it. With a population of 16.8 million (half Canada's) and a footprint the size of Lake Ontario, the Netherlands exports more of its GDP than Canada does, and the quality and diversity of Netherlands exports is better than Canada's as well.

On the UN Human Development Index, Netherlands rates fourth and Canada eighth. So – both are nice places to live. How well are the two nations managed?

An obvious difference between the two is that Canada has perhaps 1000 times the raw resources of the Netherlands. Let's factor in the land and sea area for each Country next, and see how Canada's management team compares. I like to use GDP Exports quite a bit because this is the wealth-building portion of the GDP for any nation. Canada's GDP Export is $553 billion, and the Netherlands is $707 billion. For Canada, GDP is $1.8 Trillion; and for the Netherlands – $853 Billion, but with twice the population, GDP per Capita is the same for both countries.

When a country's Imports are too high, overall GDP is impacted, so let's confirm what is happening for these two nations. Here we see that both countries have trading exports approximately 20% higher than Canada's Trade levels. Canada imports more than it exports which comes as a surprise given its vast resources.

Leading Exporters and Importers, 2013 World Merchandise Trade				
2013 Rank	2012 Rank	Exporters	2013 US$B	2013 % Share
5	5	Netherlands	664	3.5
10	8	Russia	523	2.8
13	12	Canada	458	2.4
2013 Rank	2012 Rank	Importers	2013 US$B	2013 % Share
8	7	Netherlands	590	3.1
11	12	Canada	474	2.5

Source: WTO Secretariat

Imports into Canada are 32% of GDP where Netherlands are 73%. Canada is more self-sufficient but exports only half the services that the Netherlands does. The Netherlands banned offshoring of engineering 20 years ago, so their engineering sector is thriving. The following charts are care of findthedata.com.

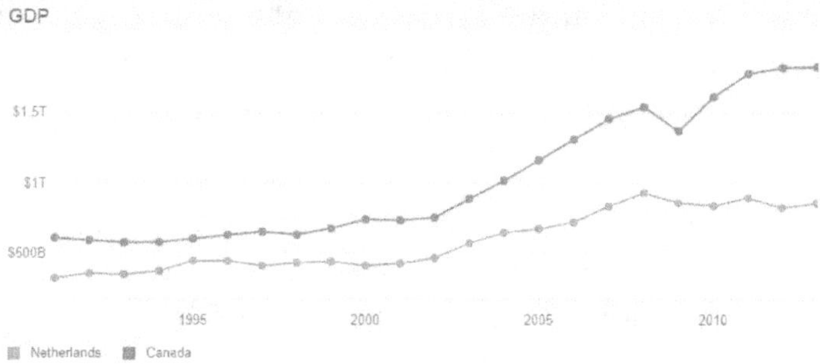

GINI Wealth Distribution is 15% better in Netherlands.

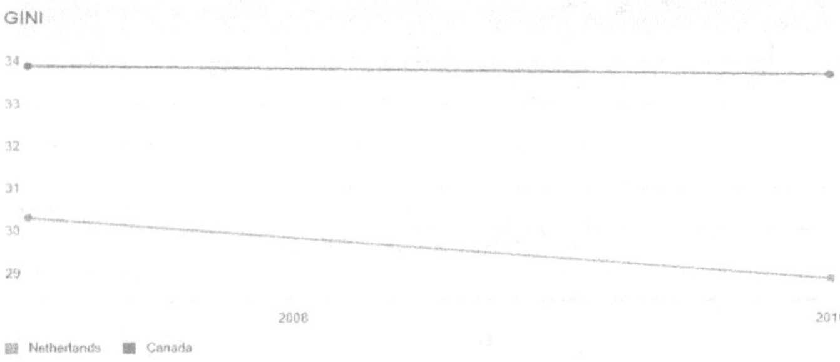

Trade is almost three times that of Canada per capita.

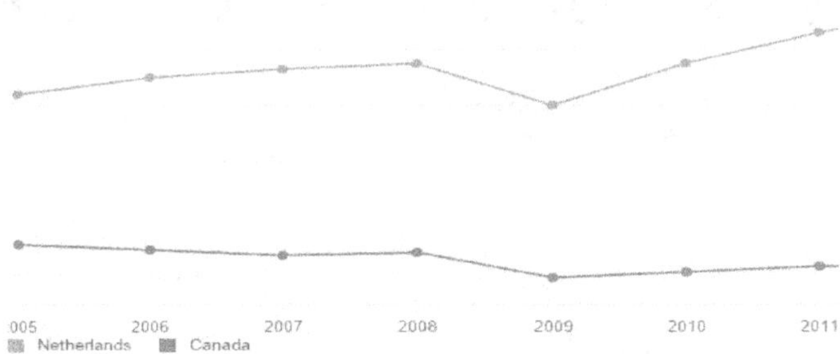

The economy appears better managed by the Netherlands government right away, but when we factor in resource management (both land and sea), we can get a sense of how efficient and inefficient is each nation. Canada has a resource base of nine million square kilometers where the Netherlands' has forty-one thousand.

Having a large GDP with high export percentage as well, is a sign of a strong economy. GDP is made up of five components, Consumption, Investment, Government spending, Exports, and Imports - C+I+G+X-M.

Let's take the example of a small country that has $1 billion in consumption, $1 billion in investment and $1 billion in government spending. GDP so far is $3 billion.

This small country happens to host a large car factory. Of course, a small country only needs so many cars, so the country exports most of these cars, making $4 billion in foreign sales. Now GDP is $7 billion.

With all that wealth coming in from the car factory, the country uses that money to import a bunch of products that it can't produce on its own, so now it has $4 billion in imports. Imports are next subtracted from GDP, so total GDP goes back to $3 billion.

In this case, exports equal 133% of GDP

$4 billion / $3 billion = 1.33 x 100

This is what happens in small countries with high productivity like Luxembourg and Singapore essentially. Due to their small size, instead of trying to be self-sufficient and produce all the products their population needs, they specialize in a few highly-profitable industries. These industries may produce more income from exports than the entire domestic economy. All that money from exports allows them to purchase imports far more than what their domestic economy could otherwise support.

Area and Export – Wealth Creation

If GDP, HDI, and GINI are the standard measures of a Good Life, let us use Exports and Land Area to decide how well-managed are the resources of each country next. Low unemployment rates are also a very good sign of good management, and I will deal with these measures in a future formula discussion.

Ranking each country's calculated results compared to neighboring countries, gives visibility to the achievements and challenges of other management teams, and useful too is the exercise of adjusting formula weightings for Area and Population Distribution. The actual values themselves are not considered important.

Other considerations enter in too. Antarctica - has massive land and resources, but relatively little of it is accessible nor populated. This will change as primary production economies automate, but for now Antarctica is an example of a very poorly managed massive land mass for legitimate reasons.

Leaving out of this formula the country GINIs presented in Chapter 6, we can plug in values for some combination of variables and call it formula "v1.3."

(GDP Export / Area) x HDI

Canada Ranks #100 in the world with a 5/10 Management Team.

($ 533 Billion / 9,984,670) x 0.902 = 5,540.10

The Netherlands Ranks #6 with a 10/10 Management Team:

($ 707 Billion / 41,850) x 0.915 = 1,691,577

We can now compare which countries are more efficient quite easily using the "GDP per country" data made widely available at Wikipedia.com and other similar websites. Formula v1.3 above produces a v1.3 Dataset with rankings of best-managed countries quite easily.

In the example formulation above, rankings will appear to penalize larger countries. Next you might realize as I did, that a majority of smaller countries and island nations have often been forced by necessity or resource scarcity to become more efficient with both their resources and economic choices.

Variations of this formula will improve rankings until a nearer-perfect formula arrives. Until then, take a look at the GDP exports of some the best twenty or thirty top management teams in the world and take a page from their lessons-learned.

Top 10 Country Management Teams

We work with these formulae until they produce a list of **Top 10 Managed Countries** who can be considered 5/10, 6/10, on up to 10/10 management teams reliably. Again, ranking is presented with the sole intension to discuss opportunities to improve and this v1.3 calculation will penalize a larger and more sparsely populated country in keeping with a greater untapped resource base.

See Rankings for the G8 countries via the v1.3 formula in the following chart. Top 10 Countries, ranked one-through-ten, are considered 10/10 management teams, and I think most will agree that, ignoring GINI, that our #1, #2, and #3 position Hong Kong, Singapore, and Luxembourg qualify as 10/10 management teams. Rankings from 11 to 20 are 90% or 9/10, from 21 to 40 are 80% or 8/10. Ranks 41 to 60 are 7/10 or 70%. Ranks 71 to 120 are just 50% efficient, and then we removed 1 point if imports were higher than exports.

Rank	Rating	Country
14	10/10	Germany
24	9/10	Italy
26	9/10	Japan
29	8/10	United Kingdom
30	8/10	France
60	6/10	United States
100	4/10	Canada
112	5/10	Russia

In the v1.3 dataset, we found that Norway sits at position 40 – and we realize that this is an example of a failing in this version of our formula that could be corrected by simply incorporating GINI (a measure of wealth distribution). V1.3 spotted Hong Kong at #1, but its GINI is the lowest at Ranking #130 (see Chapter 6) so, here again, the introduction of GINI in a v1.4 dataset would correct the rankings.

Norway's #1 ranked HDI, #6 in GINI, $20 minimum wage, and 3% unemployment are shining beacons of good management as needed to build a good life. Adding GINI in a v1.4 would benefit France, Germany, Canada, Italy, and UK, and penalize Russia, the USA, and Japan.

Rating Elected Leaders

Bench Strength (Leadership training and experience), Engineering, Accountability, adherence to CSQ Planning and Social Projects like Wealth Distribution and Wealth Generation, Safety Nets, On-shoring Engineering, Production Automation, Hitech and Sciences investment all factor into your decision as a voter.

Most democratic countries have up to four or even six parties running in most major elections. Questions for readers include:

1) What is your country's rank on the v1.3 table below? Readers will live in countries with a ?/10 (four out of ten was mine) management team. This is your country "Rank."
2) Research how many years each party had time in office since 1980 (this was the start of our most recent K-Wave Autumn). Call this value "Time."

Rank, when multiplied by the percentage of Time in Office during this past 35 years, gives you a sense of how much each party contributed to your country's current management situation. Parties tend to reward their SMEs (Subject Matter Experts) with tenure and can, therefore, be expected to continue with the team.

You might notice a recurring irony over time, in that Right-wing Conservative parties tend to be elected to lead in Good Economic Times. Left Parties, parties with more social programs, are preferred in difficult economic times more often than not

Norway is an example of an exception to this rule because a superior management team with socialistic policy leanings, was elected 20 years ago when times were good. Today, Norway is an economic powerhouse that is also widely regarded to have preserved the American Dream for its citizens - where all of the G8 countries lost theirs.

Now, compare World Peace Transition Needs to the policies and candidates of each party – and cast your vote based on:

- ▶ **Are you voting for an Engineer & Hi-tech Builder first;** then Medical; then Legal, then Sales; and then the distant last option Finance lead? When an engineer harms society they lose certification; when a business person harms society they get a bonus. It is an example of the opposite of accountability in fact.
- ▶ **Manage threats to World War III actively**. Do not permit distraction to expose your country to World War III. Comparing 1930s timelines, we could be just two to three years away from

a war right now, due to continued hardship in this latest K-Wave Winter. Russians and Americans, for two examples, are brilliant engineers of both technology and a Good Life, so commit to finding common ground and making progress together. Where will distractions of war be sponsored? By wealth inequity opponents; by profiteers of conflict-Oil; by groups who benefit when other interests suffer. "Follow the money" is a time-honored investigation tool that often leads to the truth. Who or what is paying for mercenaries? Our security personnel will find these villains and for us, a Good Life and World Peace are a Project that require our full attention.

- **Do policies reflect Wealth Distribution goals?**
  - Economic controls graduate taxes based on Wealth & Business controls "with teeth" (that work and have penalties for abuse) as Queen Elizabeth would say.
  - Get on the phone to every major company with a record breaking year and insist on rapid 100% employment nationwide – and an end to slow hiring policies.
  - Make Business Stewards Accountable. CSQ Compliance in Projects will go a long way as well.
  - Minimum wage to $20 ($16 in the U.S.) - Graduate tax with higher taxes where it can best be afforded.
  - Affordable Housing – can your children find a home in the town they grew up in?

- **Wealth Creation**
  - Reshore Engineering
  - Improve GDP Exports like Norway & the Netherlands. The Netherlands is the size of Lake Ontario yet it matches Canada's and Russia's GDP exports income. This is a gross negligence not to be proud of, so fix this.
  - Fix our broken Safety Nets and pay for this with automated profitable exports first.
  - Retraining
  - Reshore automated manufacturing

- Fund technology start-ups with real money and not empty promises. Invest in CSQ/World Peace Science & Technology Automation Projects

- **Build the Good Life**
  - Fix Broken Safety Nets that support automation of profitable exports.
  - Build the technology Projects in Chapter 7
- **Status World Peace Agenda (WPA) Projects Transparently**

Members maintain a comparison of Political Party Policy alignment versus CSQ Planning on our forums. Presently a report for North America is online, and other countries can be added to the forums as well.

Conclusions

These rankings present transparent result and are intended to be a targeted learning tool for voters, officials and country builders wishing to establish their performance KPIs and measures well. The charts are also a good chance to understand where opportunities for improvement lie.

At election time, we voters are surrounded by news of just how terrific our political parties are. These tools provide you with more background and process with which to validate those messages. Remember that everyone has an opportunity to improve their country's rank by voting for a World Peace Agenda.

More rankings, formulae, and datasets are online for discussion at CSQ1.org.

v1.3	Country	GDP	Export	GDP Export	Area
1	Hong Kong	274,027	229.6%	629,139	316
2	Singapore	295,744	190.5%	563,451	459
3	Luxembourg	60,131	203.3%	122,258	811
4	Malta	9,971	93.6%	9,334	254
5	Netherlands	853,539	82.9%	707,925	41,850
6	Belgium	524,806	82.8%	434,329	30,528

v1.3	Country	GDP	Export	GDP Export	Area
7	Qatar	202,450	75.6%	153,093	11,586
8	Switzerland	685,434	72.2%	494,541	41,284
9	Bahrain	32,898	20.0%	6,580	549
10	Maldives	2,836	111.3%	3,157	236
11	Mauritius	11,938	54.3%	6,484	765
12	Kuwait	175,831	71.6%	125,825	17,818
13	United Arab Emirates	599,000	95.2%	569,949	83,600
14	Germany	3,730,261	50.7%	90,123	357,114
15	Barbados	4,228	42.5%	1,796	308
16	Trinidad and Tobago	24,463	63.2%	15,448	2,842
17	Israel	291,567	32.9%	95,984	20,770
18	Denmark	336,701	54.3%	182,728	43,094
19	Ireland	232,077	105.3%	44,377	70,273
20	Brunei	16,111	76.2%	12,270	3,903
21	Austria	428,322	57.4%	246,028	83,871
22	Seychelles	1,445	76.3%	1,103	347
23	Lebanon	47,221	62.6%	29,537	10,452
24	Italy	2,149,485	28.6%	613,893	301,336
25	Czech	208,796	77.2%	161,191	78,865
26	Japan	4,898,532	14.7%	721,554	377,930
27	Slovenia	47,990	74.7%	35,844	20,273
28	Antigua and Barbuda	1,241	44.1%	547	316
29	United	2,678,455	42.9%	1,150,129	242,495
30	France	2,806,432	28.3%	793,659	640,679
31	Saint Kitts and Nevis	743	34.3%	254	180
32	Hungary	133,424	88.8%	118,427	93,028
33	Saint Lucia	1,336	46.0%	614	444
34	Cyprus	24,057	40.1%	9,649	9,251
35	Portugal	227,324	39.3%	89,247	92,090
36	Spain	1,393,040	31.6%	439,643	505,992
37	Poland	525,863	47.8%	251,363	312,679
38	Grenada	831	25.1%	209	260
39	Malaysia	312,434	81.7%	255,196	330,803
40	Norway	522,349	38.9%	203,089	323,802
41	Sweden	579,680	43.8%	253,842	450,295
42	Greece	242,230	30.2%	73,226	130,373
43	Saint Vincent and the Grenadines	709	27.4%	194	294
44	Lithuania	46,403	77.1%	35,791	65,300
45	Thailand	420,167	73.6%	309,117	513,120
46	Estonia	24,880	86.1%	21,417	45,227

v1.3	Country	GDP	Export	GDP Export	Area
47	Croatia	57,869	42.9%	24,849	56,594
48	Equatorial	18,532	88.5%	16,393	28,051
49	Azerbaijan	73,557	48.7%	35,837	86,600
50	Panama	40,467	71.0%	28,736	75,417
51	Jamaica	14,270	30.4%	4,342	10,991
52	Vietnam	171,222	83.9%	143,621	331,212
53	Romania	192,094	42.0%	80,641	238,391
54	Finland	267,329	38.2%	102,066	338,424
55	Bulgaria	54,481	68.4%	37,260	109,884
56	Costa Rica	49,621	35.1%	17,437	51,100
57	Dominica	498	32.8%	163	468
58	Latvia	30,886	58.8%	18,173	64,559
59	Dominican	60,612	25.5%	15,468	48,671
60	United States	16,768,05	55.7%	9,344,834	9,526,468
61	Turkey	822,149	25.7%	210,881	783,562
62	El Salvador	24,259	26.4%	6,402	21,041
63	New Zealand	189,025	29.7%	56,046	270,467
64	China	9,181,204	26.4%	2,423,838	9,572,900
65	Sri Lanka	67,203	22.5%	15,101	65,610
66	Philippines	272,067	27.9%	75,934	300,000
67	Belarus	71,710	61.2%	43,872	207,600
68	Macedonia	10,767	53.9%	5,802	25,713
69	Serbia	45,520	40.8%	18,549	88,361
70	Mexico	1,259,201	31.8%	399,796	1,964,375
71	Samoa	691	30.6%	212	964
72	Saudi Arabia	748,450	51.8%	387,622	2,149,690
73	Oman	79,656	62.7%	49,904	309,500
74	Tonga	440	17.8%	78	464
75	Jordan	33,594	42.5%	14,267	89,342
76	Cuba	78,694	20.0%	15,707	108,889
77	Bangladesh	153,505	19.5%	29,995	147,570
78	Albania	12,904	35.1%	4,523	28,748
79	Ukraine	188,350	46.9%	88,280	603,500
80	Montenegro	4,417	41.8%	1,845	13,812
81	Chile	277,043	32.6%	90,205	756,102
82	Tunisia	46,883	47.0%	22,030	163,610
83	Fiji	4,034	58.8%	2,372	18,272
84	India	1,937,797	24.8%	480,961	3,166,414
85	Bosnia and Herzegovina	17,852	32.0%	5,705	51,209
86	Venezuela	371,339	26.2%	97,179	916,445
87	Guatemala	53,797	23.7%	12,728	103,000
88	Georgia	16,127	44.7%	7,207	69,700
89	Iceland	15,330	55.7%	8,543	100,210
90	Indonesia	868,346	23.7%	206,145	1,904,569
91	Iran	492,783	32.2%	158,578	1,648,195
92	Ecuador	94,473	29.2%	27,567	276,841

v1.3	Country	GDP	Export	GDP Export	Area
93	Moldova	7,970	44.1%	3,516	33,846
94	Armenia	10,431	27.0%	2,815	29,843
95	Comoros	622	16.4%	102	751
96	South Africa	366,060	31.1%	113,991	1,221,037
97	Swaziland	3,523	55.3%	1,948	17,364
98	Uruguay	55,708	24.0%	13,370	181,034
99	Nigeria	514,965	18.0%	92,900	923,768
100	Canada	1,838,964	30.1%	553,160	9,984,670
101	Honduras	18,569	47.9%	8,900	111,369
102	Ghana	47,830	42.2%	20,165	238,533
103	Morocco	103,836	33.7%	34,941	446,550
104	Turkmenistan	41,851	73.3%	30,660	488,100
105	Colombia	378,148	17.8%	67,424	1,141,748
106	Australia	1,531,282	19.9%	304,419	7,692,024
107	Cambodia	15,250	65.7%	10,022	181,035
108	Belize	1,624	60.9%	988	22,966
109	Egypt	255,199	17.6%	44,966	1,002,450
110	Angola	121,692	55.8%	67,880	1,246,700
111	Peru	200,269	23.7%	47,544	1,285,216
112	Russia	2,096,774	28.4%	594,855	17,098,242
113	Argentina	611,726	14.3%	87,293	2,780,400
114	Gabon	16,970	58.7%	9,965	267,668
115	Brazil	2,243,854	12.6%	281,604	8,515,767
116	Paraguay	29,208	49.4%	14,423	406,752
117	Kazakhstan	224,415	38.3%	85,839	2,724,900
118	Haiti	7,691	18.2%	1,403	27,750
119	Uzbekistan	57,210	27.7%	15,824	447,400
120	Nicaragua	11,256	40.5%	4,561	120,538
121	Libya	74,597	67.4%	50,263	1,759,540
122	Rwanda	7,601	14.4%	1,095	26,338
123	Algeria	208,764	33.1%	69,184	2,381,741
124	Kiribati	175	10.5%	18	572
125	Vanuatu	800	47.8%	383	12,189
126	Pakistan	225,419	13.2%	29,800	881,912
127	Djibouti	1,456	57.1%	831	23,200
128	Lesotho	2,230	45.0%	1,003	30,355
129	Togo	4,158	39.4%	1,639	56,785
130	Sierra Leone	4,929	53.1%	2,617	71,740
131	Suriname	5,299	58.7%	3,108	163,820
132	Uganda	26,444	23.7%	6,275	241,550
133	Bhutan	1,781	40.9%	728	38,394
134	Kyrgyzstan	7,226	47.2%	3,409	199,951
135	Yemen	34,714	30.5%	10,581	527,968
136	Solomon	1,073	54.5%	585	28,896
137	Senegal	15,152	26.2%	3,970	196,722
138	Laos	10,760	37.2%	4,005	236,800
139	Botswana	14,778	55.1%	8,146	582,000
140	Kenya	54,443	17.7%	9,653	580,367

v1.3	Country	GDP	Export	GDP Export	Area
141	Malawi	5,146	46.3%	2,384	117,600
142	Papua New Guinea	15,420	51.0%	7,864	462,840
143	Bolivia	30,601	44.2%	13,520	1,098,581
144	Guyana	2,990	84.6%	2,530	214,969
145	Nepal	18,179	10.7%	1,945	147,181
146	Zambia	22,384	41.9%	9,374	752,612
147	Tajikistan	8,506	19.2%	1,631	143,100
148	Cameroon	29,568	20.7%	6,109	475,442
149	Benin	8,307	18.3%	1,518	112,492
150	Tanzania	44,698	24.7%	11,049	945,087
151	Zimbabwe	13,490	29.5%	3,978	390,757
152	Burkina Faso	12,547	27.5%	3,449	272,967
153	Namibia	12,580	43.0%	5,411	825,615
154	Guinea	7,219	28.5%	2,055	245,857
155	Madagascar	10,612	30.1%	3,191	587,041
156	Burundi	2,549	7.4%	189	27,834
157	Liberia	1,946	32.4%	630	110,879
158	Mongolia	11,516	45.1%	5,198	1,564,110
159	Eritrea	3,438	19.5%	671	112,622
160	Ethiopia	46,017	12.5%	5,748	1,104,300
161	Mauritania	5,516	66.7%	3,680	1,030,700
162	Sudan	54,595	9.6%	5,230	1,886,068
163	Mali	10,943	31.3%	3,421	1,240,192
164	Chad	10,640	32.2%	3,423	1,284,000
165	Afghanistan	21,618	6.3%	1,358	652,230
166	Niger	7,407	23.3%	1,729	1,267,000
167	Mozambique	1,105	30.2%	333	801,590
168	Central African Republic	1,585	11.7%	185	622,984

Chapter 17 - Building World Peace

World Peace - is a Project. The G8's democratically elected leaders struggle with unemployment, under-employment, and other social issues because the solutions to fixing these problems seem rarely obvious. Unemployment, Wealth Distribution, Transition to Automated Production Economies – these, and all other problems that require smart change, are just projects. Every Problem is Solvable in this way and many projects, are managed together as a program.

At the start of a project, a project manager is assigned to lead working teams through discussions of "What exists today" vs. "What is needed tomorrow". These are the project's Needs and Inventory. Next, signed-off solutions are budgeted and resourced in an implementation plan that rolls out to trained operational administrators and users. KPI monitoring confirms that the unemployment problem is improving, or that it needs to be revisited before the next planned release of the project.

Updates to the first project's deliverables might be needed every six months for application projects, or every one to four years for other projects depending on whether solutions are meeting operational targets.

In government, every Minister has a book of policy problems to manage. In this way, his portfolio can be said to have many projects and programs in addition to the policies that are status-reported while in operation.

The steps that follow describe a transparent process that has been proven to work for hundreds of thousands of projects over many years. Now all that voters have to do is ensure that smart engineers are elected to lead both the setting of strong targets for these systems and the building of solutions that meet targets for new policy.

By posting stats on projects like % Unemployment, % Unemployment under-25 years of age, % over-45 years of age, % projects that met targets, % projects that posted stats, and similar, you add transparent information that voters can reward for strong or weak performance.

World Peace too, is a large Portfolio of many programs and many sub-projects. Each country's World Peace Agenda presents status for these projects in a consistent way that can be rolled up to present a worldwide snapshot of the progress of all participating countries easily.

With adoption, a central repository will likely be administered by the United Nations. The Global World Peace Agenda Repository will reside at the CSQ1.org Content Website under Forums.

How much does it cost to build World Peace? Technology projects cost much less than social projects generally. Running to a plan is usually cheaper than running without a plan as well. Development of a budget is the first step in the creation of any Program Charter.

World Peace – the Transition uses the same Process that Hewlett-Packard Canada has used to manage 500+ complex projects a year reliably for the past fifteen years. As a consultant, I have seen similar systems evolving and maturing in Project Management offices in the U.S. and in France as well. Over time, I have seen many thousands of projects complete just like falling dominos. No drama; everyone knows their next steps; and no problem is unresolvable.

The only failings that I have seen with the system had occurred in more recent years when non-technical procurement staff and non-SME business sponsors tried to step into the leadership roles of project management teams without having sufficient skill set to know when and how to advance work through the steps.

You can hand anyone a hammer, but that will not make them a carpenter. In qualified hands, with oversight and skilled guidance, this was never a problem and the process ran like clockwork.

For me, the work of building World Peace is just another day at the office – so in this chapter, I am going to step you through this process patiently with the warning that a hi-tech or engineering background is going to serve you well. Like Alan Turning, you will need to understand how to translate real-world needs into automatable technology solutions.

For Technology Projects, a strong hi-tech and project background - with some programming, web infrastructure design, resource planning, and online collaboration tools are essential.

Technology resources vary widely in ability and productivity, so knowing how to mentor and coach past roadblocks that are real, and sometimes just imagined, will move the project forward quickly and easily without delays and unnecessary constant "fire alarms" - in a professional and even fun manner.

Document 1 - Program Charter

The Charter describes the project work assigned. This could be a Charter for Cold Fusion, Rapid Charge Batteries, Household Food Delivery, and any one of dozens of other projects both social and technical.

Program Charter - Table of Contents:

1. **Stakeholders** - Who are they? List all names and groups including authorized delegates who will participate in discussions and Approvals.
2. Can we implement this change in **phases**? Yes

2.1. A large number of Programs are needed; each will have phase 1, 2, 2.1, etc.

3. **Requirements** – a numbered list of *WHAT* needs to happen?

The following is a list of the needs of World Peace as discussed in previous chapters. If I've missed something here through the many edits needed to get this message right, feel free to add to this list as needed.

Needs include:

- A good life
- Family Friendly communities
- Equality of opportunity
- Support for non-income earning members of our families and communities through all phases of youth and old age.
- Freedom from fear and oppression.
- Maslow's Hierarchy of Needs
    - Basic Individual Needs include:
        - Food, Water, Shelter, Healthcare, Security, Family
    - Primary Family Needs
        - Living Space, love, education
    - Secondary Needs of Society
        - Art, music,
        - discovery, exploration, science, wealth

4. **Designs** describe *HOW* new solutions will be implemented. Describe an initial view of a solution.

5. **Acceptance Tests and Pilot Signoff**

6. **Timelines** and Program Structure (which are important if multiple projects are required)

7. **Budget** +- 50%

8. **Scope**

9. **Risks & Mitigations Plan**

9.1. Risk - Stakeholders must commit to timely sign offs.

10. High-Level **Goals, Deliverables and Key Measures of Success**

11. Communication Plan

 11.1. Status Reports, Working Team Meetings, Steering Team Meetings

 11.2. Town Hall meetings and workshops

12. **Working Team** & **Steering Committee** Members

 12.1. Assign a Project Manager

 12.2. More than 2 or 3 Related Projects require a Program Manager as well.

 12.3. Technical Lead

 12.4. Subject Matter Experts (SMEs) need to be assigned by Managers from all affected business sponsors.

 12.5. Program and Project **Dependencies**

  12.5.1. Which other technology or social projects must complete in order before this project can begin.

  12.5.2. Which work packages or sub-projects can begin in anticipation of dependencies

 12.6. **All Stakeholders Sign-off** Charter

13. **Taxonomy** – a fancy name for numbering Needs, Design and Implementation Deliverables so that tracing back decisions can be done easily. There should be a need for every deliverable.

When the initial Charter completes, arrange for a Project Kick-off Meeting to confirm that your Charter assumptions are correct, and then simply make revisions to points in the document based on any in-meeting discussions if needed.

Congratulations – you are started and running in the right direction with your new project.

Being a sponsor and stakeholder for Transition projects is a privilege and responsibility to be taken seriously. Stakeholders must be able to guarantee their ability to participate by reviewing and signing-off as requested within two weeks – or they must delegate a project owner with the authority to approve on their behalf. There are usually just three important signoffs per program and two per individual project.

Project Leads should always introduce meetings with an agenda, and send out written minutes taken during all meetings so that stakeholders can add or clarify discussed requirements or next steps. Meeting minutes frequently document discussions, takeaways, project requirements, and next meeting time.

2. Requirements & Inventory

Requirements Documents describe WHAT everyone's needs are, and it can take up to eight to ten weeks to assemble workshops and discovery meetings, document, and gain sign-offs from stakeholders. Often a separate Requirements document is required for each project in a larger program.

The document should contain the following:

- 2.1. Discovery - WHAT is needed? (each item should be numbered for tracking through to execution)
  - 2.1.1. Document Revision Log
  - 2.1.2. Needs By Stakeholder
  - 2.1.3. High-Level needs such as Goals, Prime Deliverables, and Key Measure of Success
  - 2.1.4. Documentation Detail & Process needs
  - 2.1.5. Business Needs - like budget and ROI (Return on Investment)
  - 2.1.6. Monitoring, Reporting, and Enforcement needs
  - 2.1.7. Operational needs
  - 2.1.8. Security needs
  - 2.1.9. Legal needs
  - 2.1.10. Technology needs
  - 2.1.11. Testing / Piloting needs
  - 2.1.12. Sustainability / Evergreen needs

2.1.13. Facility Needs
2.1.14. Service Level Needs
2.1.14.1. Incidents
2.1.14.1.1. Availability
2.1.14.1.2. Delay
2.1.14.1.3. Accuracy
2.1.14.2. Problem Management
2.1.14.3. Moves, Adds, Changes, plus limits on the number of Change Requests anticipated
2.1.15. Training / Course Requirements
2.1.15.1. for admins
2.1.15.2. for users
2.1.16. Reporting Needs
2.1.17. Documentation, Revision Control, Taxonomy Needs

2.1.18 Create Design Meeting Agendas by listing which Requirements will be discussed in each design meeting, and specify which participants and SMEs will attend each design meeting as well.

2.2. Inventory
2.2.1. People, processes, and technology
2.2.2. Reports, applicable laws, past design documents, etc.

Task: **Sign-off by all Stakeholders**

All needs are documented in detail in the Requirements Document.

As you get better at facilitating this process, and at running your projects successfully, you will begin to realize that Requirements Discovery is the time when you can and should be thinking big and aiming high - without the constraints of the physical universe – sometimes - to limit the things you might like to see.

For example, if you want a chair, or a car, to hover off the floor six inches, or if you need a robot maid, state that need here as a requirement. The project's scientists may have to ask you to modify your requirement due to budget, technical difficulty, or safety constraints after they have conducted a thorough design review, but for now, here in Discovery, this is the time when you must give scientists the opportunity to meet and think about the desired requirement fully. Do not allow SMEs, Subject Matter Experts, to constrain sponsors from stating what is the full requirement.

President John F. Kennedy commissioned the launch of a successful manned mission to the moon despite being told in advance that the mission would require the discovery of numerous new technologies and at least two metal-alloy materials that had not yet been discovered. Again, do not permit constraints and cautions to change your needs and goals. A thorough technical investigation with the right SMEs, can usually find a solution or workaround to most problems.

Important: Always stop the project until all stakeholders sign-off the Requirements document. It is the responsibility of the working team to complete this document; and it is the stakeholders' responsibility to advance the project on to the next step – the Design Phase. Ownership for any project delay lies with the stakeholders during this time. It is perfectly acceptable to require a change request to push out the end date of the project should sign-offs fail to complete within a timely manner.

3. The Design Document

Design meetings and documentation detail HOW needs will be met and often takes eight to ten weeks to gain sign-off.

The contents of the Design Document will include:

- 3.1. Document Revision Log – if required
- 3.2. High-Level Design Overview
- 3.3. Detail Designs –
    - 3.3.1. With one detail design description per requirement
- 3.4. Test Plans - Table Of Contents only here
- 3.5. Detail Implementation Plan with +- 5% final Budget Adjustments

3.6. Design Document **Sign-off by all Stakeholders**

Requirements are not always solvable in the first design meeting. Often breakout meetings are required to solve some of the items on each Design Meeting Agenda created in 2.1.19 above. Give meeting attendees four business days to investigate solutions – with vendors and external experts, and then return to complete all design meeting topics. A second breakout meeting might be needed, but most problems can be solved, or confirmed to be unsolvable, by the second breakout.

An outright failure to meet a requirement in design might be due to constraints of budget, unavailability of subject matter experts, project time considerations, currently known constraints of science, metallurgy, and so on.

When it happens that a need cannot be met at this time, the design team will look at options like a workaround. A phased approach may also be recommended to meet part of the requirement in release 1.0 of the project, and then the full requirement will be planned for delivery in a future release v2.0 or v3.0. A last option would be for the project team to return to the stakeholders and request a compromise to the original requirement altogether.

It would surprise you to see how rare that failures to meet a need are – as President Kennedy realized when we asked his scientists to solve the lunar lander's alloy requirement back in 1962. The Apollo 11 mission flew successfully in 1969 complete with two previously undiscovered alloys that did not exist at the 1962 approval date.

4. Detail Implementation Plan
Build, Test, Pilot, Release

4.1. Build all technology, facilities, processes, test plans, reports, documentation & laws as required per each requirement, goal, and success criteria.
Check online at CSQ1.org for example build Gantt charts and plans describing facilities and other build activities like

houses, mined or farmed resources, manufactured items, and so on.

4.2. Testing

4.2.1. Technology Testing

4.2.1.1. System Integration Tests – rollout the developed system into the staging environments of Development, Test, Performance, and Production

4.2.1.2. User Acceptance Test Plan

4.2.2. Structural Testing & Inspections for Facilities

4.3. User and Administrator Training

4.4. Operational Readiness Checklists

4.5. Pilot 5% of the total rollout.

Always Pilot in Production first – never implement all at once. There is always a way to run a Pilot, so stay in design until a Pilot can be configured.

4.6. Sign-off Pilot after a week or two, and then Authorize Rollout to Operating Teams.

4.7. Release Warrantee - Project Teams usually warrantee delivered work for two or four weeks after deliverables are released to production. After warrantee, project resources can be redistributed to new projects.

4.8. Close the Project by documenting the lessons that you learned during the project. Other teams can benefit from those learnings and might not suffer the same delays that your project team did, based on these lessons.

Signoffs Requested

Signoffs for project documentation is requested as follows:

- Charter – All Stakeholders
- Requirements – All Stakeholders
- Design – All Stakeholders
- Test Plans - Business Sponsors

- Change Management - Pilot & Release to Operations: Ops, Business Sponsors
- Project Close – Project Sponsor Only
- Project Review – Sponsors Only

Unlike the 4 Steps in the World Peace Transition Plan, the steps of a project MUST each complete in sequence before moving forward. Whenever a project manager assembles design meetings, all stakeholders must have already agreed on needs for availability, for example, or the Design Meeting may be a complete waste of everyone's time.

Projects become unproductive when millions are spent on infrastructure that do not meet needs. Project Management Life Cycle best-practice says that Charter is followed by Requirements, followed by Design, Implementation, Test, Training, Pilot, and Rollout.

In a project to develop application screens and forms, iterations of needs, design and test phases may repeat once or twice, and this will often add to the quality of a version 1.0 or 2.0 release. To make change more agile, you may elect to run 60-day or 90-day releases as well, but the PMLC process must maintain this sequence to ensure good quality control.

Project Management Life Cycle (PMLC)

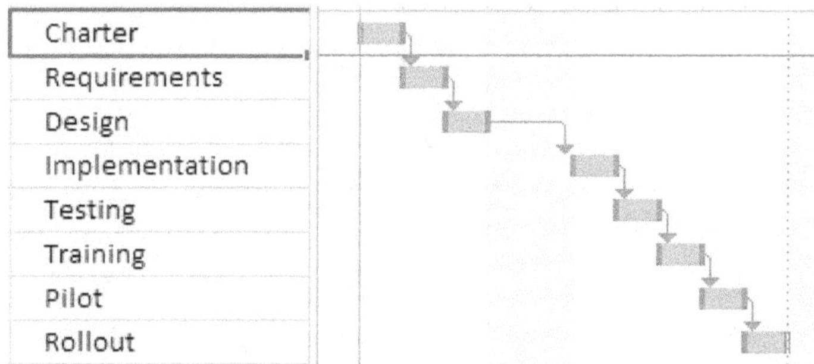

Run Book – Operators Guide

The Run Book is the Operation's guide to maintaining and configuring a sustainable solution with professional service levels while in operation.

What are Service Levels? Service Level Objectives (SLOs) are the needs of the stakeholders for an operational system. SLAs are the formal agreements promised by supporting service vendors and local service providers in support of an operation. Other important Availability KPIs include reliability measures MTTF (Mean Time To Failure), and the MTRs (Mean Time to Repair, Restore, and Recover).

1. **Availability and Reliability** - Can a busy grocery store or heart surgeon, afford to wait 4 hours for their equipment to restart after a failure. Absolutely not – redundancy by backup batteries and backup processes must be available immediately. A modern, busy grocery store might be able to check out customers manually without cash registers, but if staff were not trained by managers, in business continuity procedures, they would have to close their doors almost immediately. Some office systems are only needed Monday to Friday; other systems are needed seven days a week. Maintenance windows and backup windows can impact availability if designed in this way.

2. **Delay** – if you can connect and work with the system, but the system is too slow to keep up with volume and timely needs for information – then it as good as broken in most cases.

3. **Accuracy** – a fast, reliable system is of little value if it is not taking nor giving accurate information.

Here is an example SLA chart for one service:

	SLAs	Level 1 Support	L2	L3	Vendor Support	Cost & Staff
Availability	7/24 5 min down max.	Help Desk Ticket Logs	7/24 Knowledge base Analyst	Architect	5x12 5 hour time to repair; 5 hour time to restore service	5 L1 5 L2 2 L3 Vendor Contracts
Delay	1 sec screen refresh	Ticket	7/24 Knowledge base Analyst	Architect		
Moves/Adds /Changes (MACs)	5-day bug fix; 5-week Feature enhancement – 6 months; Next Release - 6 months	Ticket		Architect & Dev Teams		

The Service Management Industry maintains standard Service Delivery towers which include:

i. **SLAs**
   1. Availability 7/24, 5/8
   2. Performance
   3. Accuracy

ii. **Incident Management**
   1. Call/Ticket Promotion – Level 1->2->3->Vendor Support Agreements and Contact Info
   2. Tech Escalation
   3. Business Escalation

iii. **Problem Management**
   1. RCA (Root Cause Analysis), Process and Approvals

iv. **Change Management**
   1. Password Change
   2. New User
   3. New Server/Service
   4. New Functionality or Bug Fix – Application Development
   5. Maintenance Windows
   6. Change Moratoriums - Month-end/Year-End/Christmas/Peak Business Traffic Times

v. **Configuration/Asset/etc.** – deliverables based on the RFP

vi. **Service Reporting** - SLA and Other

1. Who gets these reports every month?
2. What do these reports detail (samples)?
3. Have we delivered everything that was needed?
4. Support efficiency - I like to see a report that shows what % of incidents could be resolved at Level 2 – less than 80% means documentation does not have sufficient in How-Tos

How-Tos

How-Tos are Operational Procedures that explain the documented Checklists needed to turn most outages or problems back to Running and "Green" Status.

When a project team puts together a solution, they will usually be reassigned to other projects once Operations assumes responsibility for the delivered system. This means that you will not easily find the builders once something fails, sometimes several years later.

You may have seen these documents called installation instructions, owner's manuals, and online service announcements. Mechanics receive very extensive repair manuals and videos that explain how to repair every part of a vehicle or machine.

The best reason for engineers and hi-tech developers, to document their projects well – in explaining what to do in the event of an outage, is that these support calls can often come in the smallest hours of the day, night, and holidays. For this reason, there is the real benefit of not receiving support calls for easy-to-fix problems that could have been resolved easily through documentation. The better the documentation, the fewer are the support calls. Many step-by-step instructions make extensive use of pictures and examples to avoid wordy explanations as well.

Store all How-To Procedures in a Support Knowledgebase where they can easily be found by Help Desk or Call Center Technicians.

How-Tos maintain:

i. Applications – these are SME deliverables
ii. Servers / Shared Services – Server team
iii. Databases
iv. Networks
v. Parts & Equipment
vi. Similar…

Life-cycle Managed Documentation

Each Program contains standard project documents:

1. Charter
2. Requirements / Inventory
3. Detail Design / Build Book
4. Detail Implementation Plan (DIP)
5. Run Book
6. Test Plans
7. Training Docs
8. Project/Program Status

Sub-Projects each get their own Requirements, Design, DIP, and Run Book documents (unless the sub-projects are quite small and can be grouped together), and all sub-projects within a program share one Charter.

All project documents are version controlled, and taxonomy is consistent within each project so that needs in requirements share numbering with their corresponding detail designs. This is for trace back purposes so that every design implemented can be traced back to a sponsored requirement.

Some collected needs will be addressed in future releases (2.0, 3.0, etc.) and so release version numberings are recommended – such as 1.2. 1.3…1.25, 2.0, 2.1, 3.1, and so on. Technology release management windows are often scheduled twice annually, and social project policy might be revisited every 3 to 5 years.

Sub-project documents might be versioned Requirements 1-1.1, 1-

2.1, 2-1.1 and so on – where:

- 2-1.1 is the 2nd phase of sub-project 1 version 1
- 3-6.20 is the 3rd phase of sub-project 6 version 20
- 1-6^2.1 is the 3rd phase of sub-project 2 of sub 6- v1

A Robotic Hand - Example Project

A Robotic Hand and Arm is an example of a Program with many sub-projects.

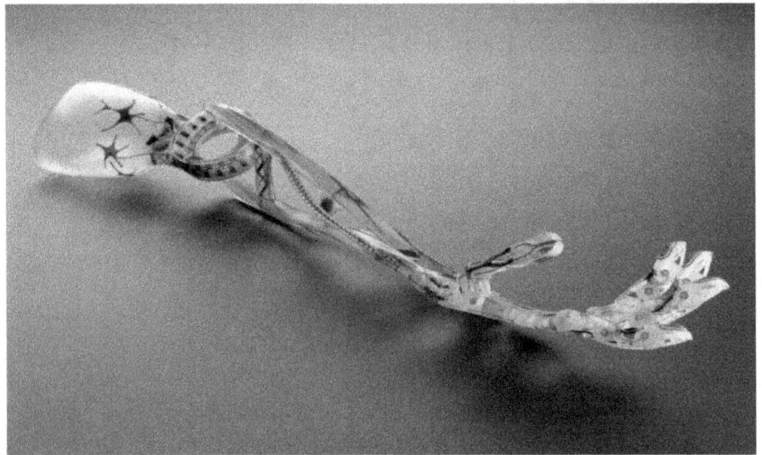

Sub-project 1 - might be a finger; Project 2 - a clasping opposable thumb; Project 3 – tools, welder, lifters, etc.; Project 4 – a hand and wrist; Project 5- elbow or elbows, Project 6 – Logic, Project 7 – Hand-off to next station or tool.

Consider that the finger may have sub-projects too – for pressure system, guidance systems. Needs like the surface friction, grip and so on of the finger might be just needs of the original finger projects, or they might need their own sub-projects for materials testing.

No matter the complexity, the project simply get broken out and then they all run the same basic PMLC Process (Project Management Life Cycle).

Performance Management Dashboards

Monitor performance of build and operations using Project and Service Management Dashboards.

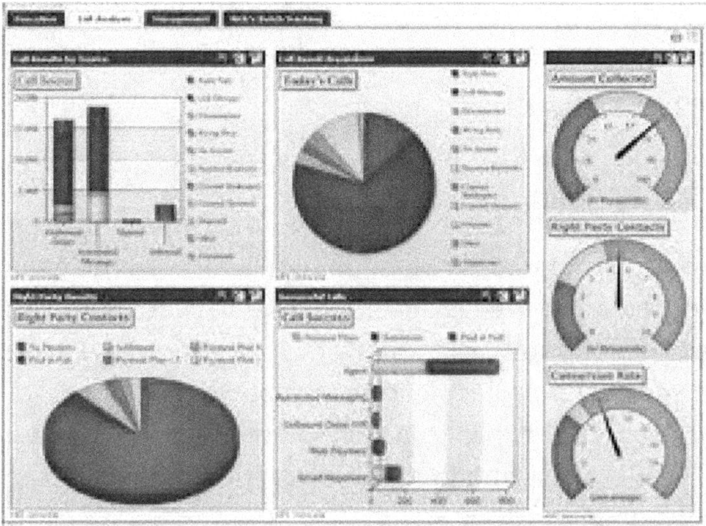

Prefer Dashboards to static reports as Dashboards add "Drill-down" and "What-if Decision Support" capabilities to performance data. If you see a Red Status, you can simply click and drill down to understand as much detail as you need.

These tools should be made available to all users and managers to ensure that the benefits, SLAs, and exceptions, of services, are well understood and communicated.

For purposes of your deliverables, monitor the KPIs of your solution on a regular basis and run regular dashboard reporting during the development of the project. Exception Alerts should focus the attention of stakeholders on issues if they occur.

Go back a year or so after each project ends - just to confirm that the costs and benefits of the projects' business case were realized properly. If yes, the project was a Success; if no, the project was a Failure in that release.

For Progress, you must be easily able to navigate through

1. Budget – approvals and funds happen quickly, and teams start work based on resource availability and not financial process delays. Corruptions in the system could arise, and the occasional mad-scientist may need to be derailed from accidentally threatening the planet – but these are fun examples extreme and easily mitigatable risks.
2. Project Status Reporting

   The working team should meet on a weekly basis and track progress against the overall plan.

   A Summary Report review meeting forces the team to look forward a few weeks and anticipate deliverables that need to get resolved now. The report itself does not have to be formal nor perfect if project leads can speak to it in Program Review meetings, but is has to be completed well in very large PMOs were PMs for hundreds of projects could never practically attend one meeting.

   The following example reports are suitable for a very large programs, like World Peace, but some teams rely on just the Executive Format for smaller portfolio review meetings where project leads can attend meetings and speak to questions directly.

   Emphasis is on the quality of information and, therefore, perfect formatting is a distant second priority.

Infrastructure Refresh – August 16th

Project Status

Initial planning to recover a 4 week delay in IBM equipment indicates that we can still implement the project by end of October.
Escalated Issues: Originally, we had hoped to accomplish the refresh without downtime. Unfortunately, **1.** the new IBM SAN required a BIOS upgrade on the switches for all Tier 1 and 2 servers; **and 2**. an SQL Server open issue required the shutdown of connected applications prior to switching to the new SAN.

	This Week	Last Week
Schedule	Y	R
Scope	G	G

Activities Overview

Server	Downtime	This Week	Next Week	Apps Affected
Tonga	1 downtime Sat 19th 11 pm to 3 am	19th – re-Zoning to new SAM	21st – Mirroring 22nd – Recabling to new switch	IRIS, Visifax
Belgium		14th - Copy / Restore 19th - Rezoning	22nd – Recabling to new switch	ICFS Custom Feed

Sub-Project Status Details

#	Prio	Project	Start Date	End Date	Comments	PM	Former Status	Current Status
1	A	Infrastructure	?	Aug 1	We are Red this week and last due to a 4 week delay in the arrival of IBM Equipment. We recommend acceptance of a project change request to adjust out the end date of infrastructure to Sept 1st. Acceptance of this new delivery date will turn this status Green.	Ed T.	Red	Red
2	A	Application Migration	?	Oct 31	Compressing 3 months of migration to 2 months adds risk but can be done : a) Database team work – average 8 db per week for 8 consecutive weeks b) Application teams – many resources will have to adjust to db team changes quickly	Greg S	Green	Yellow

Other Planned Next Week

- Finalize detail application migration plan - Due Aug 25th

Accomplished last week

- Haiti, Congo, Cuba and Canada are Migrated.

3. Program Status Reporting

World Peace Agenda - Canada
EXECUTIVE PROGRAM STATUS
December 2015

Status Options:

Green	Yellow	Red	TBD

ESCALATION ITEMS

#	*PROJECT / ITEM*	*Due*	*Comments*	*Former Status*	*Current Status*
1	PTE Readiness	Nov 18	Shakedown/Application Testing/Performance	Red	Red
2	Systems Management	Nov 18	Clean-up complete Application monitoring	Yellow	Red
3	Operational Readiness	Nov 18	Staff hiring, application knowledge transfer	Yellow	Red
4	Service Level Agreement	Nov 18	Interim SLA not accepted	Red	Red

PROJECTS & MAJOR ACTIVITIES (Non Escalation)

#	*PROJECT / ITEM*	*Due*	*Comments*	*Former Status*	*Current Status*
1	Cold Fusion	Nov 18	Shakedown/Application Testing/Performance	Green	Green
2	Rapid Charge Batteries	Nov 18	Clean-up complete Application monitoring	Green	Green
3	Pipelines – Oil & Gas	Oct 18	Staff hiring, application knowledge transfer	Green	Yellow
4	3D Food Printers	Nov 18	Interim SLA not accepted	Green	Green
5	3D Clothing	Nov 18		Green	Green
6	Tricorder	Nov 18		Green	Green
7	,,,			TBD	TBD

Also, consider adding summary info for:

- Project Manager Name
- % Complete Phase (Charter, Design, etc.)
- % Complete Total Project
- Accomplished Last Period
- Planned But Not Accomplished, and
- Accomplished but not Planned
- Risk Mitigations
- Project Issues

Status Options and Stakeholder Duties

A good Project Team always gives stakeholders a chance to correct problems before they slow down or stop the forward progress of a project.

As long as the team gives a Yellow status warning of problems looming, Stakeholders and Owners should look at Red flags as a failing of Management to provide needs of the team - as the teams needed them. When projects are green, all is well: Job well done by all. When the team hits a Red Issue that is beyond their control, the Sponsor and Stakeholders are responsible for solving problems quickly for the team.

In this way, hundreds and thousands of project teams can progress well transparently and collaboratively.

Sharing Knowledge

Individuals with a great deal of broad knowledge and capability tend to share information freely and collaboratively. They acknowledge their mistakes and opportunities for improvement easily, whenever they do not know something too. You will never be able to learn all of their knowledge and thereby reduce their value, no matter how many discussions you have with them.

Those who know little will sometimes guard that little bit they know, and perhaps even deny access to knowledge when they feel like they might not have control nor a role to play after that. These folks feel as though they would not be indispensable if they revealed that little thing they do know or control.

Never be a roadblock of information, nor permit a roadblock. Share knowledge and surround yourself with people who are smarter than you are – it rubs off.

As you proceed, report your progress well. Teams showing poor KPI performance should have a chance to appeal measures. By constantly improving the KPIs used to measure, the project builds improve as well. Measures of performance in the World Peace Agenda are transparent, straightforward, reported centrally, and are planned to be viewable worldwide on the web as well.

Policies and Projects that support the Transition Plan for World Peace are tracked in each country's World Peace Agenda. Reporting the status of these Projects in a consistently way, helps neighboring nations to compare progress, to collaborate, and to share resources and "Lessons Learned" easily.

Considerations for national security, privacy, public safety, copyright, patents, and accuracy are factored in so that minutia and detail don't sideline forward progress.

Democratic oversights like these have been working at the United Nations and the Vatican through historical bright spots - and admittedly not-so-bright spots. Chief among these lessons, is that experienced builders and engineers are effective leaders of progress – and should be called upon to do so again now.

Everything is Solvable

What is your process? What was your performance? How will you improve progress going forward? If you can't answer these questions superlatively and confidently, and show a stellar resume of progress as well, have the integrity to take a learning role for a bit until you can.

As for resource references; superb candidates rarely have so many references as the career salespeople in the crowd. Alan Turing was fairly loathed by his administrator bosses at Bletchley Park for one example.

A high-performing candidate that does brilliant work may not be well-liked and may even be thought of as very rude - unless he has authority. Once given authority, then all is well in the eyes of his or her subordinates. Britain's Prime Minister, Winston Churchill himself, had to give Mr. Turing top authority above credentialed and decorated administrators – and Alan could never have invented a working installation of the first Computer without hundreds of people supporting his work either.

This is not to say that high-performers are entitled to abuse peers and sub-ordinates. Also, high-performers without experience are not leaders either. High performers must be good bosses in keeping with the authority that they ascend to more quickly.

Alan realized that an automatable machine that could solve any problem was necessary. Given the resources, he made a machine that could solve anything and then he used it to solve a specific puzzle. Soon, engineers will build onto his universal computer, the ability to solve all the other puzzles needed to automate our production economies in just the same way.

The reasons that high-performers and geniuses seldom get good references include: i) the work and contribution is their first priority – so insecure administrators and bosses that need their ego supported, do not get what they need; and ii) human nature marginalizes genius among our peers in even our smartest gatherings.

Humans marginalize their geniuses. Spectacular examples include Alan Turing (Inventor of the binary computer) – whose chemical castration led to his death at age forty-one, Albert Einstein (Relativity, Atomic Energy) – was relegated to a patent clerk's back office for two decades after his discoveries, and Fleishman and Pons (Cold Fusion) – were outcast for twenty-five years until a next generation proved their that work was indeed accurate and repeatable. The TV Series Survivor showcased this behavior weekly as the smartest and most talented were routinely targeted as the first to go.

Embarrassing treatment of potential by academia, human resource heuristics software, and administrators within government, business and society is endemic and takes a targeted process and effort to overcome. “Hire slowly” and “team fit” are examples of this problem.

Managers who attract, nurture and produce the most high-performers within their team, need to be given authority, and managers who have no high-performers that could even replace them in their own middle-management job, should be given only limited or no responsibility over hiring and no responsibility over others.

As a Leader, who wants to accomplish great things, you want to get this right. If you are a voter, you want to see that the right plan, that plan that targets a Good Life and can even lead on to World Peace ultimately, is in place - and that performance measures for progress are met.

Look for leaders who celebrate the advances of their brilliant teams through the plan, and you are onto a winner who deserves your vote.

Funding Technology Projects

There are a few additional reasons for the unusual request for funding diversity in Technology Projects: first, some highly creative and productive thinkers work well collaboratively – and others prefer the ability to lead teams individually. First, collaborating with a University can be extremely frustrating, and is fairly unworkable when scope spans multiple faculties of science, economics, social studies and others. Second, academics are brutal on their stars and unconstrained thinkers. There is also an unfortunate tendency in many democratic settings to suffer what I sometimes think of as “Survivor Syndrome”; a phenomena showcased by the TV Series that I mentioned above, where the herd dumbs itself down as a defensive protection to being voted off the island.

In Canada, having the experience of running six hi-tech startups myself, I can assure that academic, public and even big-technology firm incubators have this exact challenge due to their leadership and administration. 99% of startups fail as a result, and none keep KPIs on companies assisted. So I can say authoritatively that this either needs fixed immediately or an alternate form of financial support is a non-negotiable must-have.

The Productive Complaint

This is not a rant nor an unfair and unproductive complaint by the way. I've described here, 70% of leadership of the technology departments of major brick and mortar corporations as seen through the eyes of a senior C-level and hands-on consultant and Hi-tech startup CEO over these past ten years.

I have used as many descriptions of shortcomings as I could possibly think of in an effort to give senior management the KPIs needed to recognize and manage against unproductive behaviors and to encourage teams to get better too.

Why is this important? Because competitors worldwide absolutely are working diligently to get this right; to be more productive and let their best and brightest rise to levels appropriate to their interests and ability.

Chandran Rajaratnam was my CEO at SoftChoice in 2001. Chandran was without equal when it came to putting in place smart processes to find and encourage top talent to ascend to the top ranks of the company. Not surprisingly, corporate performance here was nothing less than stellar during his time as a direct result.

Corporate Stewardship Cautions

Within technology organizations, a relatively small percentage of high-performers and team leads have a broad knowledge of all towers of development, infrastructure, operations, architecture, and business transformation. Engineers take on complex automation work in an immature and rapidly changing engineering area, and it takes strong processes to ensure a productive, professional workplace in which all can contribute, thrive and ascend in knowledge, seniority, and salary. Good technology leaders cross-train and cross-assign staff to all towers over years, in order to give them broad experience within each of the SME towers. Good managers will also try to advance high-performers more quickly.

At the other end of the productivity spectrum, there are corporate stewards who make themselves indispensable by hiring teams incapable of taking their management jobs from them. They stack boardrooms with subordinates or their "network" of friends and relations.

Stewards who abuse their position do not disclose conflicts of interest when voting; they overstep through their collective bargaining power; and vote to gain inequity of privilege. They say things like "you don't need to be an expert, SME, or technical to manage this"; their voice is often the voice to stop progress because "That will cost too much" or "We don't do it that way here".

Bullying; non-transparency; slow hiring; everyday there is an unanticipated alarm, these leads offshore and outsource irresponsibly, they see resources as a liability and employment is a privilege; and they rarely understand which KPIs to measure.

The worst of these managers will fire good resources quickly, whether initially once intimidated during the interview process, or later - on the job when that worker cannot or will not cow tail in day-to-day work life.

Recalling the work of Dr. Mayer of Oxford University (see Chapter 12), our society is in most peril when these managers get an MBA. An MBA is often a vehicle from which non-expert administrators can be given an express pass to legitimacy, and also the keys to the executive suite; often propelling them forwarded to undermine morale by leading teams in areas that they know almost nothing about.

Failure rates at major engineering universities are 20% to 30%

despite the most rigorous screening of any faculty save medicine. I have never heard of an eMBA that did not graduate and yet MBAs are often thrust into roles that manage Technology or Engineering staff - even within technology and engineering organizations, such as major airports, and similar.

From our discussion of Business Accountability in Chapter 12, an administrator-heavy organizational design gives strongest voice to those most able to externalize social costs in complete indifference to CSR. Boards of Directors too, are often structured with heavier admin and finance votes.

This is all to say, that my observations both professional and intuitive, warn against traditional Corporate Stewards having involvement with projects in the World Peace Agenda. If such a group were to step up to volunteer or be assigned, insist upon seeing their mitigations against all of these points before permitting distraction by their participation. Always insist that a hi-tech engineer become that organization's primary liaison to the WPA Committee.

It often takes a strong engineer at the helm, like an Elon Musk, Steve Jobs, Bill Gates, or in the case of the driverless car, Larry Page at Google, Sergey Brin, Lee Kun-Hee, and I count myself a good lead and as you are reading this book, you may have some these character traits as well.

I have hovered here for two pages in explanation already, but I will add a final example of how pervasive and endemic is this problem that warrants so much of my time here.

Let's suppose that a corporate executive was hired onto a major company almost twenty years ago. He or she went on to fire and off shore tens of thousands of jobs to cheaper, inferior engineering teams elsewhere. All socially harmful actions and events were documented credibly by credible organizations. Workers' salaries were cut in excess of $100 million while his salary increased by as much. When other CEOs took pay cuts, he took huge increases.

Fired in disgrace, with tens of millions in parachute payments - this individual is counted among one of the worst CEOs in history by several authors. In his last loss while running for a political office, the loss was accounted to he or she being potentially the most disliked person in the public eye in modern times.

If this person were running for a major election in the U.S., would you know who I was intimating and would this knowledge influence your decision to vote for or against them? Someone sponsored this individual for consideration as a candidate as well. Two fine examples of a total, unabashed failing of accountability toward sociopathic behavior in business, and within our society at its highest levels.

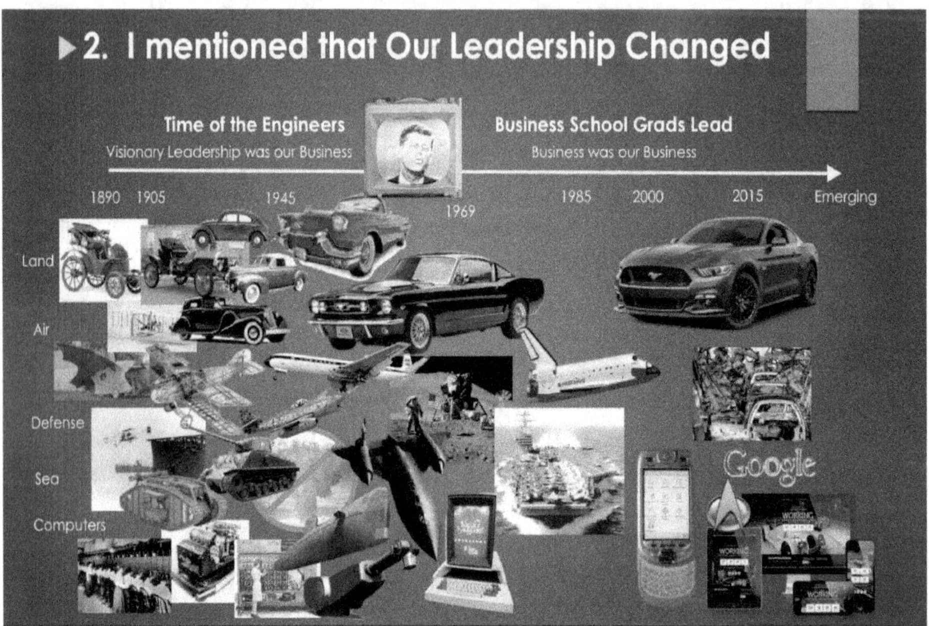

The forward progress of technology and other sciences has either crawled along, or halted in many cases, for forty-five years within the G8's major corporations. China's major cities Shanghai and Singapore, and businesses - while not perfect, are a generation ahead of those in the West in many regards and are gaining.

Realizing this, the next steps for C-Level management teams is to encourage and collect legitimate complaints, to analyze these concerns until inefficiencies are understood. In most cases, senior technology leads will replace accounting leads in engineering departments; in other cases a social accountability will be added to the acceptance criteria of major projects and management bonuses.

Chapter 18 - Diligence not Fear

In any discussion of peacekeeping and security, we are going to want to recall stories from the world's news services of the global crises that drift in and out of headlines either annually, weekly, or even daily that *might* affect our lives. Profit-driven news networks are under constant pressure to keep viewership and readership high so that news from around the globe can be piped into our homes in a steady stream.

Great pains were taken in the early years of broadcast news reporting to ensure the validity of facts presented and to protect a naïve public who assumed that what they saw on television, or read in the newspapers, was true and supported in fact. In recent years, a mature viewing audience has come to realize that everything on the television and news report is not to be believed – and yet, the sale of news must continue. In the UK, the five main newspapers are each owned by a different billionaire – and the same business forces are true for major news networks in North America by and large.

Peoples' interest in news is much more intense when there is a perceived threat to their way of life. When things are good, people are less interested in the world news but fear and poverty stimulate greater interest in the news. So notes researcher Roy Greenslade in his Sept 2007 article "The good news about bad news. It sells" (Greenslade, 2007).

Last week, a news report in our local newspaper mentioned that there might be a Soviet tank in Syria. The Russians are singled out this year in numerous accounts of stories that spin a fearful "web of intrigue" as Putin deals with what could as easily be a rogue state's sabre-rattling with nuclear weapons.

Nothing was mentioned of Mr. Putin's level-headed and consistently strategic responses to attacks both financial and military by the rest of the G20 nations in this article. Americans, Canadians, and other nations were kept fearful of a situation that had little to do with them and likely never will.

ISIS, a creation of failed security infrastructures within Iraq after the Iraq War, fills our news in 2015 with reports that these armed forces are "on their way". Scary stuff right? Well, not so much really; you have a much better chance of buying the only winning ticket in a major lottery – but fearful stories sell newspapers, and as I mentioned above, media is a business.

A FOX-News report came later and clarified that as part of President Obama's direct efforts to "decomplexify" reporting, the situation was perfectly fine in fact. U.S. and Russian forces were working cooperatively from air and land to route militants from Syria. What a breath of fresh air. Previous to this report, members of a breakfast that I had attended were very concerned about World War III boiling over.

Should we be fearful of other religions? Of course not. We should protect the good values of our society, the greater good for our communities and families, and mitigate security threats by terrorists including our own media-generated fear-mongering.

Mitigating Security Risks

Fear is not a motivator for professional security professionals and engineers asked to stand on guard against threats on a regular basis. It is true that there are many threats, but the professional response to these threats is to mitigate them with a countermeasure that eliminates or minimizes impact should a threat come to happen in real life.

Once the unthinkably horrible threat of a passenger jet being flown into three buildings was realized in 2001, mitigations were implemented on every plane flying commercially, both immediately and within the next six months, as needed to mitigate the risk of similar future hijacking situations ever happening again. But this did not prevent a fearful public from gobbling up every news article for the next eight years – and even supporting an Iraq War, which was never a measured response to the attack and later proven to have - in highest probability, had nothing to do with the attacks of Sept 11th.

Diligence is the role of security forces within our society – and make no mistake that these professionals mitigate threats by the hundreds and thousands annually.

The numerous, well-documented calls by news writers to prefer responsible and balanced reporting amounts to a waste of time, and even career suicide, as in most cases our viewing audiences greatly prefer to know what has gone wrong far more than what has gone right. We prefer to be fearful.

Civilization's great philosophers, great thinkers, and architects of rules of law discussed appropriate responses to attack for thousands of years. "Measured responses," "eye-for-an-eye," and other carefully discussed approaches are used by security professionals today just the same. These are constructs suggested by Plato, Aristotle, Socrates, Cicero and other builders of our modern universities, societies, governments, sciences, laws and ethics. The Code of Hammurabi had an eye-for-an-eye policy that differed by class in 1750 BC.

The very great value of education, reading, and the study of lessons learned in history is that without these references - we start all over, and perhaps even set back and diminish our societies.

Responding to Terrorism

The largest single night of casualties due to Terrorism in France since World War II took place on November 13th, 2015. The death toll of that terrible night was 160 at the time of this writing. September 11, 2001's New York City death toll from terror attack was 2,997. Statistics on terror are kept by the U.S. Department of State.

Seeking facts on Terrorism provides an understanding sufficient to consider what are appropriate, and inappropriate, Measured Responses.

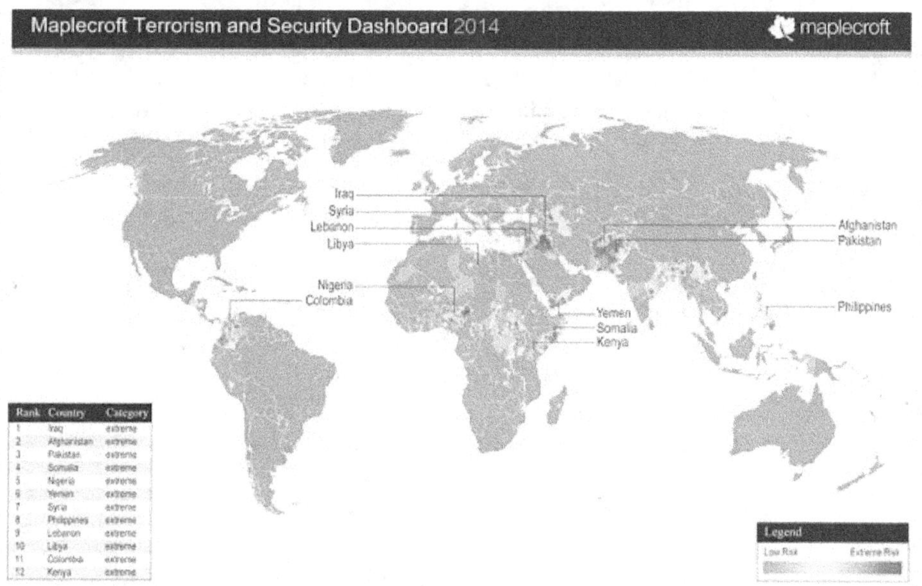

(McElroy, 2014)

2014 Highest Terrorism	Total Attacks	Total Killed	Total Wounded
Iraq	3370	9929	15137
Pakistan	1821	1757	2837
Afghanistan	1591	4505	4699
India	763	426	643
Nigeria	662	7512	2246
Syria	232	1698	1473

Year	Total Attacks	People Killed	People Injured	People Kidnapped
2007	14,415	22,720	44,103	4,980
2008	11,663	15,709	33,901	4,680
2009	10,968	15,311	32,660	10,749
2010	11,641	13,193	30,684	6,051
2011	10,283	12,533	25,903	5,554
2012	6,771	11,098	21,652	1,283
2013	9,707	17,891	32,577	2,990
2014	13,463	32,727	34,791	9,428

(CounterTerrorism, 2014)

Tactics Used in Terrorist Attacks 2014 & 2012

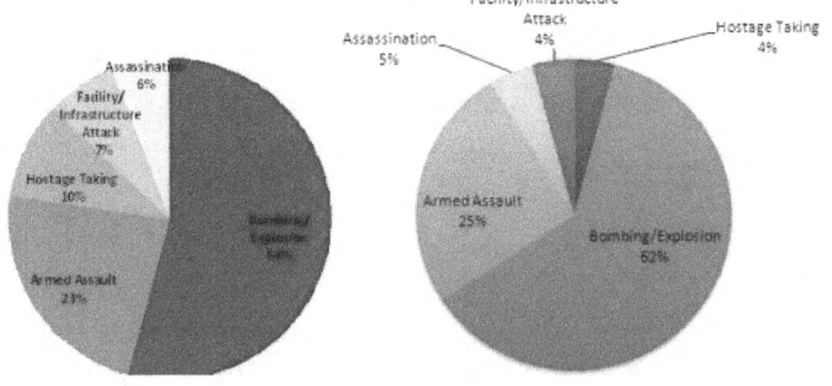

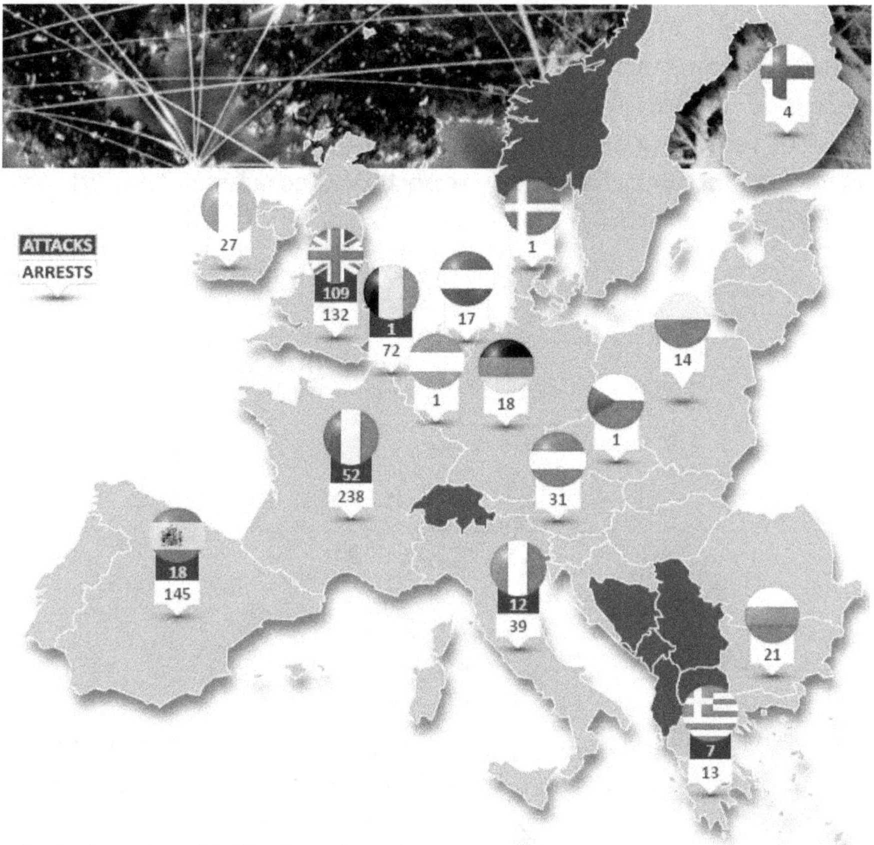

Figure 2: Terrorist attacks and arrests in the EU in 2014

(Europol, 2015)

The reasoning behind attacks.

Data source: https://www.europol.europa.eu/latest_publications/37

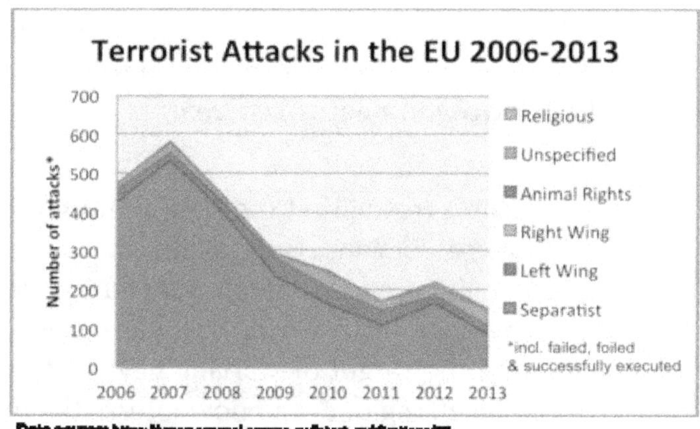

Terrorist cells and attacks require a cause, funding, armaments, security, communications, and access to an opportunity to be successful. Opportunity for terrorists came in the form of rushed refugee screening in France's humanitarian response to Syria's urgent need for relief and early reports also concluded that attackers entered France through Belgium.

2015 Gun Violence Archive in the US	
Total Number of Incidents	45,575
Number of Deaths	**12,300**
Number of Injuries	23,351
Number of Children (age 0-11) Killed/Injured	609
Number of Teens (age 12-17) Killed/Injured	2,285
Mass Shooting	309
Officer Involved Shooting	3,815
Home Invasion	1,951
Defensive Use	1,049
Accidental Shooting	1,636

Gun Lobbies in the U.S. asked, and some states even mandated, that Americans arm themselves years ago against similar threats of violence. The solution was a failed one as seen in the chart above (Archive, 2015).

Statistics reveal that U.S. deaths by these guns are much higher annually (12,300 so far in 2015) than the highest per country Terrorism death rates in the world. Iraq has the highest number of Deaths by Terrorism annually; in 2014 there were 9,920 killed. Worldwide, 25 U.S. non-combatants died from a terrorist attack in 2010. Japan had just two gun deaths last year in a country of 127 million people.

Before there were Muslims (recall that the Muslim faith is only here since 700 AD), there were Persians, Germanics, and others groups that attacked the Roman Empire consistently over hundreds of years. The problem persists through history today because islands of well-fed haves will always be the target of starving have-nots struggling to feed themselves on local, regional and now international stage.

Terrorist Militias make up a small percentage of a community. Muslim non-combatants in Iraq are terrorized by a murder rate that is ten times higher than the U.S. murder rate.

People are people; by nature - and also by nurture for the most part. People in Iraq, Syria and Somalia are educated in a way similar to Christian Americans. Fathers in the Bible belt are the head of their households and try to do the honorable thing by raising good families and educating their children not to question wisdom taught by their elders, leaders or Church. Many Americans fought against the Suffragette movement eighty years ago; and 300,000 Americans died in part, in defense of anti-slavery laws 100 years before that.

Jumping to the 100 km view of the planet mentioned in Chapter 1, many of our regions sit at different levels with their minimum levels of human rights, and this is true among the G7 countries where even America could be seen to have more challenges than other countries with high homeless rates, veteran unemployment, incarceration rates five times higher than the next highest G7 country, and the highest population of people in the world, 80%, who believe in Angels.

Even the word "Jihad" – to strive - is misunderstood and misused. The term "Jihad" was a concept to defend Good, but was hijacked by political and religious groups over the centuries to justify various forms of violence (KABBANI, 2015)(Strobel, 2010). "Jihad" is a non-violent concept that can be defended in legal, diplomatic, and economic sanctions to deliver a just outcome. If the use of force is required, female non-combatants must not be harmed, and peaceful overtures must be accepted. The word for war is "al-harb".

War too - is not a Measured Response. The Iraq War killed one million mostly non-combatants, made a handful of Oil companies and countries very modestly richer, and it also created ISIS by removing the policing infrastructure that managed militias within Iraq.

The French attack on ISIS training centers and armaments in the days that followed the Paris assault *was* a Measured Response.

ISIS is a militia of mercenaries, and relatively few in Syria, Iraq, and

Somalia are Holy Warriors. Following the money - answers important questions about who are the parties responsible for paying these mercenary salaries. When ISIS took Syrian Oil fields, who bought the oil? Proceeds from captured Syrian Oil and support by a terrorized public pumped millions of dollars into a war chest that allowed them to buy weapons. Who sold them weapons? Looting yields little usually, so which are the bank accounts from which paychecks and cash originated?

Putting an End to Terrorism

Young men get on planes from the west to join this Anarchic lifestyle because they do not feel as though they are important at home. They cannot start lives, families, find a home, food, income and other basics of life.

These problems can, however, be fixed by social projects here at home. When solving social projects, consider the following:

1. Feed a man a fish - and he can eat for a day. This is the Top-Down example project discussed in Chapter 1. It works, but benefits are often unsustainable over time.
2. Teach a man to fish, and he will eat for a lifetime. This is a Bottom-Up plan that ensures self-sufficiency and food for a family for a lifetime.
3. Build automated tools that deliver fish automatically to everyone. This plan is a Sustainable Bottom-Up plan because he will have food and he will not be a "have" among "have-nots". I mentioned in Chapter 6 that Capitalism could provide this as well, but that wealth distribution between countries requires a solid World Peace Agenda, Peace, and solid Transition Economics to make work sustainably.

First Step is to allow security forces to fix the top-down problem - as they are already working on through Measured Responses. Next, as a community, we need to get to work fixing the bottom-up projects that are detailed in Chapter 6 and 7 above, and begin automating our production economies quickly. These are all straightforward projects to run and would solve the problems once and for all.

That is what good leadership looks like; no fear, and no confusion about what are the next right steps. It's the reason that I attend meetings and write books training others to build worthwhile projects.

Instead of hearing media fear-mongering that pushes us closer to a WW3 and the end of mankind, give our engineers and communities the support to start a World Peace Agenda and worthwhile projects - to "Jihad" – to strive forward.

My deepest sympathies extend to the families of victims of terror. To the 160 killed this week in France, and to the 2,997 killed Sept 11, 2001. In Iraq, 9900 are killed annually by terrorists, in America 12,000 are killed with guns every year, and to the many more families who were born into a part of the world that does not permit enough security to know if they or their families will awake in the morning.

Taking away the root cause of Terrorism is in the providing a Good Life of importance and meaning, filled with worthwhile Top-down and Bottom-Up projects, and strong families for all. The "How" - the process that accomplishes these projects - is described in Chapter 17 and after you run a few projects successfully, you too may come to agree with the approach.

Global Crises

A wide variety of example crises fills our news reports, distract government planning, and keep young people from seeing a positive outlook and clear plan for a brighter future too. This section is from the CSQ 101 course "News vs. Reality".

Last year, the "Big" topic was Global Warming, a few years before that it was the stock markets crashing, next there was war in Iraq, and this year - the big stories are of the takeover in the Ukraine, Syria, Turkey and our Growing Global Populations.

Every year seems to bring one or two new theme-stories that keep newspapers selling and the public in an emotional state of mind. First the reports appear in the news, and then they also pop up in our movie screens, television sets, and in popular reading too. At some point, the stories may well have an underpinning of truth to them, but the overwhelming majority lack credulity for the most part.

Global Warming

Week to week, the discussion of global warming evaporates and returns - for years now. Or the crisis of rising tides? Record thaws followed by record freeze winters. Northern Ice Cap is shrinking as the South Caps Increase.

Between $900 million and $1.4 billion dollars in research grants are given to Global Warming Research in North America in 2014. Governments provided the biggest percentage by far, but unlike funds forwarded by energy companies that supported both views of the discussion, government funding only supported research in support of a Global Warming outcome.

Not surprisingly we heard quite a lot from the funded agenda – and almost nothing from the other. This summer, concern for the impacts of Global Warming were largely debunked as almost a hoax.

Pollution should always be minimized but as to the notion that man has any more impact beyond the normal trends of the earth's natural warming and cooling cycles is uncertain. I think it's safe to say that the media and specific interests in science benefit from research more than the public at this time.

This was a story that sold books, newspapers, and rallied awareness for important air quality and ozone issue discussions. More importantly, it taught us to question what we hear in politics and in the news.

A Global Energy Crisis

Any oil researcher that I have ever met has told me pointedly that there are enough already discovered oil reserves to supply the needs of the planet for 150 years or more.

Advances in Nuclear Fission and current radiation management best practices appear to be well capable of creating more than enough safe energy to supply our planet as well. Statistics of survivors at Hiroshima and Nagasaki showed radiation-induced cancer rates roughly equal to those of smoking's 20% mortality rate, so clearly radiation too is a manageable problem.

Environmentally friendly alternatives to Oil, like Audi's Blue-Crude, Coal-based Paraffin Oils and Biodiesels are potentially able to stretch or even replace our fossil fuel dependency entirely over time. Advances in battery technology are showcased on modern Tesla vehicles and in rapid-charge technologies coming online in 2016. LED energy-saving technology is now in house-lights that consume just 10% of the power required by equivalent incandescent light-bulbs.

A German Fusion Reactor based on magnetic Plasma stream containment has completed successful startup testing in December 2015 (Andrews, 2015). Cold Fusion too has been discovered to be a reliable science and the energy source and was ahead of its 2023 target schedule to begin energy production on a community scale.

The biggest obstacle to switching from Oil to other cleaner and more efficient fuels appears to be Oil Industry investment. Even the Gates foundation was found to have huge investments in fossil fuel companies in March of 2015; as did University of California, and a surprising array of other educational and technology interests.

This reporting isn't an invitation to abuse power, nor to overtly expand your carbon footprint, but as a citizens of the globe you can feel comfortable in assuming that global panics are sponsored by special interests that stand to profit from the panic. We are best served when our leaders methodically work through energy issues and solve problems as they arise.

The Global Energy Summit in Paris in the fall of 2015 resulted in a agreement by all participating countries to switch from fossil fuels altogether by no later than 2070.

An alarming environmental event took place in Canada's remote north this past summer of 2015 when four million barrels of crude oil were found to have spilled from a pipeline. The spill happened slowly without detection over months due to its remote location, but it created an environmental disaster in the area and a very expensive cleanup that could have been avoided by simply monitoring oil in, and oil out, at both ends of the pipeline. Pipelines rust, they are vandalized, things happen; and without an alarmed warning system that alerted engineers when the first 20 barrels were spilled, this was surely going to happen.

The Environment Minister's response to this glaring incompetence was "these things happen". These things don't happen actually, and clearly she had no training nor work experience in engineering to draw from before making her government's public position known on the disaster.

Fortunately for officials, and unfortunately for Canadians, the nation is too worried about their jobs, to rally a proper complaint about the public's faith misplaced in an administrator of important, dangerous resources – who has no business being in charge of anything of the sort.

Bench strength - is the measure of how well-qualified are political leaders. The Oil Pipeline spill is an especially obvious example of why I repeat the importance of voting for engineers, and career builders of complex systems, frequently in support of a nation's World Peace Agenda. Always put engineers and process-based problem solvers into elected office with your next vote.

World Population Control

In this year's popular news, and in a soon-to-be-movie book by Dan Brown, discussion of World Population controls are in the headlines. I enjoyed the Da Vinci Code's brilliant story of contemporary problem solving leveraging historically accurate clues in symbology - and so I'm sure that his next fiction "Inferno", will be compelling as well.

There certainly is a process to figure out a solution for population growth - and that solution has much more to do with good planning and management than desperate acts and round-about discussions of wars and culls, so let's take a few minutes here to talk about

Population Control.

One of my favorite reads was a book called, "Collapse: How Societies Choose to Fail or Succeed" by Doctor Jarod Diamond.

In his two excellent books, Dr. Diamond selected a handful of collapsed civilizations from around the globe, with some examples very close to home. Montana, for one example, collapsed under changes in climate, drought, and by pollution left from extensive mining and chemical contaminants that continue to leak into its rivers and crumbling dams today. Its resources became exhausted and the society and industry could no longer sustain their living there in any great numbers after costs of pickup trucks changed from two head of cattle forty years ago to twenty or thirty cows in modern times.

Civilization continues of course, it simply moves elsewhere, but the underpinning notion of collapse is that a society's resources are fixed, finite, and not necessarily renewable either.

For another example, islanders in Tahiti and neighboring Pacific islands were at risk of extinction due to overcrowding at several points in their history. The largest men in society were believed to have more Manu (God) and, as a direct result were given the power to lead. In this way, the search for the protein needed to make men larger became a very high order of priority for many parents.

Before the introduction of livestock from the West in the 1800s, Island inhabitants throughout the Pacific have been documented to resort to cannibalism in order avert starvation as well. Modern day Samoans are examples of selective breeding for great size as was King Kamehameha I, King of the first United Hawaii, who was estimated to tower over seven feet tall, weigh 300 pounds, and was capable of lifting a two-ton stone as well (if reports are accurate).

This is not to conclude that any of these cultures considered cannibalism a norm; rather I simply point out that there was consistent pressure in island nations, as an environment famously low in natural protein, to either avert starvation or attain other physical benefits by maintaining a high-protein diet.

Protein shortages and cannibalism too continued in remote parts of the world until surprisingly recently.

My father, seen in the photo above at age nineteen, served with Canadian Peacekeepers in the Congo, Africa in 1963. He told us that the local cannibals still filed their teeth to spikes and lightly pinched his arms as they spoke to him, in an effort to see if he would be a good candidate for eating at another time – as was their custom. By "filed", I mean that they took an iron file and ground their front teeth to points as in this photo of a young Mentawai Sumatran man – seen here smoking a cigar in 1890.

Livestock, and other growing crops like corn, potatoes, and other planned or engineered provisions, permitted island dwellers to sustain many multiples of population more than in pre-livestock eras as well. To this day, I'm not aware that any island nations are at risk of collapse – including Java, Indonesia - the world's most populous island with 135 million inhabitants. In all, 730 million people, or 11% of the world live in island nations that sustain themselves by desalinating fresh water, importing livestock or growing engineered crops. We mainlanders, with our relatively abundant growing places, appear to have little to worry about in comparison to island nations.

Much work has been done in recent years, to recreate the proteins found in livestock via crop-based vegetables and fruits. There may soon come a day when the protein needs of our children and adults are met by vegan science almost entirely, but generally a dairy and livestock diet supplement for infants and teenage years is the very safe building block of a healthy adult body. At adulthood, smart organic vegan lifestyles are probably a smart nutrition regiment for life, especially if your body wants to store fats - but I am no expert on this topic and I do believe in moderation.

Livestock are bred for food specifically and other technology will emerge to grow meats by themselves at a point, so there is no need to vilify meat proteins where livestock is treated humanely.

To my knowledge, no identical twins have yet been born and raised to adulthood with equal health – one by vegan and one by dairy diets, but I do know of attempts that have failed. Many parts of the world are vegan almost entirely, India is an example of a community that has recently killed people for eating meat; a practice contrary to our basic human rights.

Potatoes and corn are both important building blocks of modern civilizations – and you may not also have learned that both of these important crops did not occur naturally but had to be customized by horticulturalists over hundreds of years to reach sizes and productions that we see in modern crops today.

Too Many Monkeys

Another concern is population *density*. There is a zoo in Malaysia that became the topic of considerable study a few years ago when several neighboring zoos closed, forcing too many monkeys to come to live in one too-small enclosure together.

The normally peaceful monkeys began exhibiting traits similar to criminal acts that we see in human cities; there were muggings, gangs, rapes, and murder for the first time ever reported in ape habitats.

Humans need a little personal space and don't tolerate stress from sustained security concerns very well. Wealth seems to enter into this discussion too as people live very peacefully in Tokyo and Singapore in incredible population concentrations - but high population densities in Africa, or in poor U.S. or South American cities, where security is poor also, need much more space in order to avoid conflicts.

Moderating Birth Rates

Thirty-five years ago, Mao Zedong, Chairman of China, initiated perhaps one of history's most humane Birth Rate corrections of all time when his government enacted the Single-child Family rule to curb China's one billion citizen-strong population crisis. The rule averted an estimated 200 million births.

Countless other leaders in history had relied on war to cull human populations; Emperor Leopold II of Belgium used three competing religions to ensure that indigenous populations would kill each other off reliably for 150 years in the Congo.

Contagious European diseases and Conquistadors killed 90% of the native men and women of Mexico and South America outright in the 1600s – one of the greatest events of loss in human history; an estimated thirty-nine million people.

At this point, the world has its fair share of people and, as a result, two children families are probably a common sense population management plan that the earth will be able to sustain for several thousand years. With fewer children to care for mom and dad, society's role in caring for the elderly and infirm increases - and that is a healthy thing too.

So yes, the earth is home to billions of more humans than existed in the many centuries previous to this one. However, the problem also does appear to be resolvable fairly easily; through agriculture, or through technologies that turn wind-turbine electricity into clean drinking water; pragmatic population control, population distribution, and so on.

A Global Water Crisis

With sufficient energy, water can be extracted from seawater or air - and will eventually be synthesized directly from energy as well. The conservation of water is a short to mid-term issue for communities that have built settlements in arid regions and lack the technology to transport or create fresh water in sufficient supply. You don't hear about water shortages in Abu Dhabi nor Aruba – for example, as they simply extract clean water as needed from the sea.

I live directly adjacent to one of North America's Great Lakes, and our community draws water from the largest fresh water supply on the planet. Apparently, the Saudi's got the Oil, and Canada got the Water – but with the energy provided by oil, nuclear power plants, and other means that Oil income can buy, water can be had as needed.

When my kids came home from School deeply concerned about the importance of conserving water, I reminded my children that there is no water crisis in this area and that there won't be for many hundreds of years, if ever, in all likelihood. Why their school felt that it was important to stress this particular message of conservation is completely beyond me.

So much for artificial crisis. However, as I presented in the second TED Talk slide, water availability, and water quality is a serious concern for one-in-nine humans worldwide with developing nations carrying the brunt of the 750 million people who lack access to clean water; and the 840,000 who die from water-borne disease annually. Access to energy and to automated technology that provides clean water would solve this problem for good.

The Real Estate Super-Bubble

When I finished high-school, you could buy a modest home in a major city for around $35,000; that was in 1982. By 1985, three years later, the same house cost $90,000 and today costs roughly $350,000. When I say the same house, I really mean - the same ramshackle house – because it is now thirty years older and considerably more modest that it was back then as well.

When I was seventeen years old, my best friend's father told us the story of when he was a seventeen-year-old teenager himself, back in 1960 when he worked a summer job. The proceeds of his summer job allowed him to buy either a user car or a two acre Lake Ontario waterfront lot in a Toronto suburb. That lot would be worth millions today – but it was bought easily by a teenager's summer job income back in a K-Wave Spring Economy.

A couple of years ago, I divorced at a time just after entry level homes jumped up from $250k to $400k and now are going in the $500k and sometimes even $600k range.

There is nothing real about the price of real estate any longer. It's an alternative to the stock market at a time when investment in the stock market is becoming fool-hardy – and little else. Most expensive real-estate is being bid up and bought up by foreign business interests as well at this point.

The same story is playing out in most major western city centers. The front page of British newspapers last week headlined reports that Britain's politicians had promised the construction of 200,000 new affordable homes in the London area because their problem is bigger than ours and completely intolerable.

At the point that our younger generation can no longer launch, there are the underpinnings of revolution. Who benefits from revolution on local soil? Certainly not locals.

K-Wave theory explains that this real-estate bubble is typical of Capitalist Winter Economies. A Spring Economy historically bursts through it over the next ten years or so.

Government regulation of foreign investment will greatly reduce the pressures on real estate to balloon in the short term, as will regulation of rental markets that will try to take advantage as well. If history holds sway, Real estate prices have to plummet – if only to permit the launch of our next generation.

Who will Lead the Change

In most things, I am a moderate, so it says quite a lot when I get to the point where I read or watch the news and wonder should there be newspapers and television broadcasts at all. There is so little evidence of skill in problem solving in the evening news that all of the work of gathering and broadcasting appears largely wasted.

To get it right, I think they should gather the facts, and then relate the facts to how they will be processed and a solution found. What are the stats on success and failure for that process?

In grade eleven math, we learned that no complex problem can be fixed without a little good process – so if someone was shot, why were they shot? And who will investigate so that it never happens again? If 5% of similar crimes were resolved in the last year, the news story is that there is a serious problem with an important process and that the problem is now escalating to stakeholders properly.

The News Stories discussed in this chapter above aren't crisis to me. To me - they are just problems. Every problem has a solution and the Process needed to resolve each problem is laid out in Chapter 8 and 17 of the CSQ 101 course.

Chapter 19 – Performance & Merit

"Sometimes it's the people that no one imagines anything of, who do the things that no one can imagine."

The Imitation Game 2014 - Andrew Hodges

We can never under-estimate people nor pre-judge another person's abilities. So often it is those who you might never expect that end up changing the world - and the surest way to disincent someone from doing so is to not support or permit their work, nor to acknowledge their contribution.

This is as true for the very great majority of us, whose contributions and passions lead to raising families and building tremendous communities. Recognition and reverence are equally important for people among us who are different; who do incredible, ground-breaking work with fantastic contributions in science, technology, mathematics, medicine, and all other fields of study.

Gifted people and savants often struggle when interacting with the rest of us. "Gifts" are double-edged swords because even when a savant takes an active interest in communicating to the outside world, you can imagine that they might find that there are very few others who understand their interests with the same dexterity. A man or woman who can play any song ever heard in any key or style on a piano is not going to find very many relatively interesting others in even a hall full of concert pianists. A man who can read a play and hear the music just once before performing it perfectly - like my Anglican Minister and friend Ross Norton, will spend a lifetime waiting patiently for even the Mensa club members to catch up. This can be frustrating - so when gifted people also feel bullied or ignored by less capable others, it can be a great challenge for them to take time out for relatively unimportant social proprieties.

Alan Turing, the founder of both the computer, and the field of Computer Science - and in future he will likely be accredited as the Enabler of World Peace as well, was tried on the count of homosexuality. His sentence in 1953 was chemical castration which left him unable to work until he finally committed suicide at the age forty-one in 1954.

A mathematician at Kings College in Cambridge, England, Alan traveled to the U.S. for graduate studies at Princeton, and returned to London at the age of twenty-seven in 1939. Mr. Turing began working at the wartime cryptanalytics headquarters at Bletchley Park just north of London, England, and within a year he was decrypting Luftwaffe messages reliably. By 1943, Alan made work an electro-mechanical computer, from concept, construction, and programming, capable of deciphering the German Navy's U-Boat ciphered communications via Enigma, the most sophisticated mechanical computer ever built. He converted his initial Electromechanical Computer to an all-digital version in 1945, founded the field of Computer Science, and created the binary computer that all smartphones and computers run today.

Mr. Turing's genius and contribution are estimated to have saved the lives of twelve million people and reduced World War II by two years. The 2014 movie "The Imitation Game" told the incredible story of an outwardly very awkward young man - 'Professor shabby', nail-bitten, tie-less, sometimes halting in speech and awkward of manner.

Would Alan Turing's breakthrough work and contributions have been realized if he and fellow code breakers had not penned a letter directly to Prime Minister Winston Churchill reporting that administrators at Bletchley Park greatly hindered their efforts? What would have happened had Mr. Churchill not agreed to provide for the needs of his best and brightest scientists and engineers as well? In the end, Churchill ensured that resources were available, that administrators were checked, and the rest is history.

Albert Einstein (founder of the Theory of Relativity and Atomic Energy) was relegated to a patent clerk and ignored by academia for two decades. In 1989, Stanley Pons and Martin Fleischmann discovered Cold Fusion but academic peers condemned them to disgrace and took twenty-five years to rediscover that new formulations of nickel and palladium nano-powders created cold fusion reactions reliably.

Years lost; promise squandered; contributions ignored, and a future of clean energy delayed.

Alan Turing was chemically castrated and died alone at age forty-one; for two decades academia ignored Albert Einstein; and Pons and Fleischmann were disgraced for twenty-five years despite their world-changing discoveries in Cold Fusion technology. Years lost, promise squandered; brilliant contribution unrecognized, and worse for our society, a future of wonder delayed.

We haven't recently had a social and technology plan to measure technology and its use since the cold war ended. During the World Wars and Cold War Years from 1914 to 1969, societies had to build technology faster and better than the enemy or the Allies could suffer defeat, fascism, or much worse. Germany risked financial devastation and subjugation too. During the Cold Wars, Russia, and the U.S. averted worldwide nuclear destruction by building a compelling nuclear deterrent.

The result of having a government funded plan for technology improvement was nothing short of spectacular. By land, sea and air, engineering teams set about building and improving ships, tanks, cars, airplanes, jet airliners and fighter jets, to the world's fastest ever SR-71 and Lunar missions and Space Shuttles. City-sized nuclear-powered ships, submarines, bombs, rockets, and then the first mouse-controlled GUI (Graphics-windows-based) computer in 1970 – the Xerox Altos.

In recent years, planning fell away. Innovation was simply something "new" – not good and not bad. The market decided if something failed or succeeded – often led by non-engineers and business grads. If a spreadsheet was used by a hedge fund manager to squander one million families to joblessness, was that spreadsheet technology and the thousands of software engineers who built it, a bad investment by society?

The productivity of society over the past forty-five years was exponentially less impressive. We improved technology, computing, metallurgy, and miniaturized only because performance in these years was measured not by engineering merit, but rather by profit.

People in society were seen to be smart if they could make money with as little investment as possible, and then sustain and increase those profits year over year. Mortgage, stock, and insurance industries were good examples of products and businesses built in service of money alone.

Moneymakers were rewarded with cars and boats and houses because making stronger businesses made stronger societies – or so the business ethics classes told their graduates. With complete exemption from social accountability came the keys to the corporate washroom. Complex or labor intensive work that could be accomplished less expensively in some other part of the world went there while local engineers struggled to find work and make a living here.

When engineers and doctors did harm to society, they lost their license to practice. When Business Grads did harm to society, they were promoted and given bonuses. No doubt these same folks who lack a big-picture perspective also wonder why are "Take back Wall Street" demonstrators casting much closer scrutiny as the preventable recession and economic depression unfold today.

The points I am making are twofold; first, socially accountable Performance, is incredibly important to monitor; and second, Merit must be rewarded.

Maintaining Performance

This isn't going to be a typical performance management discussion. High IQ people can process and recall a lot of information, and High EQ can present what they know very comfortably while also being very aware of the needs of the people that they are communicating to at the same time. Put them together, however, and you often get sociopathic behaviors in a very great number of instances. Recall too the stories above when I discussed the decline of the Good Life in G8 nations over the past sixty years. So instead, we measure performance here as CSQ. Common Sense Quotient (CSQ) is the measure of the social benefit or harm that your decisions create. High-CSQ individuals make smart decisions that are expressed well, that are a benefit to all interests of business, government, individuals, and society as well.

If you are making tremendous forward progress in your car as you travel toward a cliff, ensuring that progress slows benefits you little until you finally make a change in your direction of travel.

In today's societies, funding makes the difference between the advances that you hear about – and the ones you never do. Microsoft, Apple, Google, and others are amazing collaborators to modern computing – and all got there by the funding of their sales or investors.

The countless technology companies that did not get funding, you don't know about, and you probably don't even realize that we should care about them in many cases.

As a society, we should care very deeply that Cold Fusion was delayed and denied funding for twenty-five years. Why? Because it is technology that aligns us with a brighter future and twenty-five-year stumbles like this one are preventable.

Should we be as concerned that innovations like social or video game companies like Facebook, Twitter, or Instagram are successful? If these tools improve our personal happiness, interpersonal skills, knowledge and collaboration sharing, and finally – productive progress in working to complete a Right Plan for society, then yes they are a good thing. At the point that they make our kids introverted, anti-social and unable to enjoy otherwise normal and happy lives, then they are much lower priority beta innovation. As it is, attention deficit and texting while driving is a serious problem that these apps are associated with that creates a very high accident and even death rate.

Judging Merit

Who gets hired today? At fifty years of age, with twenty-four years in senior and executive roles, with an unfinished degree, and a million lines of code written, most Human Resources' heuristics would consider me a poor candidate and therefore not hirable in a competitive job market.

Stereotypes of inflexibility, high salary, pension risk, non-relevant skills, too distant past training, ageism, racism, academic and other biases, and the dismissing of volunteerism and other community leadership experience. Preferences to female, native, minority workers without consideration of the number of incomes per household, are just a few of the many considerations that companies try to juggle.

Most important and damaging – the great majority of administrators, MBAs (Masters Business Administration), software heuristics and human resource interviewers - are not SMEs (Subject Matter Experts). They have often never worked a day in the competitive engineering fields that they are hiring for. This will tend to reduce productivity consistently in though-leadership areas like hi-tech, engineering, project delivery, product development, and strategic planning quite a lot.

Geniuses are really not incented to working for less capable people that also try to keep them on a short leash or who cannot understand how complex their automation work is. Once the high-performers get fired a time or two for working to change things for the better, you risk losing their potential forever.

For Transition Projects, a very different team than this would need to make decisions as to whether you deserve a bigger or a smaller lot in life – your merit. I have sat in a room full of 30-something Waterloo technology leads who branded one company that I had built, "unviable with no market potential". Six years later, AirBnB was worth $25 Billion market cap running that identical technology. These gatekeepers were not just wrong, they were unqualified - and consider who suffered; there were the employers and investors paying their salaries; the technology could not advance, and like ripples in a pond – these gatekeepers failed roomfuls of entrepreneurs the same. Clearly socially responsible leadership has diminished in finance, but when hi-tech Masters-level technical leads can't judge potential, our academic courses are failing us as well.

The reality that our work's potential for wealth creation and social value, are not connected to funding in any way is terrifying – but is it resolvable? We can fix this.

Chapter 20 – Human Rights

I did not realize the importance of a definition for Human Rights before I wrote this book. Over time, I have come to realize that the basic construct of establishing minimum human rights within a society is one of the most fundamentally important building blocks of a universal Good Life and World Peace.

The way that we set our "low-bar" as a society, our basic standard for Human Rights, is the difference between whether we set ourselves on a target for a Good Life – or not. There are untold thousands of variables in any economy, setting human rights to include a Good Life, gets this accomplished. This was never more clearly explained than in "Elysium."

In the 2013 Hollywood movie Elysium (Elysian Fields was "Heaven" to the Ancient Greeks), a future two-tiered civilization evolved on earth in which a fully automated production economy is created already. Automation exists in healthcare, humanoid-robotics, peacekeeping, manufacturing, construction, and others projects, but the technology has only been rolled out to the top tier of society to those 1% who could afford to be called "Elysium Citizens." The movie ends by reassigning Elysium human rights to everyone, which gives automated technology systems the direction required to promptly roll out a good life to all and quickly save the world. Does this sound familiar?

Rather than try to break new ground on a topic discussed and championed by hundreds of agencies and thesis, I will simply cover a handful of leading definitions to give the reader a high-level understanding of where we are with Universal Human Rights today.

Leading sponsors of Human Rights include International, regional and national-focus non-government agencies, the United Nations, and numerous multilateral organizations.

Criminal minds, cases where individuals can feel no remorse for actions due to a natural biological shortcoming, are born into most human societies in very small percentages. Psychiatrists realize that psychopathic personalities are those individuals who are born, or conditioned to act, without being able to discern right from wrong; good from evil, nor the value and importance of human life or human suffering. Many are unable to control their impulses to act outside of the norms of society as taught to them during the normal course of moral lessons given in childhood.

Individuals who are unable to control or modify their actions in pursuit of the things that they want may need to be institutionalized because they are legitimately unequipped to survive in our society. Cases of psychotic behavior are easy to find when obvious, and are found quickly. Those afflicted individuals that are clever enough avoid capture by hiding their condition may take a considerable amount of time and luck to discover.

The very great majority of individuals who harm their fellow man, however, do not fall into this category and do so based on nurture; based on training given to them through norms or instructions in society or in their career.

Rape cultures, gender inequality and abuse, corporate greed, endemic bullying are examples of injustices in communities, and even whole societies, who systematically engage in behaviors that disregard the human rights of others.

Respect for Life

No advanced society can exist were respect for life does not exist. When you or members of your family can be killed, there is no security, there is no peace nor liberty, and there can be no humanity.

The United Nations maintains a Universal Declaration of Human Rights which lays out thirty Articles in pursuit of eight essential human needs. Examples of needs include:

- Recognition of basic human dignity and equal rights are founded upon freedom, justice, and peace
- Where barbarous acts have outraged the conscience on mankind, human beings shall enjoy freedom of speech, belief and freedom from fear
- The right of last resort to rebel against tyranny and oppression and protection of law
- Friendly relations between nations with a common understanding of rights, social progress and better standards of life.

Rights afforded by this U.N. declaration include:

- All persons are born free and equal; endowed with reason and should act to one another in a spirit of brotherhood.
- Life, liberty and security of the person
- No man shall be held as slave nor subjected to torture or to cruel punishment
- Everyone has the right to be recognized before the law.
- Marriage, property ownership, and the pursuit of happiness & freedom are basic rights of all
- Freedom of thought, opinion and expression
- Right to peaceful assembly no-one may be compelled to belong to an association

For just a sampling...

Respect for Equality

Under the UK Equality Act of 2010, nine pieces of primary legislation and 100 pieces of secondary legislation were simplified down into one piece of law. With it, people are not allowed to discriminate, harass, or victimize another person because they have any the protected characteristics. There is also protection against discrimination where someone is perceived to have one of the protected characteristics or where they are associated with someone who has a protected characteristic of:

Age
Disability
Gender Reassignment
Marriage and Civil Partnership
Pregnancy and Maternity
Race
Religion and belief
Gender
Sexual Orientation

- Discrimination means treating one person worse than another because of a protect characteristic
- Putting in place a rule or policy or way of doing things that has a worse impact on someone with a protected characteristic, when this cannot be objectively justified (indirect discrimination)
- Harassment includes unwanted conduct related to a protected characteristic that has the purpose of effect of violating someone's dignity or that creates a hostile, degrading, humiliating, or offensive environment for someone with a protected characteristic.
- Victimization is treating someone unfavorably because they have taken (or might be taking) action under the Equality Act or support someone who is.

Who is Responsible?

- Government departments

- Service providers
- Employers
- Education Providers (Schools, Colleges)
- Providers of public functions
- Associations and membership bodies
- Transport providers

Women's Rights

To take education, and to have liberty, and to make choices that decide our destiny and happiness, are choices that not all women are permitted to have in our world.

Children's Rights

The right to be children without interference, to take a good and worthwhile education, to not be demanded to work, to have all basic needs of food, family, happiness, shelter, free from fear.

Respect for Family

Prosperity and equal rights have brought an alarming rise in divorce within the G20 countries.

In the 1950s and 1960s in Canada and the United States, it is widely believed that as many as 15% of married men hit their wives at some point in their marriage. For a short time, this embarrassing social problem was considered the responsibility of the wife and she was often asked to treat the repair of her marriage as she might treat her career at a time when a 5% divorce rate created a considerable negative stigma against divorcing her husband.

Fast forward 60 years and we see 50% to 70% divorce rates despite a very much smaller 6% wife abuse rate. Men who are weaned on lessons of integrity and honor of the family, marry women who are unwilling to continue in bad marriages in 80% of divorce cases in the United States. This means that 75% of society today are unwilling victims of a Divorce Culture as either children or adults. The cost of divorce to society is staggering. Now, two homes are needed instead of one, children must have double everything. Step-parents, family instability, poverty, anxiety, and a $12 trillion estimated annual additional cost to society in the U.S. results.

Right to Healthcare

The issue of access to Healthcare is dwarfed in many developing nations by the simpler needs of clean water and nutritious food. Once these basic needs are met, efforts to build universal access to healthcare can begin.

In the G20, all countries have access to universal healthcare plans except the United States and India. In both of these countries, private insurance, and some local state funds are set aside for healthcare, but not in a way sufficient to be considered universally accessible at the time of this writing. Documented statistics reveal that poor members of society live twelve years less than the rich on average.

Free Speech

Many nations, the United States and China among them, struggle to ensure that human rights and free speech are consistently upheld whenever a good life is missing for a large percentage of their citizens.

In Capitalist nations it takes funding to get a message heard and in this way the rich have access to a free speech that the poor never could never afford. Authors too, compete not only with volumes of competing publications, but other Media (movies, television, video games) as well until our children prefer distraction to book learning.

The Right to a Good Life builds World Peace

It is not yet considered a basic human right to be provided with a Good Life. This failing in both direction and planning, explains why we do not yet have World Peace in our time too.

In a world where so many of us struggle to find clean water, food, security and other basics of life, how could we ever think in these terms? However, once we fix this shortcoming by building autonomous delivery systems as a normal part of our automated economies, we will see the lives and human rights of seven billion change sustainably for the better.

CSR – Corporate Social Responsibility

BSR - Business Social Responsibility and CSR – Corporate Social Responsibility, go a long way toward assisting the efforts of Business and Governments to bolster a Good Life worldwide.

By not permitting investment nor lending to projects that would harm society and the earth as well, we prevent situations like Montana – where 100 year old tailing ponds from mining operations long retired, are bursting through decaying reservoir dams.

By directing investment toward sustainable automation, we spend once and realize benefit many times. In the example of tailing ponds, robotic dredgers could process contaminated soil with eco-friendly solvents separating component arsenic and other mining acids from soil. Unrecyclable chemicals can be converted to energy through a plasma-based waste elimination process and reusable chemicals can be transported to working mining installations. Once cleaned, the land can now be made-over to parkland by Spain's automated Park builder.

By monetizing while protecting the environment and society, we build strong, sustainable communities.

Chapter 21 - World War III

World War III, when or if it happens, will be a war that will dwarf our last K-Wave Winter trough war – World War II. Two nuclear bombs ended World War II, and it is that nuclear deterrent that may prevent World War III from ever starting.

Historically, every sixty years going back to 900 AD, Capitalist societies have relied on either a trough war or other event to redistribute wealth and opportunity – to begin the Monopoly Game over – if you will.

Wars are either started by competing countries or from within; and often begin over financial, resource or security concerns.

The Cuban Missile crisis almost started a war between the U.S. and USSR back in 1963 when Russia threatened to park nuclear missiles right next to the coast of Florida. The Iraq War was a financial war. Germany & USSR's invasion of Poland in 1939 was for reasons financial (to escape the Treaty of Versailles) and for territory (for *lebensraum* - living space).

So let's take a look at the world today and the current likelihood for beginning a World War III.

Countries capable of initiating a World War include the nuclear nations, USA, Russia, Japan, Germany, United Kingdom, Canada,

Italy, India, China, Israel, North Korea, the Netherlands, France, and perhaps one other - Saudi Arabia.

The following chart lists each country's Debt load versus their GDP incomes in an effort to present how difficult is their ability to repay their debts.

Country	GDP - 2010	Debt as % of GDP
United States	$14.6 trillion	92.7
China	$5.7 trillion	19.1
Japan	$5.4 trillion	225.9
Germany	$3.3 trillion	75.3
France	$2.6 trillion	84.2
United Kingdom	$2.3 trillion	76.7
Italy	$2.0 trillion	118.4
Brazil	$2.0 trillion	66.8
Canada	$1.6 trillion	81.7
Russia	$1.5 trillion	11.1

An economic trough war would be most beneficial for heavily indebted countries such as the USA, U.K., France, Japan, and Italy. Russia has negligible debt loads but will struggle to maintain loans with its presently devalued currency until current Ukrainian Crisis sanctions are lifted.

The counties with the highest probability of internal conflict are: Russia, U.K., India, and Canada. These are the countries where citizens have experienced rapid and growing gaps between rich and poor; high unemployment and under-employment rates. This also mean that young people can no longer afford the cost of living in major centers; a situation that has sparked demonstrations in the United Kingdom. Countries like France and the Netherlands control their wealth distribution and healthcare within their economies actively through law and enforcement – so they are less vulnerable.

Geography also plays a role as no one wants to start World War III in his or her own backyard either.

In world news, the Ukraine was stewarding Russian warships and nuclear warheads when it destabilized last spring. Russia's decision to de-risk its situation by removing its nuclear arsenal does not seem a random act of aggression. Putin, Russia's President, is clearly a good chess player and strategist so it looks like that conflict is well managed. Russia turned next to the defense of Syria and Iraq and the cleanup of ISIS forces on the ground.

By December, US-backed Turkey was suspected of supporting the insurgents of US-backed Syria, which was now also being defended by Russian ground and air forces. NATO (North Atlantic Treaty Organization – a military alliance between 28 G20 plus Scandinavian member nations) had not recognized Russia after the Ukrainian Crisis sanctions began in February 2014, and supported its member country Turkey in shooting down a Russian jet and in cutting off access to the Mediterranean Sea - stranding Russia's fleet in the Black Sea in completely in early December.

If Turkey has supported payments for Syrian Conflict-oil to ISIS, as Putin is suspected to believe, Turkey will likely be correcting those actions at request of NATO and Russia collaboratively and in short order. Russia's practice to "Follow the Money", and the good working relationship between both Presidents Obama and Putin, appears to be correcting a difficult situation with Turkey and ISIS very well for now.

North Korea are in the news for internal upheavals but haven't made overt threats toward South Korea or other regions in recently.

Our present K-Wave Winter is the time to avoid wars that could evolve into World War III over the next few years.

Without a financial reset and wealth distribution within our economies, the spans between rich and poor continue to increase, a few more billionaires and an exponentially greater number of poor and financially destitute people emerge. Historically these conditions result in militias, homeless, increased poverty and social problems too.

Peace therefore, would appear to require the expanded role of local government in enacting and enforcing laws that support a peaceful and proactive/reactive redistribution of wealth.

Chapter 22 – The Meaning of Life

The Meaning of Life is a topic that we have touched on anecdotally throughout this book. More than the struggle to reproduce our genetic material, the Meaning of Life is the value that every man and woman want to attain. It's the reason we make bucket lists and cry for those not given the chance to reach for a full life.

The Meaning of Life is very well discussed by scholars and philosophers the world over for millennia, and I bumped into most of the important elements during the many and varied researches needed for this book. In CSQ Common Sense 101, I stressed that living a full and interesting life, full of meaningful projects, is a very basic underpinning to Common Sense.

This brief chapter examines a few important points with a little critical thinking.

Worthwhile Projects

Remember Aristotle's views on the Right Plan of worthwhile Projects:

The Right Plan is the one whose ends, means, practical thinking and purposeful action result in a Good Life. A life full of things you need – and not necessarily a life full of everything you want. With a little luck, goods in body and soul, and by making a habit of good choices that reflect moral virtues of temperance, courage, and justice, a Good Life should be sought and found.

Abridged from Politic 322 BC (Messerly, 2013)

Aristotle's "Right Plan" represents the worthwhile projects in a society that are sustainable bottom-up plans that are scalable, flexible and reliable. They improve all societies and as such are probably the most worthwhile. Examples include the driverless car, internet search engines, computerized automation, clean energy, fuels and clean transportation - for examples based on technology. Mother Theresa's life, was a life well lived in service of building a process that still continues long after her time here has passed. These contributions deserve our recognition, respect and thanks.

Manual production projects such as training a society to build the tools needed to fish or feed themselves, so they will never need these things again – are next in importance. Whenever one group has a benefit over their neighbors, jealousy or need can create security issues that undermine a good life for this group in the long term. For this reason, these projects are worthwhile but are less so than projects that automate a solution for this community and their neighbors as well.

Care workers and philanthropists who administer top-down plans to support developing nations through tactical food and water distribution, education, and even hospitals, probably represent next worthwhile projects. Feeding the hungry is a worthwhile project.

In the same way that someone supports a swooning person from falling, someone also has to get them a chair to support the person until they can return to their feet safely. There can be no more useful purpose than a hero who administers artificial respiration to someone immediately so that they can be resuscitated to full health again in the hospital. Where we fall short is when we do not also work to minimize needs of heroic intervention, or permit the need of heroic acts to increase.

Socially accountable capitalism in a wealth-distributed society is the traditional foundation of the American Dream and a Good life. Our Chapter 6 - Social Projects are designed to restore capitalism and a good life; therefore, these are worthwhile projects too.

Buying a fine home and latest and greatest luxury car in support of worthwhile projects cannot be a bad thing. When your socially accountable workers have Good Lives and retire with full pensions, your life of socially responsible projects have merit.

Buying a fine life from proceeds of not worthwhile projects, probably should be a bad thing. Employing, and treating a team well that operates in a socially irresponsible way, simply compounds the harm unleashed on society, and so that should be seen as a bad thing as well.

Less-Than Worthwhile Projects

Is a life spent in support of no projects or projects that are not worthwhile, a life lived with less meaning?

Charles Dickens' character of Ebenezer Scrooge is an example of an extreme capitalist persona; successful financially – at the cost of personal misery for himself and great hardship for others. There are comparably extreme examples of good and giving individuals just the same. We all fall in between the two extremes to many and varying degrees.

Some individuals are conscious of their contribution and also harms done, but I have quoted great thinkers in this book who observed that many other people are oblivious to impacts beyond their personal benefit and line of sight. People surround themselves with peers, to blot out the inequities of others too.

What would a list of worthwhile projects seek to exclude?

Debt is an obvious start. Lending to support families and new business ventures in the construction of a Good Life is a good thing. However, lending to support usury and wealth inequality, warrant laws to protect us against.

Socially responsible behavior in business includes: CFO and Human Resource hiring problems, benefits (for pharma, dental and health, pension, and daycare, maternity leave, and so on), and rapid procurement. COOs have business targets to meet while guarding against high workloads, worker shortages, and high-travel assignments. CIOs report to CEOs and must manage engineering work locally as much as possible. They must differentiate the business through high-value service offerings and automation, and they manage technology investment. For example, spending hundreds of millions of dollars on ERP solutions or simple databases is incompetence.

All executive members are managing automation and safety net needs for worthwhile external social projects at the same time. The executive team often have no members with social accountability credentials, and this should change at a board level too. Until it does, more weight should be given to smart, socially responsible voices, which add to the success of the business. Use the process in Chapter 17 to work through this discussion and revisit annually as well.

Hiring problems include slow hiring policies, policies that build herds and shut out high-performers, Non-SME interviewers, the discount of volunteerism and number of household incomes, avoiding senior hires, pension risk aversion, and other socially irresponsible behaviors.

Procurement professionals too, are non-SME administrators that administer the process, but decide nothing. Forcing adherence to Procurement Professionals in an effort to ensure propriety, often boils down to the tail wagging the dog, with no improvement in propriety either. An example was a friend whose seven-member procurement team were released one by one in separate incidents for inappropriate business practices or bribery.

Make Religion a "Good" Thing

Religion is probably best examined as we suggested in Chapter 1 at a blades of grass detail level first, to the 100,000 foot forest view next, and then finally, to the 100 km view from space.

At the detail level...

Is belief in a deity, the definition of religion? Some might say yes, some no, but I think probably most of us believe, in the importance of goodness and in good works. The word Good sounds like God (in English & German) but is Good therefore "God" as well?

Belief in Good - is the underpinning of every Religion - whether you call your religion Spirituality, Christianity, Muslim, Buddhist, or other. Jesus (a human representation of Good) - was made human only by a decision of the Nicaean Council of 325 CE - and that decision could have gone either way during much debate by scholars at the time.

It may be difficult for many of us who are programmed since childhood, to believe that God is a grey-haired old man, but God is actually a metaphor for Good.

To understand why "God" had to be made into a personage, you only need to remember that 2000 years ago - when these constructs of Good were needed and first taught, peasants had no formal education. People had forty-year life spans on average and could not easily grasp concepts nor abstractions. A billion years is a concept; have you ever seen a billion years? No. So instead, Creationism in the bible taught that God made the seas, land, man, etc. in "a day".

To test my logic, simply replace the word "Day" with "a billion years" and you may see that Creationism and Evolution have more in common than you previously thought.

Alarmingly, 40% of Americans are taught Creationism still – there are even five types: Young Earth Creationism, Gap Creationism, Progressive Creationism, Intelligent Design, Theistic Evolution (Evolutionary Creationism). 77% of Americans also believe in Angels as well.

I do want to send out a note to teachers that Creationism is equivalent to training kids to not ask questions about explanations that don't make sense and have no foundation in science nor reality. Teaching not to question religion is more about control than good.

Teachers and parents disrespect the freedoms won by much sacrifice when they fail to teach kids that religious scripture is full of euphemism, metaphor, and analogy for a reason. Social lessons like the Golden Rule, Ten Commandments, Evolution, living meaningful lives, and the wisdom to know ourselves and make good decisions - can be taught more scientifically and accurately now because we educate our children today and have less need of the over-simplifications mandated by societies 2000 years ago.

Teaching lessons from old bible stories less-literally means that "Blood" no longer describes only the red stuff in our veins, "God & Jesus" represent "Good" and they are a somethings, and not someone. Angels, and even fate too, are not living beings for us to depend upon but metaphors for our power to defend ourselves and help others.

Belief in a deity - is not the point of religion. Over the centuries, politicians and special interests shaped messages to make it appear so – for many reasons - but belief in Good is the more important message.

At the forest level ...

To the important question - is there room for Religion in life today? There is more room for Good today than ever before. It is definitely worth figuring out how to teach our kids about what it is to be and act for good as well.

Religious groups the world over do amazing work in thousands of communities as do well managed secular charities like MSF (Medecins Sans Frontières – Doctors without Frontiers), Salvation Army, and UNICEF.

The fact that we sometimes prefer to see the religion's negatives is part human nature; and partly the media selling newspapers through fear, and it also says that you have been very fortunate so as to never need the charitable support of religious groups, and so you forget that it exists for those who do need it.

On the whole, supporting religions that bring good work to communities is a good thing for our society. We should all - every religion - have the good sense to ignore the negative messages that got dragged into the Good word over the centuries – and even work to slowly remove the negatives too.

The 100 km Summary view...

Christians had a supreme Emperor Constantine at the Nicaean Council in 325 AD who commanded all groups to unite and scrub their scriptures - including the violent passages – and still there was flagrant abuse of those scrubbed scriptures over time. The Muslim faith has not had this leadership and millions have died over centuries, and appear certain to die in future in-fighting if nothing changes.

Until all religions can represent Good consistently, and embrace contemporary human rights, they risk making themselves irrelevant.

Without Good in Religion, we are better off with well governed secular charities for the poor; and Philosophy Classes that educate our kids about Good, the Golden Rule and Ten Commandments.

I would dearly love to pass on the good part of my religious heritage on to my children - but I would not in good conscience sacrifice the safety of other children to do it. Religion will probably vanish, and Atheism will replace it if that is what it takes to remove the violent scriptures from 2000 year old scripts that continue to kill people.

Finally, consider that the only reason that we sit around discussing differences in religion in the first place, is because we don't know what other discussions need to be addressed next in our own society. World Peace Transition Projects assign every country a technology deliverable that builds the American Dream sustainably for everyone; so maybe direct your attention to those projects in

Chapter 6 and 7 instead and join in online at csq1.org/mag or csq1.org/wpprojects.

The Meaning of Life

Early on in this book, I introduced the Meaning of Life web library of academic papers administered by Dr. L. Messerly. Messerley summarizes the life's work of Dr. R. Hepburn (Hepburn, 2015a) on the subject and I paraphrase both men's notes below.

The following brief paragraphs are the summary of many lifetimes of research in what is the Meaning of Life.

1. Meaningful lives are purposeful and are spent pursuing valuable ends.
2. Meaning - is not found but created – which removes actionless values discovered in religion for one example.
3. To be unreflectively happy - is neither to find, nor to fail to find, meaning.
4. If we are lucky enough to participate in a meaningful life - through good health and opportunity, meaningful lives manage to combine good coping skills with their worthwhile projects.
5. It is not egoistic to want worthwhile projects to be compelling nor interesting.
6. We should work to live both morally, and as well as possible, as we ensure the progress of our worthwhile projects throughout our lives.

This is a terrific library that I encourage you to take a look at when seeking to broaden your understanding of a meaningful life.

All You Need Is Love

John Lennon's famous lyric is All You Need Is Love. How does this wisdom hold up after taking into consideration what we know about the Meaning of Life?

Building a family is certainly a worthwhile project; our children need stability in adolescent years and the love and support of a family provides that.

Western Christians and Chinese couples exchange wedding vows which are very similar. Through good times and in bad, in sickness and in health, beauty or plain, I will love you, comfort, respect, and protect you all the days of our lives. Others offer obedience and helpfulness to the other.

I mentioned above how unreflective happiness is not considered meaningful in itself and certainly most marriages go through ups and downs. A great majority of couples will openly admit battling against strong feelings of unhappiness in bad years and then there are good years too. Remember that happiness is something we give or deny ourselves too; consider that some people live with excess in life and are routinely unhappy, and others live with the basics of a Good Life and are very happy.

Infidelity in an Era of AIDs and Herpes, and cases of Spousal Abuse too, are valid reasons to accept that a marriage is no longer safe for the bride nor groom. The same can be said for substance abuse cases, extreme alcoholism, drugs, and other. For these conditions, most societies have granted annulments and divorce requests in approximately 10% of cases in even strongly anti-divorce cultures like the former Soviet Union and America through the 1950s and 1960s.

A 70% divorce rate means that 75% of our society are victims of family instability as children or adults. Divorce drives the need for 50% more jobs and 50% more households, as it drives poverty and burdens social safety nets and courts.

In a meaningful life, one of society's most worthwhile projects is your family. In family, therefore, clearly All You Need is Love. We love and support our friends as well, but there are always individuals who you will not like personally, and they won't like you either, and that is natural.

Love, therefore, clearly has to look beyond "liking" someone else, nor even knowing them. The Golden Rule or Ethic of Reciprocity, "Love your neighbor as yourself" acknowledges this fact. Every religion in every land through antiquity voices this exact same rule (Flew, 1979).

World Peace requires more than love of self - and family can be thought of as ourselves as well. World Peace requires action in support of people we do not like, or do not know, by building projects that sustain Human Rights that give us all a Good Life.

The Right Plan is the one whose ends, means, practical thinking and purposeful action result in a Good Life.

Aristotle, Politic 322 BC

So in summary, "All you Need is Love" by itself, is a vapid notion. When we love each other, as we love ourselves, and begin to action a Right Plan that builds a Good Life for everyone – then; by way of the Golden Rule, and all that we've learned about the meaning of life - All You Need is Love.

Chapter 23 – Causal Conclusion Case Studies

We live in a content-rich age with TV and Print News services, Internet and Flipboard Smartphone Newsstands and websites, Wikipedia, and excellent search tools for most publically available knowledge out there today.

We also have ready access to email, Facebook, Google and Twitter, Pictogram, and other tools which easily permit us to share interesting news and information from throughout the world, to friends, neighbors and family. When articles arrive at my inbox, I have to dismiss most of them because there are just too many. I also realize that quality has been lost to a very large degree, and is replaced by quantity.

When people ask me to review online letters that discuss leadership and economics, I notice a consistent problem with the author's ability to string together facts that support the conclusion credibly. These authors appear to consistently make invalid conclusions based on the facts that they present. Other articles are spot on and quite surprisingly well researched. How do we tell the difference?

Case 1 – Working the Numbers

The TV Series "The Newsroom – Season 1, Episode 1 – We Just Decided To" called on statistics (and not rhetoric) to explain that America was, and that it could be again, but that "America is Not the Greatest Country in the World Anymore". Check out the video at https://www.youtube.com/watch?v=dOwMdUyLw1k

It's NOT the greatest country in the world, Professor. That's my answer.... And with a straight face, you're gonna sit there and tell students that America is so star-spangled awesome that we're the only ones in the world who have freedom? Canada has freedom. Japan has freedom. The U.K. France. Italy. Germany. Spain. Australia. BELGIUM has freedom. Two hundred and seven sovereign states in the world, like, a hundred and eighty of them have freedom....And you, Sorority Girl, just in case you accidentally wander into a voting booth one day, there's some things you should know. One of them is there's absolutely no evidence to support the statement that we're the greatest country in the world. We're seventh in literacy. Twenty-seventh in math. Twenty-second in science. Forty-ninth in life expectancy. A hundred and seventy-eighth in infant mortality. Third in median household income. Number four in labor force and number four in exports. We lead the world in only three categories: Number of incarcerated citizens per capita, number of adults who believe angels are real, and defense spending, where we spend more than the next twenty-six countries combined, twenty-five of whom are allies. Now none of this is the fault of a twenty-year-old college student, but you nonetheless are without a doubt a member of the worst, period, generation, period, ever, period. So when you ask what makes us the greatest country in the world, I dunno what you're talkin' about. Yosemite? (We) Sure used to be. We stood up for what was right. We fought for moral reasons. We passed laws, struck down laws, for moral reasons. We waged wars on poverty, not poor people. We sacrificed. We cared about our neighbors. We put our money where our mouths were. And we never beat our chest. We built great big things, made ungodly technological advances, explored the universe, cured diseases, and we cultivated the world's greatest artists and the world's greatest economy. We reached for the stars. Acted like men. We aspired to intelligence. We didn't belittle it, it didn't make us feel inferior. We didn't identify ourselves by who we voted for in the last election, and we didn't, oh, we didn't scare so easy. Ha. We were able to be all these things and do all these things because we were informed. By great men. Men who were revered. First step in solving any problem is recognizing there is one. America is not the greatest country in the world anymore.

This was a response made by actor Jeff Daniels' News Anchor-character Will McAvoy, to a commentator's insistence that Will explain his truthful opinion to a student's question "Why is America the Greatest Country in the World?" after other panelists sited rhetoric in answer to the question.

I would want readers to note the difference in approach presented by writer Aaron Sorkin - as Will McAvoy responded with logic, fact, and accurate historical references regarding great advances in technology and civilization - rather than rhetoric. Even to this academic audience, in the YouTube video, it will be more apparent that Will was very hesitant to disclose his political alignment Right or Left, Good or Bad, Socialistic or Capitalistic – in order to avoid the audience's, the media's, and the voter's, agenda of branding and too-easy dismissal of the facts. How often have you heard figures in the media prefer an emotional dismissal like "He is a Socialist"? The practice is so widespread that it even blocks out free speech.

Summaries like "That is Good" or "That is Bad" are slippery slope conclusions. A "slippery slope" argument says that we cannot erect speed limits because soon everyone will be travelling 5 miles/kms per hour. Too often, these are used to invoke an emotional and even illogical dismissal of discussion on a policy based on its merit alone – and this IS a bad thing. Every four years a billion dollars are spent to reinforce this thinking in the U.S. and you will see examples of this mindset daily in Political reports and on Facebook.

Policies are Capitalistic, Socialistic, and Communistic is a little more socialistic. Right or Left - and that's it really. "Democratic" is something different; Democratic is a system of voting used by capitalists, socialists and communists as well.

"Democratic" is also a word often misused by extreme Capitalists or Communists, when they do not want to tell you this is extreme policy – as in, this is a policy that benefits the rich and ignores the poor – or vice versa, but that is fine, because after all - the current majority will vote for it "democratically". That means it must be "Good"; and that means further measured consideration is "Bad".

Every government manages a book of hundreds of policies and every Policy Book has policies that could be called Right or Left. Are the American Government communists because they only permit 100 years leases to cottagers in national parks? No - this Communistic Policy is simply the most common sense policy for this need. Are there risks associated with taking on any one specific solution or policy? Yes - Often there are, so simply mitigate the list of risks – with No Emotion, No Drama, and move on to the Next Policy ...

Risk Lists and Mitigations are important, because if we want the revenue of cottage renters, but there is a risk that cottages might litter, start fires, hunt or fish illegally, or other risks – then we can simply i) list the mitigations for those risks, ii) monitor, and then iii) check back to see if mitigations are working as hoped every three to five years. In this way, we set speed limits that save lives, but we will also not reduce limits below the safe design thresholds of the roadway, vehicles nor drivers.

Avoid marginalizing by branding good nor bad; Right or Left; Socialist or Capitalist – both have their failings. Rely instead on smart, considered, common sense solutions that are proven to meet needs, and improve society with a productive projects and momentum.

Case 2 – Building HOW back into our Schools

In September of 2005, on the first day of school, Martha Cothren, a History teacher at Joe T. Robinson High School in Little Rock, Arkansas, did something not to be forgotten. On the first day of school, with the permission of the school superintendent, the principal and the building supervisor, she removed all of the desks in her classroom. When the first period kids entered the room they discovered that there were no desks. 'Ms. Cothren, where are our desks?'

She replied, 'You can't have a desk until you tell me how you earn the right to sit at a desk.'

They thought, 'Well, maybe it's our grades.' 'No,' she said.

'Maybe it's our behaviour.' She told them, 'No, it's not even your behaviour.'

And so, they came and went, the first period, second period, third period. Still no desks in the classroom. Kids called their parents to tell them what was happening and by early afternoon television news crews had started gathering at the school to report about this crazy teacher who had taken all the desks out of her room.

The final period of the day came and as the puzzled students found no seats on the floor of the desk-less classroom. Martha Cothren said, 'Throughout the day no one has been able to tell me just what he or she has done to earn the right to sit at the desks that are ordinarily found in this classroom. Now I am going to tell you.'

At this point, Martha Cothren went over to the door of her classroom and opened it. Twenty-seven (27) Veterans, all in uniform, walked into that classroom, each one carrying a school desk. The Vets began placing the school desks in rows, and then they would walk over and stand alongside the wall. By the time the last soldier had set the final desk in place those kids started to understand, perhaps for the first time in their lives, just how the right to sit at those desks had been earned.

Martha said, 'You didn't earn the right to sit at these desks. These heroes did it for you. They placed the desks here for you. They went halfway around the world, giving up their education and interrupting their careers and families so you could have the freedom you have.

Now, it's up to you to sit in them. It is your responsibility to learn, to be good students, to be good citizens. They paid the price so that you could have the freedom to get an education. Don't ever forget it.'

By the way, this is a true story. And this teacher was awarded Veterans of Foreign Wars Teacher of the Year in 2006. She is the daughter of a WWII POW.

Do you think this is worth passing along so others won't forget either, that the freedoms we have in this great country were earned by our Veterans?

Let us always remember the men and women of our military and the rights they have won for us.

Veterans risk everything to defend the values and citizens of their countries and there can be no more impressive and respectable role in our society. We owe it to them to make the home country of veterans the best that it can be for each of their families and children. I take great exception to stewards and administrators who abuse the charge of these great men and women when they fail to make our society a great one.

This Case Study FaceBook post exposes an unfortunate gap between i) understanding “What is the Goal” and ii) “How to Build it”. This GAP exists in most high schools today. Consider that teachers and Veterans instruct us all in what are the most important goals of our society. This article reminds us “What” is important clearly.

Our education system is next charged with giving students the explanation of “How”. Instructions in how to build that better place for our society. Since Aristotle’s time “Teachers teach and Doers do” has been a problem because Aristotle’s curriculum is the core of all modern universities and it taught us how to solve real problems in an around about way.

Some problems needed Math to solve them, so Math was taught. Some needed language, chemistry, history, biology and so on. The Application of how to utilize all of these skills to build meaningful projects that solve important problems, got lost in the implementation of his curriculum in most of our schools however.

Does telling kids that “they have much to be thankful and appreciative for”, give them the tools they need to understand how to vote for good economic policy that sustains a good society. Does it teach them how to build the other social and technology projects needed to build a sustainably better life? Put another way, does this explain how to honor the sacrifices of great men and patriates by building a better home for us all?

No, of course it does not. Change Management and Project Management does this. And yet, we teach almost everything *but* these subjects in High School. We teach kids skills - in Math, Language, Science, Technology, Physics and MetaPhysics, but we don't teach them good change process, good project management, the implementation of critical thinking, until they get into university or college.

At the point in College when they are introduced to Change, Project, or Transformation Management, it is from within a silo – a faculty like Economics or Business, Technology, Engineering, Science and so on.

Planning and building a Right Plan for a Good Life, as needed to respect this promise to our veterans, combines solutions from each of these faculties, and so it needs training our kids in HOW to do this at around Grade 9 when kids are still in general studies and not in siloes or faculties. Corporate Social Responsibility CSR is a Business silo solution, wealth distribution is Economics, the Technology Projects are Health, Science, Physics, Computer, Robotics and Engineering solutions, and the reasoning for treating projects with priority is found in a Philosophy course.

Chapter 17 teaches readers the process to build social and technology projects that build a Good Life and then a sustainable World Peace. CSQ 101 and the 100 Year Plan expand on this with a course for high-school level kids, and their parents, that teaches them "How" to do great things.

It is important, but not enough, to teach our kids "What" is important only; we must teach them "How" to achieve important goals every time too.

Case 3 – An Easy Case Study

A Danish School Teacher introduced herself in a popular Facebook posting earlier this year.

She explains: I make about $61,000 a year. We get free education, you don't have to pay for going to the doctor, nor the hospital, and also we keep paying our students for getting the education that they want.

We can afford to have our own house, a car, get 1-year maternity leave when she has a baby, every time... fathers get paternity leave too, just not as long. Everybody gets a pension from the age of 65.

I have lived in the USA, but I prefer the Danish way of taking care of everyone. Some may say we pay high taxes, maybe, but we get so much more than you get in the USA.

You pay an average of $8,500 per year in deductibles and many fees for healthcare. Your higher education is over $21,000 a year. We pay our students to learn. You have no job security, vacations, or any other benefits that we do.

Everyone in the world wants the American dream. Every Dane has that opportunity.

I have little to add here really. I can mention again that only India and the United States do not have universal healthcare within the G20 nations; and that the poor in the U.S. live up to twelve years shorter on average than their wealthier countrymen. The rest is pretty self-explanatory and absolutely true and verifiable online as near as I can tell. Unemployment rates in Denmark are approximately 3% to 5%.

Statistics show that because everyone has a Good Life here, that wealth creation efforts are higher per capita than in all other G8 countries as well – and not only in Denmark, but also in most other neighboring socialist countries as well. The Dutch generate three times the wealth per capita that Canadians do and 50% more than do American's.

The Danish, are people called Danes from Denmark; the Dutch are called Dutch from the Netherlands; Finns are from Finland and referred to as the Finnish; Norsemen are from Norway and are Norwegian; Swedes are Swedish and hail from Sweden, and I hate to think about all of the many variations in translation. Together,

38 million people live in these countries that are among the most successful and humanistic places to live in the world.

So what can be learned by a causal analysis of messages that are backed in reality and fact? Everything actually. Once you do your research to confirm the facts, make sure that that you are open to the learning.

Learning from fact asks us to rethink our basic programming – and that too is an important education that is not easy for many among us.

Case 4 – Not as easy

Here is an example letter, sent to me by a friend who forwards things that contain good facts - if even conclusions don't match. The letter begins with a creditable list of researched statistics – but fails to make valid use of them when it comes time to conclude in the summary. I analyze why, at the end of the letter.

Check this out. Donald is right! GOD, I'M GLAD I LIVE IN HAWAII

California....LARGEST INSANE ASYLUM IN THE WORLD

It is interesting that the LA Times did this. Lou Dobbs reported this on CNN and it cost him his job. The only network we would see this on would be FOX. All the others are staying away from it. Whether you are a Democrat or Republican this should be of great interest to you!

Just One State - be sure and read the last part...try for 3 times.

This is only one State....If this doesn't open eyes, nothing will!

From the L. A. Times:

1. 40% of all workers in L. A. County (L. A. County has 10.2 million people) are working for cash and not paying taxes. This is because they are predominantly illegal immigrants working without a green card.

(Donald Trump was right)!

2. 95% of warrants for murder in Los Angeles are for illegal aliens.

3. 75% of people on the most wanted list in Los Angeles are illegal aliens.

4. Over 2/3 of all births in Los Angeles County are to illegal alien Mexicans on Medi-Cal, whose births were paid for by taxpayers.

5. Nearly 35% of all inmates in California detention centers are Mexican nationals here illegally.

6. Over 300,000 illegal aliens in Los Angeles County are living in garages.

7. The FBI reports half of all gang members in Los Angeles are most likely illegal aliens from south of the border.

8. Nearly 60% of all occupants of HUD properties are illegal.

9. 21 radio stations in L. A. are Spanish speaking.

10. in L.A. County 5.1 million people speak English, 3.9 million speak Spanish. (There are 10.2 million people in L. A. County.)

(All 10 of the above facts were published in the Los Angeles Times)

Less than 2% of illegal aliens are picking our crops, but 29% are on welfare. Over 70% of the United States 'annual population growth (and over 90% of California, Florida, and New York) results from immigration. 29% of inmates in federal prisons are illegal aliens.

We are fools for letting this continue.

HOW CAN YOU HELP?

Send copies of this letter to at least two other people. 100 would be even better.

This is only one State...If this doesn't open your eyes nothing will, and you wonder why Nancy Pelosi wants them to become voters!

IF YOU DON'T AGREE JUST DELETE--IF YOU DO--PASS IT ON! WHERE DO WE GET THESE MORONS?

Windfall Tax on Retirement Income Adding a tax to your retirement is simply another way of saying to the American people, you're so darn stupid that we're going to keep doing this until we drain every cent from you. Nancy Pelosi wants a Windfall Tax on Retirement Income. In other words tax what you have made by investing toward your retirement. This woman is a nut case! You aren't going to believe this.

Nancy Pelosi wants to put a Windfall Tax on all stock market profits (including Retirement fund, 401K and Mutual Funds)!

Alas, it is true -- all to help the 12 Million Illegal Immigrants and other unemployed Minorities!

This woman is frightening. She quotes...' We need to work toward the goal of equalizing income, (didn't Marx say something like this?), and in our country and at the same time limiting the amount the rich can invest. (I am not rich, are you?)

When asked how these new tax dollars would be spent, she replied:

'We need to raise the standard of living of our poor, unemployed and minorities. For example, we have an estimated twelve million illegal immigrants in our country who need our help along with millions of unemployed minorities. Stock market windfall profit taxes could go a long way to guarantee these people the standard of living they

would like to have as 'Americans''.

(Read that quote again and again and let it sink in.) 'Lower your retirement; give it to others who have not worked as you have for it. 'Send it on to your friends. I just did! This lady is out of her mind!!!!! PASS IT ON

Case 4 Conclusions

Few Causal Conclusions can be found in this letter.

Donald Trump has expressed views on Immigration that sound very well supported by well researched statistics – that conclusion is supported by statistics. Nancy Palosi has expressed the correct need for wealth and income distribution and then she is marginalized here as a "nut-job" based on a "tax the investor (rich)" strategy.

In World Peace above, I discuss the importance of Graduating Tax after "K-Wave Summer" begins, so that 51% of those with a vote, pay the lion share as they are able to, and distribute wealth ongoing. This group have 99% of wealth with instruction to trickle it down, so practically speaking, where else can it come from? It's not clear here that a Windfall Tax is targeted or if it affects everyone, so more clarification is needed as that is not well explained. If the Tax distributes Wealth, it's a good thing and she is far from being a "Nut Job" for suggesting it.

Militia and Organized Crime

Note that people living within a country with no security will seek to form militia (called "gangs" in this letter). Readers can take a page right out of American History as confirmation. America has the highest incarceration rate in the world next to Seychelles, Africa; their incarceration rate is five times higher than then next G8 country, the United Kingdom.

Militia and organized crime groups are a very expensive social problem – just ask King George III of England in 1770. His view of American militia would have been to consider them criminals in actions against the crown certainly; and he would have sought to incarcerate offenders at the very least. Had finances, arms and reinforcements not arrived from the French and Spanish, he likely would have succeeded in his efforts to return the British colony's 95% of non-combatants to the more productive pursuit of commerce as loyal British colonists.

Support for Organized Crime, War Lords and Gangs exist in places in the world where a Good Life cannot be found through legitimate employment and social safety nets. It is true that there are a percentage of criminal minds within human communities consistently, but more often, support for gangs are based on economic need. "Soldiers" within these groups are left with few options and without security, for themselves and their families, when they do not cooperate.

Regarding points made by Ms. Palosi on wealth distribution:

Wealth Distribution in the U.S.

- 40% of the U.S. population control 0.3% of the wealth - that's 160,000,000 people without universal medical care and without starvation wages. They have no vote because 51% are "fine" so things don't change.
- A Berkeley economics professor wrote a paper in 2013 that documented 95% of all new wealth into the US since the 2009's bailouts have gone to the 1%. This prevents economic recovery through trickle-down as it was meant to - and that is the reason that we continue in the depression that we are supposed to be in - according to Longwave (or K-Wave) Economic Theory.

K-Wave have more than 18 documented cycles in recent history and can be considered a fact for purposes of making good conclusions.

Level-setting examples of Good vs. Bad Causal Conclusions include:

- If the glove doesn't fit, you can't commit - Certainly you can; one has nothing to do with the other especially and there are many other pieces of evidence to rely upon.

- If you concentrate on Wealth Distribution, you are ignoring Wealth Creation. The Economic facts are that any country and party left or right MUST concentrate on both. John McCain criticized Obama for doing both during 2008's election. McCain, J., & GONZALEZ, J. (2008)

Case 4 Conclusion

Letters like this lose their value when they ask us to make conclusions based on unrelated facts. Look to the stated conclusion and ask – is it supported well – or at all? Sleep on the subject for a night and look at it again in the morning if you think there is enough subtle truth in it. Nine times out of ten, I find that the expressed conclusion or opinion of the sender is not supported by facts presented in any way – as in this Case 4 example.

Generally, if you cannot find a valid conclusion, dismiss the letter and make good decisions based on your well-educated analysis instead.

Chapter 24- Getting the Word Out

In December 2015, a link to the first release of the electronic version of World Peace – The Transition, went out to all World Leaders in every country. A WP Magazine was also released at csq1.org/mag, csq1.org/wpprojects, #wpprojects and CSQ Research forums were built out to support the launch as well.

As the CEO of six Hi-tech Startup companies, I learned one very clear lesson years ago; that without investment, marketing, sales and strong distribution, even the greatest ideas, technologies, and products are doomed to flounder, fail, and are never to be heard of after that.

The technologies, software apps, celebrities, and politicians that we all come to know are just the products that we get a chance to see several times, and that succeed in garnering a significant market share of support through direct and indirect advertising.

This chapter of the book is dedicated to reminding us to spread the word; to take an interest in and commit a little time to finding sponsors for a plan to build a sustainable Good Life in your community and beyond.

Are the projects that make it to market the best technologies ever conceived? Rarely. Do the smartest leaders and engineers front them? Sometimes Yes – but not as often as not.

Overnight success stories are the very great exception and even "viral sensations" seldom find the management team able to sustain the tremendous luck that gave these products an initial introduction into our lives. This book, the Transition, was written on a shoe-string budget with little left for professional editing and that will impact easy reading and understanding which also adds risk to its adoption.

According to CNN News, President Obama raised $750 million in support of his 2008 campaign; his 2012 campaign raised and spent $1 billion, and estimates today put the cost of an election at $1 to $2 billion U.S. Dollars to win an American Presidential Election Campaign. These are the costs needed to stay ahead in the media wars on TV, radio, online, in the mail, and on-the-ground grassroots outreach.

The viability of a candidate for President of the United States, therefore, is predominantly based upon their ability to raise campaign funds. This puts candidates like John E Bush and Hilary Clinton in the lead long before formal election campaigns begin. Others who fail to secure funding commitments will fall away realizing that they cannot possibly compete for the popular vote of a 350 million member electorate that bases its decision on the messages they receive on TV, print, or on the Web.

Elections that require enormous financial support, will tend to build Oligarchy leaderships because only the very rich can support whatever message that finally gets out to voters. For this reason, many democratic nations will call for caps on election spending and this gives the best and brightest a chance at the leadership positions.

Naturally the support of a book with an important message such as this is a simple thing. Surely I won't have to solicit dozens of Literary Agents, the President and Deans of Harvard University, broadcasters from the CBC and similar TV Production companies. I won't have to discuss it with Members of Parliament, Captains of Industry, and media and book reviewers either. That fulfills the satire requirement for the entire book; I will have to do all of these things and much more too in hopes of getting the word out. Online Magazines, speaking engagements, applications to appear as TED Talk Presenters, on and on.

Some of these folks who routinely turn away book inquiries are official gatekeepers whose job is to ensure that your messages don't move forward unless referred through internal industry channels. Celebrities of industry are just way too busy to ever hope to parse through the hundreds and thousands of email requests that hit their desks every day too.

A "Solution to every problem" is something that all of my writings have worked toward, and yet I struggle for a solution to this problem of getting the word out just a little. Here is an example problem where luck and tenacity are not necessarily guarantors of success. Inside-sales best-practice tells me to build a sales funnel and make 100 sales calls a day. But there it is, I do not have a passion for sales.

The message to all like-minded sales-averse colleagues is - see that this particular medicine tastes bad; and then take the medicine. If you are pursuing a Right Plan, or worthwhile project, measure your tenacity by valid measures and then effort the work pragmatically using good time management (described in CSQ 101) and work your funnel leads. You cannot be lazy about this step and a little luck will go a long way too – as Aristotle

Beginning a Good Life in Law

Building World Peace is a straight forward engineering deliverable whose underpinnings lie in i) a high-school course curriculum for Project Management that most teenagers could master with just a little practice. And ii) in the basic understanding of why it is important to build our society based on a careful plan that grows to achieve nothing less than World Peace.

Most major nations and cities failed to run according to a strategic long-term plan this past 30 years. This has meant that the basic needs of young families to own a home and start a family in their hometown is often unattainable until they are thirty or forty years of age.

Our capitalist monetary system is troughing in recession and depression for the eighteenth time in since 900 AD - without government recognition nor pre-emptive economic controls - nor management "with teeth", as Queen Elizabeth put it. Enacting "teeth", or economic controls, requires that lawmakers recognize an important need, and then they must create and defend a bill according to a right plan. Next they must enact those laws, and then empower them with balanced punitive measures.

Making changes in law requires many steps that can be expensive and time consuming. Forward progress in a democratic process can also be easily derailed by many voices both well-intended and self-motivated, at any point prior to a World Peace Agenda bill becoming a law.

See "I'm Just a Bill" by Schoolhouse Rock! On YouTube

What is not at all clear, to most people, is the whole story – the Where, What, How, Who, When and Why as follows:

1) **Where** to start. With Charters followed by Needs, Designs and Implementation as discussed in Chapter 17.
2) **What** projects should be pursued? We have not established a right plan of what are our society's projects, so we could as easily support socially negative projects without this.

3) **How** – it takes financial support and a good process to keep the wheels of positive change turning. See Chapter 17.
4) **Who** – Bills - proposals for new Laws - in many democracies must be sponsored by a Member of Parliament and then defended before at least two houses of review and final approval.
5) **When** – a K-Wave Winter is a good time to start a plan for World Peace because:

   i) This is a preferable alternative to trough wars, revolutions, and hardship; and
   ii) Wealth distribution projects create a thirty-year window of capitalism-fueled opportunity with which to automate our production economy.

6) **Why** – for a sustainable Good Life at Home and World Peace abroad.

For these reasons, it often happens that only sponsors who are capable of financially affording the support of a lengthy and expensive legal change process, can be the only ones to enact laws. As the rich are able to afford to make laws and media announcements, where the poor cannot, a freedom of speech problem in Capitalist countries is created that is also seen in communist countries.

The best chance of building a sustainable Good Life, is to enact a Right Plan into law that embeds a Good Life definition in everyone's basic Human Rights - and then require a voter referendum to update that planning if needed during federal elections. Look for CSQ Certifications to confirm program alignment with necessary World Peace Transition Projects in technology and economic controls.

Difficult Summaries

During the first release of this plan, eight or ten topics needed introduction in Chapter 1. The weight of these important problems proved a considerable challenge to create strictly positive summary messages for and I failed to find a strictly positive telling.

Some are downright grim but I will leave their important messages here in hopes that perhaps you will have more luck with this jigsaw puzzle of summaries than I did.

Fixing problems for our Young People

Would it surprise you to hear that our university-aged young adults and teenagers now regard our society as a failure? I just watched another Hollywood movie last night, "Fantastic Four" which expressed this sentiment again; and last month it was TomorrowLand.

Is it upsetting to you that our kids regard our world as one that needs urgent repair and one that doesn't align with their own personal values either? I think it very well should upset.

No war veteran ever risked his or her life to protect the needs of a handful in order to make life worse for his children and nation. Instead, he or she upheld values that ensured that his family and children could live in freedom, security, and opportunity.

In North America, the retiring baby-boomers are now seen to have handed down a greatly diminished place to live in the eyes of our young adults, and our current leaders are not communicating a credible plan to turn that around either. My own five kids are entering into a society where they could never afford to purchase a house in their hometown.

In my home just outside Toronto, Canada, good jobs are as thin as hen's teeth, and opportunities to complete requisite higher education, start families, raise children and find affordable housing before the age of twenty-five, and even thirty-five, are all but impossible goals for the overwhelming majority of young people.

Productive Safety Nets

In capitalist countries, individuals that have it good, have it very good – but their numbers are shrinking with rising unemployment rates, underemployment, and big-business' pragmatic squandering of pension commitments. Those that do not have a good life are unproductive, and unable to contribute to their country's economy - often through lack of opportunity and insufficient social safety nets.

Socialistic policies distribute wealth and are statistically proven to create more new wealth per capita too, so two safety nets are needed:

The First is the **Engineering Project Safety Net**. This Safety Net funds World Peace Transition Project resources not already funded by other companies or government. This would likely include a great number of senior engineers displaced by layoffs at major corporations seeking to reduce their pension risks.

Second is a catch-all **Guaranteed Income Safety Net** for all of society designed to meet the basic human right needs of a Good Life Similar to the Pilot Program that Sweden rolled out in December 2015 but adjusted per region to permit people to live in the place where they grew up. Remember, no-one needs a car once driverless taxis are freely available via an Uber-like app or phone call. This and several other considerations change as we roll-out the Transition as well.

Context in Great Thinking

Remember that today's computers and automation were difficult imaginings 2500 years ago. Today, the ancient's extrapolations of reasoned thought are realities and very achievable technology projects.

The Do Nothing Option

What if we do not follow a Right Plan? What if we continue down our current road? Well, in some honesty - World War III potentially; either an extinction-level war that could bring an end to mankind as a species altogether, or a marginally trimmed down non-nuclear equivalent. If we follow the same timelines that governed World War II's rise after the last great depression in 1929 , the next world war is just three to five years away.

If Capitalism persists without Wealth Distribution, depression will persist and deepen until the poor have sufficient vote to effect socialistic policies that Distribute Wealth.

Anyone who watched Tony Stark deploy thirty independently navigated bomb-payloads from a single missile in the movie Ironman 2, can imagine that a single ICBM (Intercontinental Ballistic Missile) could rain down from space upon a city and completely destroy both its offensive and defensive military capabilities without the inconvenience of nuclear fallout. New anti-satellite missiles; and anti-ship and submarine missile developments too, are well capable of redefining who holds a military upper hand.

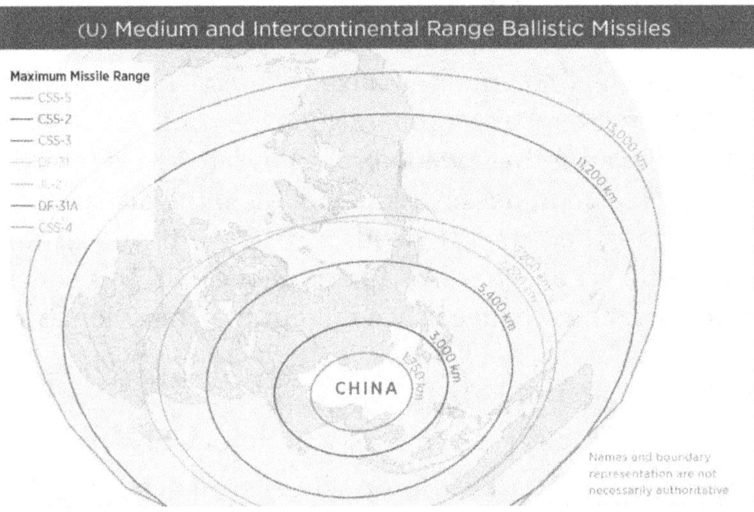

Once financial conditions become bad enough, and the two causes for the start of World War II are needed again (debt relief and living space), a motive for war exists and a World War III that no-one can win, could begin.

World War III will take just one country misbelieving that their first-strike capabilities are sufficient to eradicate today's active deterrent of 1800 high-alert nuclear missiles scattered worldwide during every hour of every day.

Social Planning

World Peace – The Transition, explains a plan that solves this problem at the same time that it provides a platform for a Good Life for everyone right here at home.

In Canada, unemployment, under-employment, and time to find work is on the rise; businesses routinely layoff twenty-plus-year employees to avoid pension obligations; and divorce is at 70% which means that more than 75% of our population will be unwilling victims of family instability as a child or as an adult.

In America, 40% of citizens control just 0.3% of the wealth in the country while the top 20% control as much wealth as all the G20 nations combined (excluding Japan and China). Democracy ensures that 51% fair very well enough, but the other 49% endure social problems often consistent with third world countries with 20% shorter lifespans, militia's arising from security concern, and income inequity that brings many other social problems.

Why do we work for thirty or forty years without a social plan that actively avoids arriving at this outcome? What does that say about us as a society, as parents, and as thoughtful individuals?

It says that our schools, government systems, and businesses too, let us down, because while we were busy struggling to make a life - and following the only plan we were taught, no-one stood in a position to plan for important social needs; no-one was preparing for our future at the same time.

Are these serious problems correctable now? Of course they are; every problem is solvable. Is a Good Life possible within three years to five years? Yes. Is World Peace possible within twenty years? Yes, it absolutely is.

This book lays out a method to get government leaders correcting things immediately, and leading technology companies are working toward many of the projects that what we need to sustain a world peace too. In many cases, projects continue without even realizing their potential – which showcases the value of planning once again.

Business & Government

Did you think that big business was going to lead the charge? The last forty-five years should be evidence enough - to prove conclusively that a future worth having is going to take good leadership within our governments and society's engineers to work toward a plan of social and technology projects that make a better future for us all; to build the projects and programs that lead to real World Peace.

I suggest that there are two primary reasons for this – rightly or wrongly. First, Business Ethics courses in every business school on the planet fail society when they validate the irresponsible notion that everything good for business must also be good for society; this is true only in a strong trickle-down that stopped being the norm in the 1990s. Second, no certification program exists for business leaders - and, therefore, social accountability cannot exist. When an engineer or a doctor do harm to society, they lose their license and can never work in the industry again. When a business lead does harm to society, they are bonused, promoted and golden parachuted until they can even run for President of the United States.

Corporate Stewardship and good leadership are polar opposites as well because not only are leadership often not engineering leads, but they are not accountable for their ongoing failures to society either.

Chairmen must, therefore, look closely at their boards, and balance the voices for profit with leaders in engineering, technology, business transformation, and also must look for licensed individuals who can "Do no harm" as they solve business problems and consider new opportunities.

Leading the Cheer

If you are ready to turn your personal, family or society's prospects around with a little good planning now; to work toward a Right Plan of worthwhile projects as needed to deliver World Peace within the next forty years, then you have come to the right place, and you have come to the right book.

My father's favorite wisdom still applies - but it needs updating now. He would say: Ignore everything you hear; believe half of what you read and everything you see with your own eyes and experience.

Today, in the internet era in which we live, this needs modifying to say: Ignore everything you hear and see on entertainment TV and web; believe half of what you read from properly researched and statistically valid reports, and believe everything that you see as long as it's aligned with the Right Plan.

Take your favorite chat site for example; does it provide a good life? Does it lay out fundamental lessons in wealth distribution and responsible national wealth creation – and a plan to move on to the next step in our human journey? Probably not. Does it automate and safety net self-sustaining communities? No again. Does it distract both kids and voters, from reading, learning, and building a self-sustaining community? A case could be made to say yes actually.

The fact that ADD (Attention Deficit Disorder) sufferers will tend to hang out on distraction sites quite a bit does not mean that it causes the condition I suppose. But it does seem to exaggerate the time wasting problem.

There are so many extreme and superficial reports on social networks that I have come to ignore them except to rely on them to create case studies as in Chapter 23.

Get the Word Out

In summary, "Get the Word Out", and for heaven's sakes find more upbeat summaries than these when you do. Speak about the importance of building a World Peace Agenda within your country - to politicians; to teachers in schools, in community events, and in business too.

Chapter 25 – Transition Economics

According to the AEA (American Economic Association) website, when observation shows phenomena that are absent in, or inconsistent with, available theories, Economic theorists look for new theories. Transition Economics falls into this "new theory" category in that no previous work has taken a direct look for an "automation to export to safety-net" and "government investment vs. spending" equation in previous theory. Housing Transition too must be considered and transition algorithms developed.

As the technical field of automation is advancing at a rapid rate and no economic theory is presented to manage it, Transition Economics is suggested to set guidelines on how to weather the storm of the next 20 to 30 years of consistent automation change within our societies.

Although I will spend probably one complete book fully covering the topic, Transition Economics is discussed in two parts. Those developing core skills and those explaining the application of the skills in specific examples.

A criticism of mine for economic theorists, is the burden of mathematical proofs as a measure of deductive reasoning. Any theory may be true in unrealistic isolation, similar to the Fibonacci sequence's description of plant growth, but observation remains the only valid measure of a forest. The importance of the forest's many thousands of other ecosystems cannot easily be modelled one on top of the other and many systems can be too easily forgotten as well.

Similarly, in Economics, Mathematics works well to model very specific and isolated systems like Supply and Demand or Interest Rate reactions, but it can rarely accurately predict the effect of hundreds and perhaps thousands of external economic influences that each relies upon probability and even luck (exceptions to highest-probability outcomes). In macroeconomics, invariably, mathematical modeling emerges purposefully and then proceeds in a direction of diminishing return on investment until we can dismiss theory and return to observation.

Longwave, or K-Wave Theory, is a superlative example of an economic theory divined through observation of repeating patterns alone; it is to economists what geysers are to geologists - if you can follow my analogy. K-Wave Theory is far more useful in determining which economic controls are appropriate for our capitalism's cyclic seasons, than any other mathematically modelled system that we presently know of.

For this reason, I attempt to keep just enough pattern recognition math as needed to ensure useful tools for deductive reasoning here.

Considerable work was done years ago when the Soviet Union switched to a monetized economic model back in the 1980s, the work created a reverse scenario and economy that was later called the "Transition Economy". I will refer to some of those constructs to reverse engineer a Transition Economics model in my next book "Transition Economics".

Economists organize by field - from agriculture and business, to urban planning, and I foresee much work arising in this new field as World Peace – The Transition emerges.

Important Data

Each country is going to want to consider the following Ingredients in calculating their own Transition Economic Plan:

- GDP Export
- GDP Import
- Highest Value GDP Industries World Wide
- # of Workers per Industry
- Number of Production Economy Automations
- Rate of Automation per Industry
- Cost of New Safety Nets per Automation
- ROI (Return on Investment) Optimization
- GINI and HDI
- Unemployment and Underemployment
- Minimum Wage
- Housing, Mortgage, and Usury Debt Limits
- Military Spending
- Underfunded Retirements
- Debt
- Real Estate
- Social Problems
  - Divorce Rate
  - Poverty
  - Violence
  - Depression
  - Anxiety
  - Fear
  - Loneliness
  - Alcoholism
  - Child abuse
  - Suicide
  - Gun Death Rate
  - Incarceration Rate

To name just a few measures.

Optimal Rate of Transition

With the basic understanding of which industry automations benefit the country first, and return investment most quickly, engineers should be enlisted and projects started in earnest.

The following few sections suggest a number of charting tools for determining our current inventory and trends in economy. I will introduce a number of those charts here and then work through specific case studies in the full transition economics book due in mid-2016.

Accelerating Trends in Industry Automation

Automations have started already in many industries and some of the safety net costs may need to be offset by revenue generating initiatives quickly now - in order to avoid national debt problems from becoming unmanageable in just the next few years.

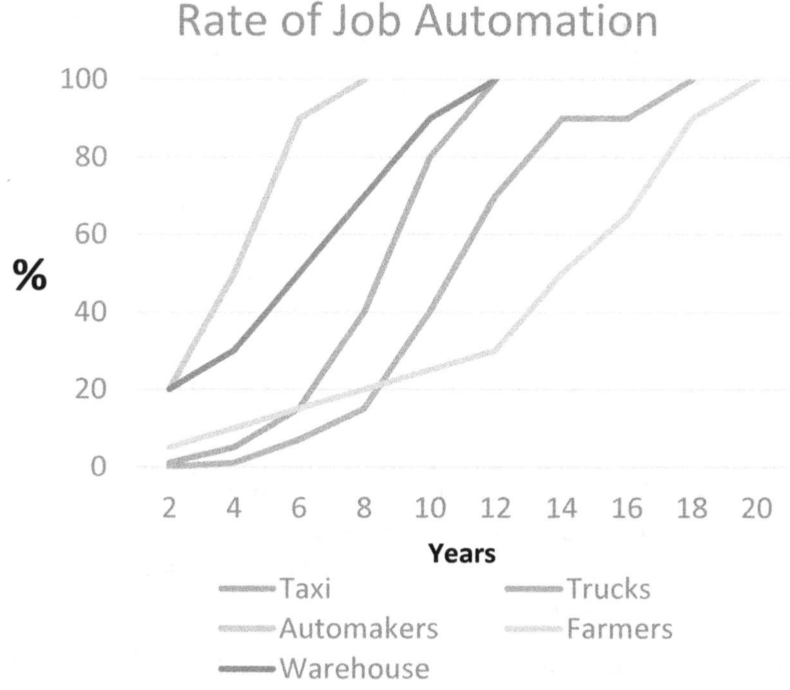

Unemployment and Underemployment

Consumption is the biggest percentage component of the GDP and economy. Consumers consume when they're working, growing income, and are confident about their short-term economic circumstances. Not only do we have chronically high unemployment, it's not going down, and the number of workers defined as "long-term unemployed" is at record levels.

The next chart shows the change in unemployment in this recession versus the prior ten recessions. Note that seven years in, we show unemployment comparatively higher and more consistent than in any other recession since the last great depression, which this certainly is as well.

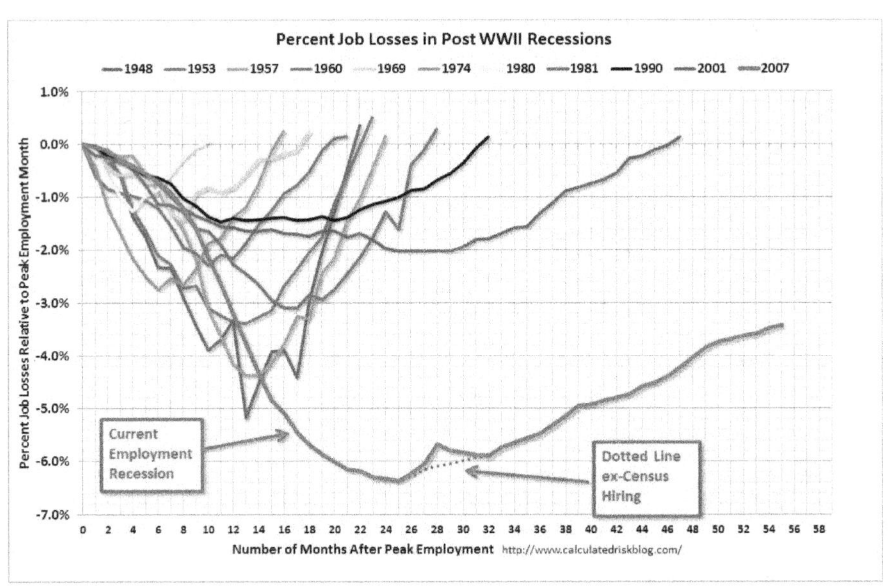

Underfunded Retirements

Corporate layoffs of twenty-year workers, unfortunate investment instability and choices, and inadequate saving habits are creating very real concern for long term. Our corporations are failing us and they have a lot to answer for now.

According to the Employee Benefit Research Institute, 47% of workers age fifty-six to sixty-two are probably going to come up short in meeting all the expenses that retirement will throw at them. That's half of the working population!

According to Public Integrity, the number of pensions at risk inside failing companies more than tripled during the recession. But not to be outdone, the public sector has even bigger problems. Related to the Municipal Finance Crunch referenced above, and poor investment results since 2000 have put many public retirement plans in dire straits.

Funded Ratio Map

The following map shows funded ratios Morningstar compiled, with green states reporting the highest ratios. Click a state to display 2011 figures:

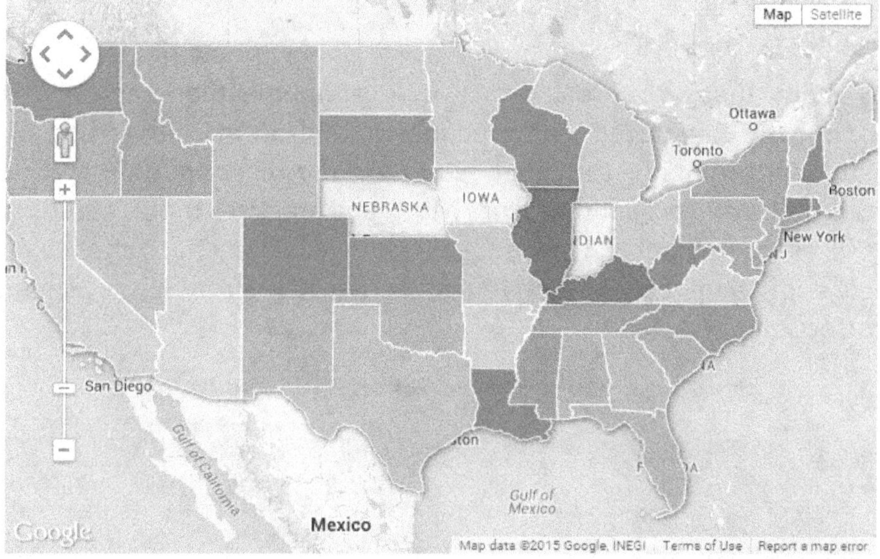

This is a chart showing the state-by-state comparison of retirement fund status. To these statistics we also need to add Social Assistance and Medicare/Universal Healthcare as well.

Debt

The USA has record debt at $18.1 trillion by today's count, which is approximately a debt to equity ratio of 15%. At the current rate, we'll hit a point where we'll be issuing debt to pay the interest on our debt. Fortunately the world has not lost interest in buying US debt. As borrowing rates begin to climb, debt will accelerate and we have to manage currency debasing leading to inflation and hyperinflation.

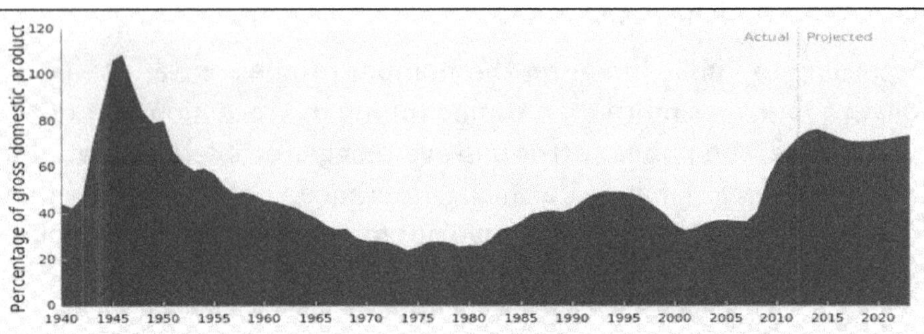

This is a look at the trajectory in debt accumulation.

Real Estate

By many measures, real estate is in a bubble that has become completely separated from economic indicators altogether. All Federal attempts to turn real estate around failed in 2010 and we stand poised to repeat the Usury Mortgage practices that created the 2008 as those that lost their houses then are coming back after seven years of bankruptcy forgiveness. Here at least we're seeing correction pressure.

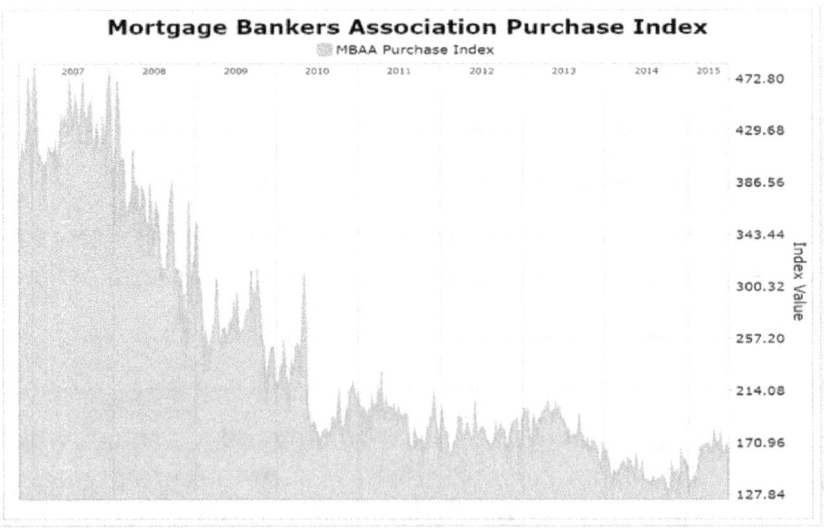

This shows the real estate mortgage purchase application index.

Note the decline since the end of 2005.

I discussed Bubblenomics in Chapter 13 – Land Ownership. We will expand on managing these KPIs throughout the Transition in the next book.

Worldwide Markets

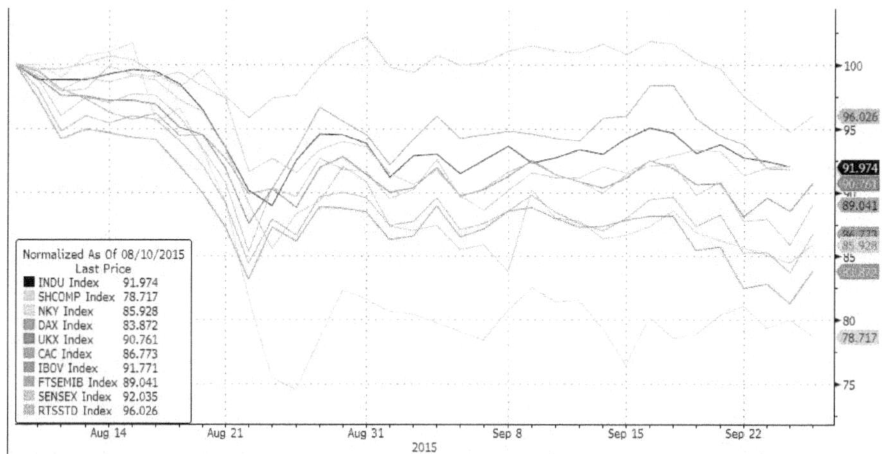

This chart shows the Top 10 Markets in the world crashing in response to China's "Black Monday" meltdown in August 2015. The United States was down 2000 points from highs, with two consecutive daily drops of 500 points. Japan's Nikkei is extremely volatile and down more than 3000 points as well.

China plummeted 40% from highs earlier this year. Germany has lost one-fourth of the value of all German stocks. The UK, down 16% and their economy is on shaky ground. France's stock are down 18%. Brazil plunged 12,000 points and is officially in recession as is Canada. Italy is down 15% with shaky economy as well. India stocks dropped 4000 points and finally Russia, is doing better than others, but half of their exports are oil and will suffer as long as oil prices stay low as they are today.

Financial Indicators Summary

Financial stats are valuable but are just data as well. The economy is most felt when we are denied access to income through

Employment. Don't stress over financial stats. Instead, at a country level, work toward wealth distribution, increase foreign trade and act vigilantly to automate both profitable exports and self-sufficiency.

On a personal level, manage your personal stress well as you make your way through this ten year trough until brighter news prevails.

Chapter 26 - Conclusion

Five months ago, I set out to create the "Right Plan", as Aristotle called the solution to a sustainable Good Life and an inevitable, unavoidable World Peace. There can be vapid attempts to refute it - amounting to acute voices raised high in support of undefendable objections or fears - but this plan checks out as supported in the facts and critical thinking collected from sufficient significant independent sources and presented here in a way that can be validated by any academic or credible audit review body.

Whenever a beauty pageant contestant is asked how will she help build World Peace? From this day forward she can confidently assure her interviewers of her commitment to pursuing a Global World Peace Agenda that builds worthwhile projects and an automated production economy that serves all mankind - in every corner of the globe. What more big-picture, honorable and noble pursuit can there be?

World Peace – the Transition, is my second book this year - written over the past eight months. Earlier this year, I wrote 'CSQ Common Sense 101' with no notion that the explanation of a system to improve the value of our decisions would have to lead to a credible and even obvious Project for World Peace as well.

At some point in your life, you may have wondered why World Peace seems so complex. The reason is that we cannot reasonably look at the problems of others until we ourselves can easily take care of our needs individually, and for those of our families, first. Once we have a sustainable Good Life throughout our own community, we can look to extend self-sufficient communities globally.

"World Peace – The Transition" is a Proof; a Case Study consolidating evidence to establish the truth of a statement. Leveraging planning, approach, and lessons learned from CSQ's 100 Year Plan applied to one of our biggest problem solving needs on a global scale.

For me, it is never enough to stop at explaining "What" the future holds, it is important to explain "How" to achieve those things too, and that is what made it possible to solve this most important of all problems too. By explaining how, you will find this book and story different from all the other movies and TV shows that you nor I have yet seen.

Building futuristic innovation, robots, and excellent lives take smart planning, good project process, brilliant engineers and hard work – and with these ingredients plus strong leadership, anything is possible. Whether a trip to the moon, or a robot crop picker or miner, or an entirely automated production economy, anything is buildable. World Peace - is just a Project.

CSQ 101 made me realize, as Aristotle did, that a better world started right here at home, and that the right technology and social plan for Transitioning to World Peace, could be built with PMLC process when put in the hands of enough sponsored scientists, engineers, governments and citizens as needed to get it done reliably without undue burden on any one group.

A Good Life, a robot maid in every home, two day, three day or even three-hour work weeks, and one day - an anti-gravity car in every driveway one day perhaps as well. Family friendly communities where equality, human rights and liberty are assured to all, right around the world.

So many books set out to inform and entertain, and then once in a very long while, a book really changes your life and makes a difference too. This was that book for me as I don't think I will ever look at the world in the same way again.

This year, I was absolutely thrilled to see that a Hollywood Movie looked to the future for adventure and not just drama again. How long has it been? The latest Star Trek movie won a heck of a battle against an alien invader of course – but really, they could use less special-effect cannon fire; I just wanted the science and adventure to continue.

Leadership with Results

I hope that a few of my specific conclusions did not offend. When I say that, historically, the weak links in leadership are rarely the engineers; and that the obstacles to progress are consistently the stewards and administrators who are given authority over engineers and scientists. I am basing my calls on facts as presented here in the TED Talk summary in Chapter 9 so feel free to get online with the forums and explain your opinions as well. Step up and be an engineer – you don't need a tin ring. You need project experience and social accountability, so go out and get that experience.

Since the 1970s, the ranks of our administrators has grown to included politicians, business leaders, and even business academia – in another correctable bit of history, so let's get that corrected too, with your vote and participation with things political.

I mention that the rate of mankind's technical advancement from 1900 to 1969 was staggering. Since 1969 however, advances stopped fairly abruptly. I am going to end the book with this story and rollup.

The Dodge Charger automobile is a good case study in what went wrong. During the 1970s, salaries started to drop and off-shoring automotive parts and whole vehicles began to ramp up. Profit-focused CEOs worked to reduce the costs of labor and retooling assembly lines. Small changes, incremental improvements, happened at a rate that ensured profitability. 50 years past and the same car sits on our new car lots.

So many people aspire to earn as much as possible, for as little contribution as possible within our society; now a plan like this one is overdue. Reject a mediocre life, plan for humanity's next great advance, and let's engage our young people and best and brightest engineers to take our civilization to the next level.

We are so close and it only takes your vote for leadership who will get us over any hurdle and make it happen.

Life is what we make it so let's build a good one. Be a Leader of results, and make your projects, and life, meaningful at the same time. If Common Sense is that voice from within that tells you to make yours a great life; to make it the best, most interesting and most fulfilling life you possibly can, I hope you will feel compelled to chip in and help build this plan and transition to an American Dream and World Peace of your own; both for yourself and for our children too.

An automated society is our generation's Moon Launch and legacy. So as Gene Roddenberry's fictional character Captain Picard might like to say – Engage.

ABOUT THE AUTHOR

The process to build and transition to a sustainable World Peace heralds from learnings in longwave and transitional economics, history, leadership, government, technology, engineering, physics and social science. I certainly hope that readers take my example and are inspired to keep their studies wide and goals important.

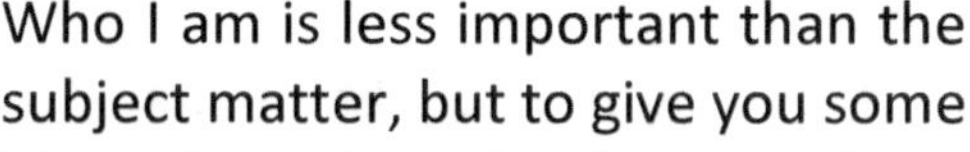

Who I am is less important than the subject matter, but to give you some idea where these books come from; I have raised five terrific kids, built six high-tech startup companies, I learned something new every day of a 25 year high-tech engineering career after four years of post-secondary study.

I am a Lecturer, CEO; CIO, CTO, CFO, COO and I have led 300+ complex projects in dozens of major programs with budgets up to $100+ million, 100,000 staff, and 200+ project team members in organizations that ran more than 500+ projects year after year. I have terrific coping skills that I exercise and practice often; I have worked with terrible bosses and terrific bosses; I know how to make a dance partner shine. I ran a full marathon a couple of years ago, I mountain bike - well, read widely and watch a lot of movies. I was a lifeguard, canoeing instructor and took Judo, Wrestling and Karate as a teen alongside eight years of horseback riding lessons at a Christian Ranch where I attended my second weekly religious service; and all of this was great learning. Travel, language, and degree studies, cultures, Religions, Academia, Business, Technology, Geography and History mattered too.

I can say easily that I am a capable big-picture, process-minded strategic thinker with a well-balanced resume for someone who thinks they know enough about World Peace to write a book that figures it all out too. I exceed a PhD equivalence in Engineering and Technology in career; and I greatly enjoy research in facts, history, economics, business and political science.

Family and society are important to me. My forefathers were Puritans and Founding Fathers, and so my history connects me to the importance of leveraging lessons from the past.

In community life, I founded one of the largest local Minor Football organization in Toronto, and that volunteer work gave me the chance to hire eight management teams and to get to know 800 young people and their parents every year as well. This volunteerism connected me to the next generation and to their hopes for a bright future.

World Peace is an important topic and book and I hope that I've provided valuable insights. Writing it has changed the way that I look at the world.

I choose to write books that are prodigious and maybe even epic, about how to build projects that are worthwhile and can change the world for the better. I hope that your life, is filled with successful worthwhile projects and family too.

Anything is Buildable - just keep working the problem.

Wishing you All the Best.

Edward Tilley

PS.

Look for my new books Transition Economics and Teaching Doers in 2016 and 2017.

Bibliography

Aldrick, P. (2015). Bubblenomics is back. Markets have become completely detached from economic reality – and it's going to get ugly – Telegraph Blogs. Retrieved November 23, 2015, from http://blogs.telegraph.co.uk/finance/philipaldrick/100024624/bubblenomics-is-back-markets-have-become-completely-detached-from-economic-reality-and-its-going-to-get-ugly/

Allied_army_positions_on_10_May_1945.png (1216×769). (n.d.). Retrieved November 8, 2015, from https://upload.wikimedia.org/wikipedia/commons/1/11/Allied_army_positions_on_10_May_1945.png

American Press. (1995). A 120-Year Lease on Life Outlasts Apartment Heir - NYTimes.com. Retrieved November 23, 2015, from http://www.nytimes.com/1995/12/29/world/a-120-year-lease-on-life-outlasts-apartment-heir.html

Andrews, R. (2015). Germany Just Successfully Fired Up A Nuclear Fusion Reactor | IFLScience. Retrieved December 13, 2015, from http://www.iflscience.com/technology/germany-just-successfully-fired-their-nuclear-fusion-reactor

Archive. (2015). Gun Violence Archive. Retrieved November 14, 2015, from http://www.gunviolencearchive.org/

Bartz, G. (2005). The Code of Hammurabi. Retrieved from https://en.wikipedia.org/wiki/Code_of_Hammurabi

Blodget, H. (2011). TRUTH ABOUT TAXES: Are Today's Rates High? Retrieved November 8, 2015, from http://www.businessinsider.com/history-of-tax-rates

Bryce, T., & Klengel, H. (2015). Egyptian–Hittite peace treaty. Retrieved from https://en.wikipedia.org/wiki/Egyptian%E2%80%93Hittite_peace_treaty

Calkins, D. (n.d.). How to make CO2 (Carbon Dioxide). Retrieved November 8, 2015, from https://www.youtube.com/watch?v=b_qdkKnftt8

CounterTerrorism, B. of. (2014). Country Reports on Terrorism 2014National Consortium for the Study of Terrorism and Responses to Terrorism: Annex of Statistical Information. Retrieved November 14, 2015, from http://www.state.gov/j/ct/rls/crt/2014/239416.htm

Davis, J. (2015). Diesel From Water And Carbon Dioxide | IFLScience. Retrieved November 8, 2015, from http://www.iflscience.com/chemistry/audi-make-diesel-water-and-carbon-dioxide

Diggs, C. (2015). Russian Floating Nuclear Power Plant. Retrieved November 8, 2015, from http://bellona.org/news/nuclear-issues/2015-05-new-documents-show-cost-russian-nuclear-power-plant-skyrockets

Drugs.com. (2014). Top 100 Drugs for Q4 2013 by Sales - U.S. Pharmaceutical Statistics. Retrieved November 18, 2015, from http://www.drugs.com/stats/top100/sales

Elia, J. A., Baliban, R. C., & Floudas, C. A. (2012). Nationwide energy supply chain analysis for hybrid feedstock processes with significant CO2 emissions reduction. *AIChE Journal, 58*(7), 2142–2154. http://doi.org/10.1002/aic.13842

EPA. (2001). AMENDMENTS TO THE CALIFORNIA ZERO EMISSION VEHICLE PROGRAM REGULATIONS: December 2001. Retrieved November 8, 2015, from http://www.arb.ca.gov/regact/zev2001/fsor.pdf

Europol. (2015). *European Union Terrorism Situation and Trend Report.* Retrieved from https://www.europol.europa.eu/sites/default/files/publications/p_europol_tsat15_09jun15_low-rev.pdf

FAS.com. (2015). Nuclear Disarmament Resource Collection |

Reports/Books | NTI Analysis | NTI. Retrieved November 26, 2015, from http://www.nti.org/analysis/reports/nuclear-disarmament/

Flew, A. (1979). The Golden Rule.

Giles, M. (2015). Inside The Z Machine, Where Scientists Turned Hydrogen Into Metal | Popular Science. Retrieved November 9, 2015, from http://www.popsci.com/z-machines-hydrogen-gambit?dom=tw&src=SOC

Gordon, I. (2009). Fourth K-Wave Winter. *Longwave Group.*

Gray, R. (2015). Audi creates DIESEL from air and water and its already powering a car | Daily Mail Online. Retrieved November 8, 2015, from http://www.dailymail.co.uk/sciencetech/article-3059025/Audi-creates-DIESEL-air-water-fuel-future-powering-car-driven-German-minister.html

Greenslade, R. (2007). The good news about bad news - it sells | Media | The Guardian. Retrieved December 12, 2015, from http://www.theguardian.com/media/greenslade/2007/sep/04/thegoodnewsaboutbadnewsi

Grossman, S. (2014). These Maps Show Every Country's Most Valuable Exports | TIME. Retrieved November 18, 2015, from http://time.com/106666/world-export-maps/

Hepburn, R. W. (2015a). Summary of R. W. Hepburn's, "Questions about the meaning of life." Retrieved November 12, 2015, from http://reasonandmeaning.com/2015/11/09/

Hepburn, R. W. (2015b). The Meaning of Life | philosophy, evolution, death, transhumanism. Retrieved November 9, 2015, from http://reasonandmeaning.com/

Herring, D. (2003). *The Puritans and Education.*

Ho, A. (2015). The unlivable dwellings in Hong Kong and the minimum living space | Hong Kong Free Press. Retrieved November 10, 2015, from https://www.hongkongfp.com/2015/07/27/the-unlivable-dwellings-in-hong-kong-and-the-minimum-living-space/

Jackson, M. O. ., & Morelli, M. (2009). *The Reasons for War. Elgar Publishing.* Retrieved from http://web.stanford.edu/~jacksonm/war-overview.pdf

Jaide, D. (2013). King Leopold's Agenda: The historical origin of Christianity in Africa by Professor Chinweizu Ibekwe | Rasta Livewire. Retrieved November 24, 2015, from http://www.africaresource.com/rasta/sesostris-the-great-the-egyptian-hercules/the-historical-origin-of-christianity-in-africa-by-professor-chinweizu-ibekwe/

Jowett, B., & Aristotle. (2015). The Internet Classics Archive | Politics by Aristotle. Retrieved November 8, 2015, from http://classics.mit.edu/Aristotle/politics.html

KABBANI, S. M. H. (2015). Jihad: A Misunderstood Concept from Islam - What Jihad is, and is not. Retrieved November 16, 2015, from http://islamicsupremecouncil.org/understanding-islam/legal-rulings/5-jihad-a-misunderstood-concept-from-islam.html?start=9

Kassel, B. (2012). The Great Transportation Conspiracy. Retrieved November 8, 2015, from http://www.brooklynrail.net/NationalCityLinesConspiracy.html

Kwitny, J. (1981). The Great Transportation Conspiracy. Retrieved November 8, 2015, from http://www.brooklynrail.net/images/NationalCityLinesConspiracy/Harpers_Magazine_Feb_1981.pdf

Latimer, H. (1858). *The sermons and life of ... Hugh Latimer, some time bishop of Worcester, Volume 2.* Aylott. Retrieved from https://books.google.com/books?id=eJZQAAAAYAAJ&pgis=1

Many. (2015). United States, Presidential Election 2008. Retrieved from https://en.wikipedia.org/wiki/United_States_presidential_election,_

2008

McCain, J., & GONZALEZ, J. (2008). McCain Campaign Calls Obama a "Socialist" -- But Why is That a Smear? | Democracy Now! Retrieved from http://www.democracynow.org/2008/10/24/mccain_campaign_calls_obama_a_socialist

McElroy, D. (2014). Global terrorist death toll soars as attacks become deadlier. Retrieved November 14, 2015, from http://www.telegraph.co.uk/news/worldnews/middleeast/iraq/10985470/Global-terrorist-death-toll-soars-as-attacks-become-deadlier.html

Messerly, J. G. (2013). Aristotle on the Good Life | The Meaning of Life. Retrieved November 9, 2015, from http://reasonandmeaning.com/2013/12/19/aristotle-on-the-good-and-meaningful-life/

Norton, M. I., & Ariely, D. (2011). Building a Better America--One Wealth Quintile at a Time. *Perspectives on Psychological Science, 6*(1), 9–12. http://doi.org/10.1177/1745691610393524

OKBM. (n.d.). Reactor Plants. Retrieved November 9, 2015, from http://www.okbm.nnov.ru/english/npp

Quigley, C. (2012). Kondratieff Waves and the Greater Depression of 2013 - 2020 | Christopher Quigley | FINANCIAL SENSE. Retrieved November 8, 2015, from http://www.financialsense.com/contributors/christopher-quigley/kondratieff-waves-and-the-greater-depression-of-2013-2020

Ryken, L. (2015). A Puritan's Mind » That Which God Hath Lent Thee: The Puritans and Money. Retrieved December 10, 2015, from http://www.apuritansmind.com/stewardship/rykenlelandpuritansandmoney/

Strobel, W. (2010). McClatchy blog: Nukes & Spooks. Retrieved November

16, 2015, from http://blogs.mcclatchydc.com/nationalsecurity/2010/08/terrorism-in-2009.html#ixzz1WQlpO0tP

Tesla. (2013). How the Tesla Model S is Made | Tesla Motors Part 1 (WIRED). Retrieved November 8, 2015, from https://www.youtube.com/watch?v=8_lfxPI5ObM

Thompson, D. (2014). Get Rich, Live Longer: The Ultimate Consequence of Income Inequality - The Atlantic. Retrieved November 8, 2015, from http://www.theatlantic.com/business/archive/2014/04/more-money-more-life-the-depressing-reality-of-inequality-in-america/360895/

Tilley, E. (2015). CSQ 100 Year Plan | CSQ Common Sense 101. Retrieved November 8, 2015, from http://csq1.org/csq-100-year-plan/

Toussaint, E. (2012). Debt Cancellation in Mesopotamia and Egypt from 3000 to 1000 BC | Global Research - Centre for Research on Globalization. Retrieved November 8, 2015, from http://www.globalresearch.ca/debt-cancellation-in-mesopotamia-and-egypt-from-3000-to-1000-bc/5303136

UN. (2015a). No Title. Retrieved from https://en.wikipedia.org/wiki/List_of_countries_by_Human_Development_Index

UN. (2015b). The Global Goals. Retrieved November 26, 2015, from http://www.globalgoals.org/

US Census Bureau, D. I. D. (2015). Income - US Census. *US Government*. Retrieved from http://www.census.gov/hhes/www/income/data/historical/household/

Various. (n.d.). Communist Manifesto 10 Planks. Retrieved November 26, 2015, from http://www.libertyzone.com/Communist-Manifesto-Planks.html

Wilkinson, R. (2011). Richard Wilkinson: How economic inequality harms societies | TED Talk | TED.com. Retrieved November 8, 2015, from https://www.ted.com/talks/richard_wilkinson?language=en

Wolf, E. N. (2015). Wealth Inequity in the United States. Retrieved from https://en.wikipedia.org/wiki/Wealth_inequality_in_the_United_States#cite_note-levyinstitute.org-12

Zarlenga, S. (2010). The Usury Problem Remains | AMI. Retrieved November 8, 2015, from http://www.monetary.org/the-usury-problem-remains/2010/12

www.ingramcontent.com/pod-product-compliance
Lightning Source LLC
LaVergne TN
LVHW020647110826
845149LV00012B/1940

* 9 7 8 1 9 8 7 9 6 4 0 4 2 *